Understanding Batch Chemical Processes

Modelling and Case Studies

Understanding Batch Chemical Processes

Modelling and Case Studies

Thokozani Majozi
Esmael R. Seid
Jui-Yuan Lee

CRC Press
Taylor & Francis Group
Boca Raton London New York

CRC Press is an imprint of the
Taylor & Francis Group, an **informa** business

CRC Press
Taylor & Francis Group
6000 Broken Sound Parkway NW, Suite 300
Boca Raton, FL 33487-2742

First issued in paperback 2019

ISBN-13: 978-1-4987-7317-1 (hbk)
ISBN-13: 978-0-367-87863-4 (pbk)

Library of Congress Cataloging-in-Publication Data

Names: Majozi, Thokozani, author. | Seid, Esmael Reshid, author. | Lee, Jui-Yuan, author.
Title: Understanding batch chemical processes : modelling and case studies / Thokozani Majozi, E R Seid, and J-Y Lee.
Description: Boca Raton : Taylor & Francis, a CRC title, part of the Taylor & Francis imprint, a member of the Taylor & Francis Group, the academic division of T&F Informa, plc, [2017]
Identifiers: LCCN 2016034450 (print) | LCCN 2016053916 (ebook) | ISBN 9781498773171 (acid-free paper) | ISBN 9781498773188 (eBook)
Subjects: LCSH: Chemical processes.
Classification: LCC TP155.7 .M33 2017 (print) | LCC TP155.7 (ebook) | DDC 660/.2812--dc23
LC record available at https://lccn.loc.gov/2016034450

Visit the Taylor & Francis Web site at
http://www.taylorandfrancis.com

and the CRC Press Web site at
http://www.crcpress.com

Contents

Foreword

Batch chemical processes are always ideal in the manufacture of small-volume, high-value-added products, as traditionally encountered in pharmaceutical, agrochemical and food and beverage industries. The main distinction between these processes and their continuous counterparts pertains to the handling of time. Continuous operations can attain steady state, wherein time is completely overridden, whereas batch operations are never at steady state. This simply implies that time can never be ignored in the analysis of batch processes. Consequently, the synthesis, design and optimization of these operations should embed time as a critical dimension.

An additional complexity arises from the presence of binary variables, which are necessary for indicating the activity of a particular task in a particular unit. Unfortunately, these variables also entail significant mathematical challenges, since the computational time of concomitant mathematical models strongly depends on the number of binary variables – the more the binary variables, the longer the CPU time. Consequently, the major part of this book concerns the reduction of the number of binary variables without compromising the accuracy of the final solution.

This book covers most of the key aspects of batch plants, particularly those that are characterized by complex recipes, the so-called multipurpose batch plants. The authors start by giving a broad description of these processes to bring the reader closer to the subject. This is followed by detailed mathematical modelling ranging from scheduling to design and synthesis with hot water optimization considerations. The authors have also provided the necessary background for most of the complex aspects and further substantiate the performance of the presented models with illustrative examples and case studies. The aims of the book are summarized as follows:

1. Introduce the reader to the key or critical aspects in the mathematical modelling of batch processes.
2. Present techniques that overcome the computational complexity in order to yield models that are solvable in near real time.
3. Demonstrate how batch processes could be analyzed, synthesized and designed optimally using proven mathematical formulations.
4. Demonstrate how water and energy aspects could be incorporated within the scheduling framework that seeks to capture the essence of time.
5. Present real-life case studies where the mathematical modelling of batch plants have been successfully applied.

It is my view that the authors have succeeded in achieving this goal. The book is likely to close a lingering knowledge gap in this area of chemical engineering.

Worthy of mention is the fact that the authors are recognized experts in this area of research, having published a similar treatise on the subject in 2010 and recently via an edited book in 2015.

Professor Dominic C. Y. Foo, PEng, CEng, FHEA, MIEM
University of Nottingham, Malaysia Campus
Selangor, Malaysia

Acknowledgements

This work would not have been accomplished without the support from various funding bodies and inspiration from some of the leading figures in the field of process integration. Foremost among these are

- The National Research Foundation (NRF) and the Department of Science and Technology (DST) through the NRF/DST Chair in Sustainable Process Integration at Wits University, South Africa.
- Water Research Commission (WRC), South Africa.
- Professor Mahmoud El-Halwagi at Texas A&M, USA, who has always been a reliable sounding board on various ideas that are presented in this book.
- The late Professor Toshko Zhelev who was a friend, a mentor and one of the pioneers of batch chemical process integration. His influence is everlasting. May his soul rest in peace!

Authors

Thokozani Majozi is a full professor in the School of Chemical and Metallurgical Engineering at Wits University, Johannesburg, where he also holds the NRF/DST chair in sustainable process engineering. His main research interest is batch chemical process integration, where he has made significant scientific contributions that have earned him international recognition. Some of these contributions have been adopted by the industry. Prior to joining Wits, he spent almost 10 years at the University of Pretoria, South Africa, initially as an associate professor and then as a full professor of chemical engineering. He was also an associate professor in computer science at the University of Pannonia in Hungary from 2005 to 2009. Professor Majozi earned a PhD in process integration at the University of Manchester Institute of Science and Technology in the United Kingdom. He is a member of the Academy of Sciences of South Africa and a fellow of the Academy of Engineering of South Africa. He has received numerous awards for his research, including the Burianec Memorial Award (Italy), S2A3 British Association Medal (Silver) and the South African Institution of Chemical Engineers Bill Neal-May Gold Medal. He was also twice a recipient of the NSTF Award and twice a recipient of the NRF President's Award. Professor Majozi is the author or coauthor of more than 150 scientific publications, including two books in batch chemical process integration published by Springer and CRC Press in 2010 and 2015, respectively.

Esmael Seid earned a BSc in chemical engineering at Bahir Dar University, Ethiopia. He then worked in the process industry for three years before joining the University of Pretoria, Pretoria, South Africa, in 2009, where he earned an MSc Eng and a PhD in chemical engineering. Dr. Seid has several publications in international refereed journals on design, synthesis, scheduling and resource conservation, with particular emphasis on water and energy for multipurpose batch plants. He is currently a visiting scholar at the Chemical Engineering Department, Texas A&M University, working on sustainable process design through process integration.

Jui-Yuan Lee is an assistant professor at the Department of Chemical Engineering and Biotechnology, National Taipei University of Technology, Taiwan. He earned a PhD in chemical engineering at National Taiwan University (NTU) in 2011 and conducted postdoctoral research at NTU (2011–2013) and at the University of the Witwatersrand, Johannesburg, South Africa (2013–2014). His research centres on process integration for energy savings and waste reduction, and his interests include low-carbon energy system planning and hybrid power system design. He has published more than 30 journal papers and 40 conference papers and coedited a book on batch process integration. Dr. Lee works closely with several collaborators in Southeast Asia and Africa.

1 Introduction to Batch Processes

1.1 INTRODUCTION

Batch processes differ from continuous processes in several ways. The main difference is that time is inherent in batch processes. In batch processes, every task has a definite duration with starting and finishing times, whereas in continuous processes, time is important during non-steady-state operation. As a result, scheduling of batch processes is vital to the operation of any batch facility. Furthermore, in batch plants, detailed requirements for the various products may be specified on a day-to-day basis. A production schedule must indicate the sequence and manner in which the products are to be produced and specify the times at which the process operations are to be carried out. It is clear that the overall productivity and economic effectiveness of batch plants depend critically on the production schedule as it harmonizes the entire plant operation to attain production goals. While flexibility of batch plants improves productivity, it also makes plant scheduling a challenging task. Much research has focused on developing optimization techniques for scheduling batch plants with the aim of reducing the CPU time required to attain the optimal objective value.

This chapter provides a detailed literature review on various scheduling techniques. The review covers work which has been done in the field of mathematical models used for scheduling of batch plants. Papers presented in the last two decades are considered in the review, with focus on models based on unit-specific event-point continuous-time representation. The chapter is systematically divided into six major sections. Section 1.1 discusses recipe, State Task Network (STN) and State Sequence Network (SSN) representations for batch plants. One of the major characteristics that differentiates batch plants from continuous plants is the existence of intermediate storage in batch plants in order to separate operations in multipurpose equipment and to free the equipment for subsequent processes. The different intermediate storage operational philosophies that exist in batch plants are discussed in Section 1.2. Section 1.3 discusses the different types of batch plants under the main division of multiproduct and multipurpose batch plants.

Section 1.4 discusses the different types of models developed for scheduling of multiproduct batch plants. In this section, the models are categorized under the main division of graphical technique and mathematical technique. In the mathematical section, the different models are grouped into models based on sequence precedence and slot time representations. A brief of scheduling techniques for continuous and semi-continuous plants is given in Section 1.5. Section 1.6 details scheduling techniques developed for multipurpose batch plants. Finally, conclusions and limitations of the current scheduling techniques are briefly discussed in Section 1.7.

1.2   BATCH PROCESS REPRESENTATION

Batch plant processes were first represented in terms of 'recipe networks', as used by Reklaitis (1991). This is analogous to the flow sheet representation of continuous plants, but proposed to describe the process itself rather than a specific plant. Each node on a recipe network corresponds to a task, with directed arcs between nodes representing task precedence. Although recipe networks are certainly adequate for several processing structures, they often involve ambiguities when applied to more complex ones. Kondili et al. (1993) showed that a recipe network was not enough to represent batch processes without vagueness, as illustrated in Figure 1.1. It is not clear from this representation whether task 1 produces two different products later used as inputs of tasks 2 and 3, respectively, or whether it produces one type of product which is then shared between unit operations 2 and 3. Similarly, it is also impossible to determine from Figure 1.2 whether task 4 requires two different types of feedstock, respectively produced by tasks 2 and 5, or whether it only needs one type of feedstock which can be produced by either 2 or 5. Both interpretations are equally plausible. The former could be the case if, say, task 4 is a catalytic reaction requiring a main feedstock produced by task 2 and a catalyst which is then recovered from the reaction products by the separation task 5. The latter case could arise if task 4 were an ordinary reaction task with a single feedstock produced by 2, with task 5 separating the product from the unreacted material which is then recycled to 4.

In order to remove these ambiguities in a systematic fashion, Kondili et al. (1993) proposed a new representation for chemical processes called State Task Network representation. The distinctive characteristic of the STN was that it had two types of nodes: namely, the state nodes, representing the feeds, intermediate and final products; and the task nodes, representing the processing operations which transform materials from one or more input states to one or more output states. State and task nodes were denoted by circles and rectangles, respectively. State Task Networks are free from the ambiguities associated with recipe networks. Figure 1.2 shows two different STN representations, both of which correspond to the recipe network of Figure 1.1. The STN representation is equally suitable for networks of all types of processing tasks, such as continuous, semi-continuous and batch.

Majozi and Zhu (2001) introduced a new concept called the State Sequence Network representation for scheduling of batch plants. The SSN was a graphical network representation of all the states that exist in batch plants, and the network was

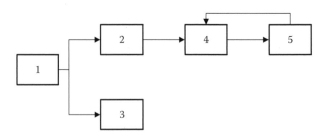

FIGURE 1.1   Recipe representation of chemical processes.

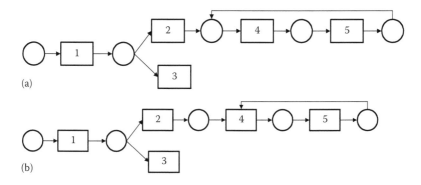

FIGURE 1.2 STN representations for recipe representation.

formulated based on the production recipe. A state changes from one state to another state when it undergoes a unit operation such as mixing, separation or reaction. The building blocks of the SSN are shown in Figures 1.3 through 1.5. From these building blocks, it is easy to construct a SSN for any process recipe. The SSN representation was developed by realizing that (1) the capacity of a unit in which a particular state is used sets an upper limit on the amount of state used or produced by the

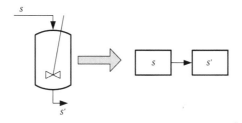

FIGURE 1.3 Simple unit operation.

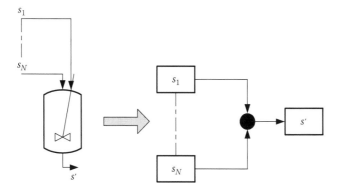

FIGURE 1.4 Unit operation with mixing/reaction.

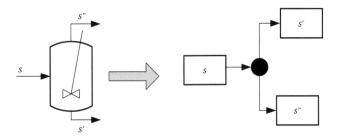

FIGURE 1.5   Unit operation with splitting.

corresponding task, (2) the presence of a particular state in an operation corresponds to the existence of a corresponding task and (3) the usage of state s corresponds to the production of another state s'.

1.3   DIFFERENT STORAGE OPERATIONAL POLICIES IN BATCH PLANT

According to Reklaitis (1982), the multistage nature of a batch processing network allows several different storage and waiting options:

- Unlimited intermediate storage (UIS)
- Finite intermediate storage (FIS)
- No intermediate storage (NIS)
- Central intermediate storage (CIS)
- Unlimited wait (UW)
- Finite wait (FW)
- Zero wait (ZW)

The ZW or FW policy is used where unstable intermediate material must be processed immediately or within a short time after the previous step has been completed. In the NIS policy, there is no intermediate storage between process units, but a product can be stored in a process unit before moving to the next available processing unit. A more practical operational storage policy is that of FIS, where there is limited storage capacity between process units. The UIS policy is more of a theoretical one. If the intermediate products are compatible, the CIS policy is recommended. After materials are processed in a unit, they may be held in the unit temporarily under the UIS, FIS, UW, FW or NIS modes, but they must be transferred to the downstream unit immediately in the ZW mode. Some batch plants have a mixed intermediate storage (MIS) policy with a combination of the FIS, NIS and ZW modes.

1.4   MULTIPURPOSE AND MULTIPRODUCT BATCH PLANTS

Chemical batch plants can be divided into multipurpose and multiproduct. In a multiproduct batch plant, different products are processed in a sequence of single product campaigns. Each product only follows a single route through the plant. They are serial multistage (single unit in each stage) plants, single-stage plants with parallel units and multistage plants with parallel units (Liu 2006), as shown in Figures 1.6 through 1.8, respectively. In a multipurpose batch plant, different batch sizes of products are produced through different production paths, as shown in Figure 1.9. A product in this plant is produced in campaigns where each campaign involves one or more batches.

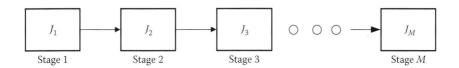

FIGURE 1.6 Schematic of a single-stage process with parallel units.

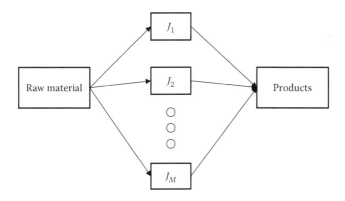

FIGURE 1.7 Schematic of a serial multistage process.

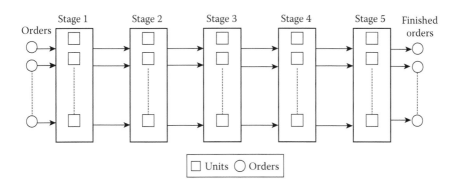

FIGURE 1.8 Schematic of a multistage multiproduct batch process with parallel units.

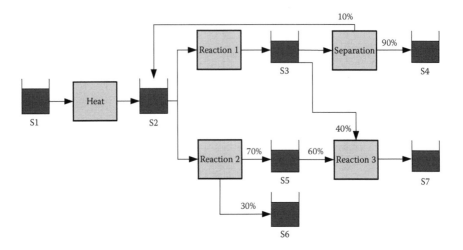

FIGURE 1.9   Schematic of a multipurpose process.

1.5   SCHEDULING TECHNIQUES FOR BATCH PROCESSES

The scheduling techniques for batch processes are broadly categorized into graphical and mathematical techniques as detailed here.

1.5.1   GRAPHICAL TECHNIQUES

Sanmarti et al. (1998, 2002) proposed a novel graph representation that takes into consideration the specific features of chemical processes in scheduling. The advantage of this work was its capability to exploit the problem structure directly, with drastic reduction in computational intensity. Once all the processing tasks had been represented in the recipe, an optimal schedule could be generated using the S-graph framework. The features of S-graph approach are it does not have binary variables which increase the computational complexity of a problem, and it does not require the discretization of the time horizon so it can truly be called continuous.

The S-graph was constructed with nodes representing the production tasks and arcs representing the precedence relationships among tasks. The processing orders of two consecutive tasks were given by a weighted arc (recipe-arc) established between the nodes of tasks. Furthermore, additional recipe-arcs were established from the nodes of tasks generating products to the corresponding product node. The weight of a recipe-arc was specified by the processing time of the task. Figure 1.10 illustrates the conventional, while Figure 1.11 shows a graphical representation of the recipe of an example with three products.

Depending on the operational philosophy (UIS or NIS), the precedence arcs were set. For UIS, the arcs started at the node of the same unit and terminated at the same unit completing the next task. However, for the NIS policy the arcs started at the following node and terminated as before. Figure 1.12 gives an example of an S-graph, from the many possible S-graphs of master recipes with NIS policy.

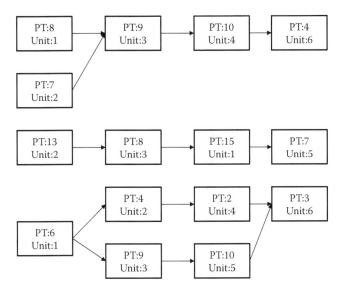

FIGURE 1.10 Conventional representations of the master recipes for three products.

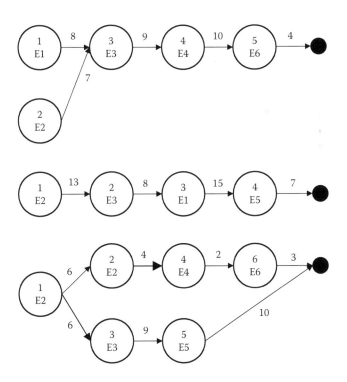

FIGURE 1.11 Graphical representation of the recipe.

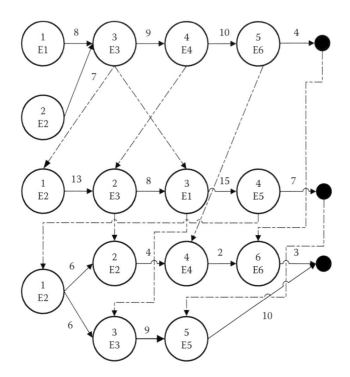

FIGURE 1.12   S-graph for NIS policy for minimization of makespan.

Sanmarti et al. (1998, 2002) introduced the longest path algorithm that used the schedule graph to calculate the makespan of a schedule. Since the resulting formulation of this algorithm was polynomial, it could be applied effectively in solving scheduling problems. The longest path algorithm traced the graph backwards starting from the end node and terminating at the source. The path with the maximal length expressed the makespan of the schedule. The start time of each task in the schedule could be determined by the longest path of this task. Thus, the longest path algorithm also provided the timing of the whole schedule.

Further advances on the S-graph approach for scheduling have been made by Majozi and Friedler (2006). Their formulation extended the work of Sanmarti et al. (2002), where instead of minimizing makespan they maximized throughput over a given time horizon of interest. In their formulation, a new concept, which was a novel node-cutting algorithm which improved the computational efficiency, was introduced. Although this method seems good compared to the mathematical models, the generation of S-graphs is complex and is suited to the case where tasks such as storage are not definite.

1.5.2   MATHEMATICAL TECHNIQUES

The different continuous time models proposed in the literature can be broadly classified into three distinct categories: slot-based, event-based and precedence-based (sequence-based) time representation (Méndez et al., 2006; Shaik et al., 2006; Shaik and Floudas, 2007, 2008).

1.5.2.1 Models Based on Precedence Sequencing

Hui and Gupta (2000) presented a continuous-time mixed integer linear programming (MILP) model for short-term scheduling of multistage multiproduct batch plants based on immediate precedence sequencing. The model determined the optimal sequencing and the allocation of customer orders to non-identical processing units by minimizing the earliness and tardiness of order completion. The previous approach to this scheduling problem relied on the application of tetra-index binary variables (i.e., order, order, stage and unit) to represent all the combinations of order sequences and assignments to units in the various stages. This generated a huge number of binary variables and, as a consequence, much time was required for solutions. The developed formulation by Hui and Gupta (2000) replaced the tetra-index binary variables by one set of tri-index binary variables (order, order and stage) without losing the model's generality. Consequently, the computational time was reduced.

Méndez and Cerdá (2000) proposed a new continuous-time MILP framework for the short-term scheduling of resource constrained multistage batch plants supplying intermediates to end product facilities at specified time intervals. The model was based on immediate precedence sequencing. The proposed approach was applied to the scheduling of a real-world two-stage batch plant involving seven parallel units and six storage tanks manufacturing eight intermediate products. The proposed MILP model gave much smaller size than previous approaches.

Méndez et al. (2001) proposed a new MILP mathematical formulation for the resource-constrained short-term scheduling of flow shop batch facilities with a known topology and limited supplies of discrete resources. They presented new general sequencing constraints. If a pair of orders $(i, i') \in I$ has been assigned to the same equipment $j \in J_{ii's}$ at stage $s \in S$, then order $i \in I$ can be processed either before or after order $i' \in I$. Order i may be a direct predecessor (if it is manufactured immediately before order i') or a non-direct predecessor of order i'. Therefore, an order $i' \in I$ is not constrained to have a single predecessor, but it may feature several depending on its position in the processing sequence. The notion of a predecessor of order i' adopted in their work is more general than that used by Cerdá et al. (1997) and Méndez et al. (2000) since it does not only apply to the order previously processed but to any one produced before in the unit assigned to order i. Their formulation achieved a better objective value and lower computational effort when compared to the model of Pinto and Grossmann (1995).

Lin et al. (2002) proposed medium-term production scheduling for a multiproduct batch facility. The methodology was based on the decomposition of the whole scheduling period into successive short time horizons. A two-level decomposition model was proposed to determine the current time horizon and identify those products to be included. Then a continuous-time formulation for short-term scheduling of batch processes with multiple intermediate due dates was introduced. An iterative procedure was used until the whole time horizon was completed.

Gupta and Karimi (2003a) developed a two-step MILP approach based on immediate precedence sequencing for multistage facilities in non-continuous chemical industries. They incorporated in the model the real and complex features, such as limited shelf lives of intermediate products, batch splitting at the storage, a batch filling multiple orders and general product specifications. The proposed formulation was industrially more realistic and computationally more efficient when compared to previous models in literature. The objective of the formulation was to minimize tardiness subject to common operational considerations such as both sequence- and unit-dependent setup times, the initial plant state and order-unit release times. The developed model was able to solve 90 batches, which made the model an important and significant contribution towards solving industrially important scheduling problems.

Gupta and Karimi (2003b) presented an improved MILP formulation for scheduling multiproduct multistage batch plants. The formulation allowed both sequence-dependent and unit-dependent setup times and common operational considerations such as initial plant state and order/unit release times. In contrast to the previous work, the model minimized tardiness as the scheduling objective because of reducing equipment ideal times in the beginning of the scheduling period and due to the computational superiority when compared to minimizing earliness as the objective function. The proposed model was tested for the multiproduct batch plant with five stages, 25 units and up to 10 orders. When compared to the previous models, their formulation required roughly 30% fewer constraints, which resulted in superior schedules and reduced the computational time roughly by 60%.

Liu and Karimi (2007a) contributed by developing a novel continuous-time formulation for scheduling multistage batch plants with identical parallel units based on immediate precedence sequencing. In contrast to the existing work, they increased the solution efficiency by considering each stage as a block of multiple identical units, thereby eliminating numerous binary variables for assigning batches to specific units. They demonstrated that a multistage process with identical parallel units could be scheduled much more efficiently than a process with non-identical parallel units using a MILP formulation that avoided assigning batches to individual units.

Ferrer-Nadal et al. (2008) proposed a scheduling technique for multistage multiproduct batch plants based on general precedence sequencing. In their formulation, they assumed non-zero transfer time between processing stages. Most of the existing mixed-integer linear programming optimization approaches have traditionally dealt with the batch scheduling problem assuming zero transfer times, and consequently no synchronization between consecutive processing stages. Synchronization implies that during the execution of the transfer task, one unit will be supplying the material whereas the other one will be receiving it, and consequently, no other task can be simultaneously performed in both units. They demonstrated in their work that ignoring the important role of transfer times may seriously compromise the feasibility of the scheduling whenever shared units and storage tanks, material recycles, or bidirectional flows of products are to be considered, as it is depicted in Figures 1.13 and 1.14.

1.5.2.2 Models Based on Slot Representation

Liu and Karimi (2007b) presented a scheduling technique based on slot representation where the task starts and ends at the same slot. They compared and contrasted the different scheduling models for scheduling of multiproduct multistage batch plants.

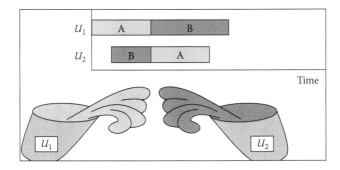

FIGURE 1.13   Infeasible schedule resulting from ignoring transfer time between units.

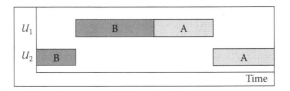

FIGURE 1.14   Feasible schedule by considering transfer time between units.

The formulations were based on unit-slots, stage-slots, process-slots, a variety of slot arrangements and sequence-modelling techniques and four-index and three-index binary variables. The authors demonstrated that their models give better initial RMIP values and were generally faster than the two existing models by Castro and Grossmann (2005) and Gupta and Karimi (2003a,b). However, more importantly, they found that no single model was a clear winner in all test problems, and the idea of using a few competitively best models simultaneously was advised. They also suggested that having several novel formulations and using them simultaneously was an excellent idea, if somehow all of them could be stopped, once one of them achieved the desired solution.

Liu and Karimi (2008) developed scheduling models for multistage batch plants with parallel units and no interstage storage. Little work existed on the scheduling of multistage multiproduct batch plants with no interstage storage in contrast to plants with unlimited storage. Furthermore, most of this work had assumed non-identical parallel units and single (uniform) interstage storage configuration. Many real plants avoid interstage storage, employ mixes of interstage wait policies such as unlimited and zero wait, and possess mixes of stages with identical and non-identical parallel units. Their paper made an important contribution to the batch plantscheduling literature. Furthermore, it has also addressed several important scheduling objectives such as makespan, weighted tardiness, weighted earliness and weighted just-in-time. Their models employed an appropriate and justified mix of slot-based and sequence-based modelling approaches. This was in contrast to most batch plant scheduling models that employed monolithic modelling approaches. In particular, it was found that a four-index slot-based approach worked best for sequencing batches on stages with non-identical parallel units, while a sequence-based approach was the best for stages with identical parallel units.

Their work also supported a frequent observation that models with fewer binary variables or those with tighter MILP relaxations do not always perform better. The performances of models also differ significantly for different objectives. In particular, they found four-index models to be better than three-index models for makespan minimization, but the latter seemed better than the former for weighted just-in-time scheduling.

Erdirik-Dogan and Grossmann (2008) proposed short-term scheduling for multiproduct batch plant for identical/non-identical parallel units. The scheduling technique allowed a task to continue processing on multiple slots. Their motivation was a real case study at the Dow chemical company. Their formulation catered for sequence-dependent changeover, mixed intermediate storage and batch splitting. In order to handle this problem, a MILP model was presented based on asynchronous time slots. Since the formulation needed prepostulation of slots, it was computationally expensive for large problems. Therefore, they developed a bi-level decomposition algorithm where the original problem was decomposed into upper level sequencing and lower level scheduling and sequencing models. The upper-level model was a recent planning model where mass balances were aggregated over time periods and detailed timing constraints were dropped. Solving the upper-level model gave an upper limit for the objective value and a good prediction for the number of slots required. The lower-level model was solved using the predicted number of slots and a subset of products to get the lower bound. The problem was solved iteratively until the bounds converged.

1.6 SCHEDULING TECHNIQUES FOR CONTINUOUS AND SEMI-CONTINUOUS MULTISTAGE MULTIPRODUCT PLANTS

Sahinidis and Grossmann (1991) presented cyclic scheduling for multiproduct continuous parallel line plants. The proposed model resulted in a large-scale mixed integer nonlinear program (MINLP). A reformulation technique was applied to linearize it in the space of the integer variables. The authors applied their formulation to the real-world case study of a polymer production plant. In contrast to previous works, transitions were treated as being sequence dependent, processing times were discretized and nonlinear costs and inventory considerations were accounted for in the formulation.

Pinto and Grossmann (1994) presented optimal cyclic scheduling of multistage continuous multiproduct plants. The method consists of an MINLP sub-problem in which cycle times and inventory levels are optimized for a fixed sequence, and an MILP master problem that determines the optimal sequence of production.

McDonald and Karimi (1997) proposed production planning for semi-continuous parallel line processes. The models were formulated as MILP problems. The objective function in the model was to minimize the production, inventory and transition costs. They tested their model for a plant that had two facilities and two units producing 34 products. Karimi and McDonald (1997) proposed short-term scheduling for semi-continuous parallel production lines. The developed models were used for a single stage multiproduct facility. The two developed models were represented as M1 and M2. The main differences between the two models were how the slots were represented in the time horizon. For model (M1), there was no a priori assignment of slots to time periods. A slot could be of any length and could continue over multiple time periods. For model (M2), each time period had a pre-designated number of slots, and each slot was therefore fixed to a particular time period. Through the different case studies it was shown that the model (M2) outperformed the model (M1).

Lamba and Karimi (2002a) developed short-term scheduling of semi-continuous plants with a composite objective of minimizing transitions and maximizing productivity. The model catered for resource constraints, such as the structure and capacity of upstream and/or downstream material handling facilities and the availability of common resources. They considered synchronized slots across all processing units in the formulation. The formulation resulted in a complex combinatorial problem with a moderate integrality gap. The solution time drastically increased as the problem size increased. Lamba and Karimi (2002b) proposed a branch and bound technique to improve the computational performance of the model they presented (2002a). It was shown in the paper that the developed algorithm gave an optimal solution in reasonable CPU time when it was applied to an industrial case study of a detergent plant.

Lim and Karimi (2003) presented scheduling model for semi-continuous plants with parallel multistage production line. They considered in their models sequence-dependent transitions, minimum campaign lengths, inventory costs, safety stock penalties and backorder penalties which have not been addressed in the previous literatures. The models are based on the continuous-time domains of parallel lines using slots that are asynchronous across lines and have variable lengths. It was demonstrated that the use of asynchronous slots can reduce the number of binary variables significantly and improve the performance tremendously.

Reddy et al. (2004) presented a new continuous-time representation for scheduling of crude oil operations. The formulation resulted in a MILP problem with fewer constraints and variables compared to the discrete-time representation. They tested their models for a weekly schedule with a refinery that had eight tanks, three distillation units and two classes of crude oil. It was shown that the formulation gave a better objective value and better computational performance compared to the previous models used to schedule refinery operations.

1.7 SCHEDULING TECHNIQUES FOR MULTIPURPOSE BATCH PLANTS

This section presents a brief overview of the time representation methods that have been developed and presented in literature in recent years.

1.7.1 MODELS BASED ON DISCRETE-TIME REPRESENTATION

In early formulations, the time horizon in the mathematical models was divided into a number of time intervals of uniform duration. Literature models based on

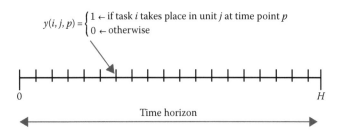

FIGURE 1.15   Discrete-time representation.

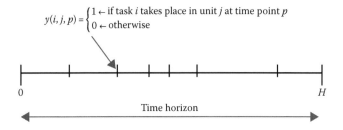

FIGURE 1.16   Continuous-time representation.

this approach include those presented by Kondili et al. (1993), Shah et al. (1993), Pantelides (1993), Dedopoulos and Shah (1995) and Pekny and Zentner (1993). Events such as the beginning and ending of tasks are associated with the boundaries of these time intervals. To attain appropriate approximation of the original problem, it was necessary to use a time interval that was sufficiently small, for example, the greatest common factor (GCF) of the processing times. This generally leads to very large combinatorial problems of intractable size, especially for real-world problems, and hence limits its applications. A detailed comparison of the discrete- and continuous-time representation was presented by Floudas and Lin (2004). They showed that discrete approximation of the time horizon leads to suboptimal solutions and unnecessary increase of the overall size of the resulting mathematical formulation due to the introduction of a large number of binary variables associated with each discrete time interval. The basic concept of the discrete-time approach is illustrated in Figure 1.15.

The main benefit of the discrete-time representation is that it provides a reference grid of time for all operations competing for shared resources, such as equipment items. This provides the chance of formulating the various constraints in the scheduling problem in a relatively easy and simple manner (Floudas and Lin, 2004). The continuous-time representation is given in Figure 1.16.

1.7.2   MODELS BASED ON GLOBAL EVENT POINT REPRESENTATIONS

Due to the drawbacks of the discrete-time approach, researchers started developing continuous-time models in the past decade. In these models, events are potentially allowed to take place at any point in the continuous domain of time. The different continuous-time models proposed in the literature for scheduling of multipurpose batch plants can be broadly classified into three distinct categories: slot-based, global event–based and unit-specific event-based time representation (Shaik et al., 2006; Shaik and Floudas, 2007, 2008). In the global event point–time representation, the starting time associated to tasks at the same event point are similar.

Zhang and Sargent (1996) presented a scheduling formulation based on the Resource Task Network (RTN) representation. In the RTN, the resources include feeds, intermediates, utilities, equipment, storage facility, manpower, transportation facility and cleaning. The resource may be temporally engaged on a task and generated at the end, or permanently consumed and generated as in the case of feeds and products. The formulation gave a large MINLP problem later linearized to give a MILP model. They proposed a technique to linearize the nonlinear terms, which resulted in a model difficult to solve for complex scheduling problems.

Pinto and Grossman (1997) formulated a scheduling technique that considered resource constraints such as manpower, electricity and utilities. In order to cater for the resource constraints, binary variables were assigned and as a result gave a large subset of big-M constraints, which was difficult to solve simultaneously as a MILP. To circumvent this problem, they proposed a combined LP-based branch and bound procedure with disjunctive programming. For a smaller problem of a plastic compounding plant, their formulation reduced the number of enumerated nodes by up to two orders of magnitude when compared to the direct MILP formulation.

Maravelias and Grossmann (2003) proposed a MILP formulation for scheduling multipurpose batch plants. In the formulation, the utilities were also considered as resource constraints. Their model was general and addressed changeover time and costs, utility limitations and storage used for multiple states. A new class of tightening inequalities was proposed for tightening the relaxed LP solutions. Castro et al. (2004) presented a scheduling technique for batch and semi-continuous processes. The formulation was based on the RTN representation. They considered two case studies, a continuous and a batch plant. For the case of finite intermediate storage for a semi-continuous plant, they found a new optimal objective value. The work extended the work of Castro et al. (2001). The formulation performed better than the previous models in literature.

1.7.3 MODELS BASED ON SLOT TIME REPRESENTATION

Schilling and Pantelides (1996) proposed a MILP formulation for short-term scheduling of multipurpose batch plants based on the RTN representation. They proposed a novel branch and bound algorithm that branched both on discrete and continuous variables to reduce the integrality gap of the formulation.

Sundaramoorthy and Karimi (2005) proposed a MILP formulation for scheduling of multipurpose batch plants. In their formulation, they used a continuous-time representation whereby the time horizon was divided into unknown slot lengths. Their motivation was to show that some attempts (Ierapetritou and Floudas, 1998, 1999; Giannelos and Georgiadis, 2002; Maravelias and Grossmann, 2003) at scheduling multipurpose batch plants to decouple the task and unit binary variables from the three-index binary variable (task on unit at event point or slot) in order to reduce the binary variables required in the formulations did not lead to the intended reduction in binary variables. Although reducing the number of binary variables in a formulation is generally a desirable modelling objective, it is well known that this does not guarantee improved solution times. Their model performed better than previous models in literature. They also commented on the importance of different solvers, software versions and the types of computers used in evaluating the performance of the different models in the literature.

Susarla et al. (2010) presented models that used unit specific slots that allowed tasks to span over multiple slots. Their models also allowed non-simultaneous transfer of material into a unit to get a better schedule. They compared their models with the different models in literature. They demonstrated that it was difficult to compare the computational time required by their model with that of Shaik and Floudas (2009) since the model by Shaik and Floudas (2009) needed to specify different values for the maximum number of time points a task needs to span over to get the global optimal solution. The computational time required increased with the increasing value of the maximum number of time points a task needs to span over. Their model performed better than the previous models.

1.7.4 MODELS BASED ON UNIT-SPECIFIC EVENT-POINT REPRESENTATION

In the unit-specific models, the time horizon is divided into unknown lengths, where the starting times of tasks at the same event point can take different values. The sequencing of different tasks is achieved through special big-M constraints. Ierapetritou and Fluodas (1998) were the first to introduce the concept of the unit-specific event-point time representation. In the paper, the three-index binary variable $y(i, j, n)$ was separated into unit $yv(j, n)$ and task $wv(i, n)$ binary variables. Although the model originally claimed its superiority due to both decoupling of tasks and unit events and non-uniform time grid, later it was shown that the model efficiency was due to the reduction of events as a result of the non-uniform time grid representation. The model resulted in a MILP problem which was simpler to solve than the previous models.

Majozi and Zhu (2001) introduced a new batch plant representation called the State Sequence Network representation. The SSN was a graphical network representation of all the states that exist in the batch plant, and a network was formulated based on the production recipe. Only states were considered in this network, thereby eliminating the need for task and unit binary variables as required by the STN formulation. This made it possible to define a single unit binary variable $y(s, p)$ throughout the formulation. This reduction of binary variable was a great achievement in solving industrial scale problems in reasonable computational time.

Lin and Floudas (2001) presented simultaneous design, synthesis and scheduling of multipurpose batch plants. The model considered the trade-offs between capital costs, revenues and operational flexibility. The formulation resulted in both MILP and MINLP problems, where a globally optimal solution was achieved for the nonconvex MINLP problems based on a key property that arose due to the special structure of the resulting models. In the previous models, design of batch plants did not incorporate scheduling strategies very well, which resulted in over-design or under-design. In the design phase of the plant, for resources that are involved in the design are to be used as efficiently as possible, detailed considerations of the plant scheduling must be examined. Therefore, it was necessary to consider design, synthesis and scheduling simultaneously. They considered two case studies to demonstrate the performance of their model in terms of design, synthesis and scheduling with that of Xia and Macchietto (1997). A better objective value and CPU time was obtained as compared to the previous models.

Janak et al. (2004) presented a short-term scheduling technique for multipurpose batch plant that considered resource constraints and different storage policies. The formulation was based on the enhancement of the unit-specific model of Ierapetritou and Floudas (1998). Tasks were allowed to continue over several consecutive event points in order to monitor resource and storage accurately between the specified limits. Two sets of binary variables were used to cater for the situation of tasks spanning over multiple time points, one that indicated whether a task starts at each event point $ws(i, n)$ and another that indicated whether a task ends at each event point $wf(i, n)$. A continuous variable $w(i, n)$ with 0 and 1 bounds was also employed to indicate if a task was active at each event point. In the formulation, states and utilization of resources were considered as tasks, where the sequence and timing of these new tasks were related to the processing tasks in order to maintain the timing change in resource level and amount of states. Their model performed well when compared to the global event model of Maravelias and Grossmann (2003). It was also shown in the paper that the unit-specific event-driven model required fewer event points when compared to the global event–point models.

Shaik et al. (2006) presented a comparative study on continuous-time models for short-term scheduling of multipurpose batch plants. They categorized the various continuous-time models into slot-based, global event–based and unit-specific event-based formulations. In the paper, several literature examples were taken to compare the performance of the various models with the two different objective functions, maximization of profit and minimization of makespan. The different continuous-time representations are depicted in Figure 1.17. In the slot-based continuous-time representation of Figure 1.17a the time horizon is divided into time intervals of unequal and unknown lengths, and typically tasks need to start and finish at an event. In the global event–based continuous-time representation, only the start time of the tasks needs to be at an event point and the events considered are common across all units (Figure 1.17b). In the unit-specific event-based time representation, only the start time of each task in a unit has to be an event point, and the time associated to an event can be different across different units (Figure 1.17c).

For the specific instance of the four tasks on the three units as depicted in Figure 1.17, the slot-based representation required five slots (or six events), the global event–based representation required four events, while the unit-specific event-based representation required only two events. They concluded that the unit-specific event-based models resulted in smaller problem sizes compared to both slot-based and global event–based models and were computationally superior.

Janak et al. (2007) developed a new robust optimization approach for scheduling of batch plants under uncertainty in processing time, demand and price of a product.

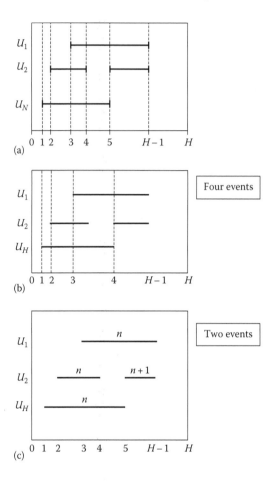

FIGURE 1.17 Different continuous-time representations. (a) Slot-based continuous time. (b) Globalevent–based continuous-time. (c) Unit-specific event-based continuous-time.

The uncertainty was considered by a known probability distribution function. The work extended the robust optimization approach by Lin et al. (2004) for scheduling under bounded uncertainty and bounded and symmetric uncertainty to consider uncertainty by a known probability distribution function. As it was shown by Lin et al. (2004), the optimal solution of an MILP program may be severely infeasible, that is, one or more constraints are violated substantially if the nominal data are slightly perturbed. This makes the nominal optimal solution questionable.

Different probability distribution functions including uniform distribution, normal distribution, difference of normal distributions, general discrete distribution, binomial distribution and Poisson distribution were considered. The general discrete distribution was less conservative, which gave the best objective value. The binomial and Poisson distribution always gave the most conservative results with the worst objective function values. The uniform distribution was less conservative than the bounded and normal distribution. Their formulation results showed that the approach provided an effective way to address scheduling problems under uncertainty, giving a reliable schedule.

Janak and Floudas (2008) presented an improved unit-specific event-based continuous-time approach that addressed closing the integrality gap. A new set of rigorous constraints was developed to reduce the complexity and highly combinatorial nature of scheduling problems. The method was based on four distinct steps. First, the state task network was used to examine the problem under consideration to determine practical limitations, as well as relationships inherent to the problem. For instance, tasks that were continued at the first or the last event point could be fixed and tasks that could not take place at these event points could be excluded. Also, constraints could be added in the model to represent one or more of a group of tasks that must occur at the beginning or end of the time horizon. Tasks that were inherent in the process could be fixed, such as tasks that must occur at consecutive event points due to storage limitations.

In the second step, in order to tighten the relaxed LP solution, the following tightening constraints were developed: (1) The summation of the processing times of the tasks assigned to a specific unit should be less than or equal to the time horizon. (2) Constraints that are expressed in (1) were further tightened by defining relevant minimum and maximum time horizons for each processing unit. (3) The summation of processing times of all tasks starting in a unit at event point (n) or greater must be less than or equal to the amount of time remaining. (4) The summation of the processing times of all tasks finishing in a unit before event point (n) must be less than or equal to the finishing time of the unit at event point (n) − 1.

At the third step, the following bounding constraints were developed: (1) In every unit, the sum of the binary variables over all event points must be between some lower and upper bounds. (2) The sum of durations of processing times for each task in each unit over the entire time horizon is placed in between lower and upper bounds. (3) The amount of material of a state s used in the entire time horizon is also placed in between lower and upper bounds. In order to get rigorous bounds, supporting schedule problems were solved. These problems sought to minimize and maximize the sum of the variables represented in the bounding constraints alternatively, subjected to bounds on the scheduling objective functions. In order to solve these supporting problems, a valid lower and upper bound for the scheduling problems objective function needed to be determined. For the case of profit maximization, a valid lower bound could be found from any feasible solution. Once these bounds were obtained, the original scheduling problem, without any timing and sequence constraints, was solved subjected to lower and upper bounds on the scheduling objective function, where the new objective function was changed in each supporting problem to either minimization or maximization of a different sum of variables.

At the last step, new constraints were generated using a reformulation linearization technique (RLT) developed by Sherali and Adams (1994). The technique could be used to generate a tight LP representation and strong valid inequalities for the MILP. The computational results obtained showed that the proposed formulation was more effective and efficient than the model proposed by Wang and Guignard (2002).

Shaik and Floudas (2008) presented a unit-specific continuous-time approach for short-term scheduling of batch plants using the RTN representation. The model addressed the FIS storage policies without considering storage as a separate task. For unit-specific event-based models, the RTN had not been explored in the literature. In their paper, they used the RTN representation for short-term scheduling based on an improved version of the model of Ierapetritou and Floudas (1998). The performance of the model was evaluated along with the continuous models of Ierapetritou and Floudas (1998), Giannelos and Georgiadis (2002), Castro et al. (2001, 2004) and Sundaramoorthy and Karimi (2005). Benchmark examples using two different objective functions, maximization of profit and minimization of makespan, were considered. It was observed that the slot-based and global event–based models always required the same number of event points while the unit-specific event-based model required fewer event points to solve the problem. Their model required fewer event points and as a result gave smaller problem sizes and was computationally superior. The unit-specific event-based models by Shaik and Floudas (2008) sometimes required more event points for finite storage case compared to the unlimited storage case, but the number of events required was still fewer compared to the slot-based and global event–based models. Their model handles the FIS operational philosophy accurately, which is overlooked by the pervious unit-specific event-based models.

Shaik and Floudas (2009) developed a novel unified modelling approach for short-term scheduling of multipurpose batch plants. In the paper, the necessity of allowing tasks to take place over multiple event points to get a better schedule was shown. In order to understand the necessity of tasks spanning over multiple time points, the sequence constraints for different tasks in different units by Ierapetritou and Floudas (1998) was examined.

$$tu(i, j, n+1) \geq tf(i', j', n) - H(1 - w(i', n))$$

This constraint states that the consuming task at the current event should start after the end time of the consuming task at the current event point that processes the same state, which need not be true if there, is sufficient material for the consuming task to start production. The model by Janak et al. (2004) also allowed tasks to span over multiple time points but it resulted in poor LP relaxations and required a large number of constraints, non-zeros and CPU time. The reason is that the model was originally developed for resource constraints; it did not reduce well in terms of problem statistics to the case of no resources. The model by Shaik and Floudas (2009) had the following features: (1) It could handle problems with resource constraints by allowing task to take place over multiple event points and (2) merged both scheduling problems with and without resource constraints into a common framework and effectively reduced to the simple case of no resources.

1.8   CONCLUSIONS

From this literature survey, it is clear that there are two broad categories of scheduling techniques, graphical and mathematical programming. Graphical methods are used to schedule batch plants running only on the NIS or UIS operational policies. Although this method seems to be good compared to the mathematical models, the generation of an S-graph is complex and is suited to the case where storage is not definite and only one batch of a product is produced in the given time horizon. The computational time of mathematical techniques depends on how the time is represented in the model. In early formulation, the time horizon was divided into a number of time intervals of uniform duration, which led to very large combinatorial problems of intractable size when modelling practical problems.

In this literature review, the performance of different continuous-time models proposed in the literature were compared. It is observed that both the slot-based and global event–based models always require the same number of event points, while the unit-specific event-based models require fewer event points to solve a problem to global optimality. Thus, the unit-specific event-based models result in smaller problem sizes compared to both slot-based and global event–based models and are computationally superior. The most recent and advanced scheduling technique based on unit-specific event-based continuous-time representation is that of Shaik and Floudas (2009). They showed a need to allow tasks to take place over multiple events in order to find the global optimal solutions. Their model is also general and handles the different operational philosophies.

The models by Ierapetritou and Floudas (1998), Majozi and Zhu (2001) and Shaik and Floudas (2007) are capable of solving scheduling problems for batch plants without utility resources and for the unlimited intermediate storage case. The RTN-based model of Shaik and Floudas (2008) is suitable for batch plants without resource and with unlimited and dedicated finite storage policies. However, both these models do not allow task to take place over multiple events, which has been demonstrated to be necessary for obtaining a global optimal solution, by Shaik and Floudas (2009).

The model of Janak et al. (2004, 2008) is relevant for batch plants with resource constraints. The model allows tasks to span over multiple time points, but computationally it does not perform well for problems without resource constraints. The proposed model by Shaik and Floudas (2009) bridges this gap in literature and unifies both problems with and without resources into a generic common framework and efficiently reduces to the simple case of no resources. The formulation uses a three-index binary variable and has fewer big-M terms when compared to Janak et al. (2004, 2008) hence results to improved LP relaxations.

The recent model developed by Susarla et al. (2010) is based on unit-specific slots where special sequence constraints are involved in the models in order to take care of timings when the material balance around a unit and storage is performed, which makes the model similar to unit-specific events. The model does not require specifying the maximum number of event points a task should span over to get the global optimal objective value, unlike the model of Shaik and Floudas (2009). Consequently, it avoids unnecessary iterations. The added feature of their model is in handling non-simultaneous transfer of materials into a unit in order to get a globally optimal solution.

The approach of allowing a task to continue processing on multiple event points results in an increase of the number of binary variables and event points required. As a result, the models require excessive computational time and it was difficult to solve scheduling problems for a long time horizon. All the mathematical models used for scheduling of multipurpose batch plants require pre-specification of the number of time points required and the number of time points related to the optimal solution is found after a number of iterations. This results in an increase in the computational time, limiting the practical application, especially for a long time horizon. These drawbacks necessitate the development of a new scheduling technique and introduction of a new method for the prediction of the optimal number of time points for scheduling of multipurpose batch plants.

REFERENCES

Castro, P., Barbosa-Póvoa, A.P.F.D., Matos, H., 2001. An improved RTN continuous-time formulation for the short-term scheduling of multipurpose batch plants. *Industrial and Engineering Chemistry Research.* 40, 2059–2068.

Castro, P.M., Barbosa-Povóa, A.P., Matos, H.A., Novais, A.Q., 2004. Simple continuous-time formulation for short-term scheduling of batch and continuous processes. *Industrial and Engineering Chemistry Research.* 43, 105–118.

Castro, P.M., Grossmann, I.E., 2005. New continuous-time MILP model for the short-term scheduling of multistage batch plants. *Industrial and Engineering Chemistry Research.* 44, 9175–9190.

Cerdá, J., Henning, G.P., Grossmann, I.E., 1997. A mixed-integer linear programming model for short-term scheduling of single-stage multiproduct batch plants with parallel lines. *Industrial and Engineering Chemistry Research.* 36, 1695–1707.

Dedopoulos, I.T., Shah, N., 1995. Optimal short-term scheduling of maintenance and production for multipurpose plants. *Industrial and Engineering Chemistry Research.* 34, 192–201.

Erdirik-Dogan, M., Grossmann, I.E., 2008. Slot-based formulation for the short-term scheduling of multistage, multiproduct batch plants with sequence-dependent changeovers. *Industrial and Engineering Chemistry Research.* 47, 1159–1163.

Ferrer-Nadal, S., Capón-Garćia, E., Méndez, C.A., Puigjaner, L., 2008. Material transfer operations in batch scheduling. A critical modeling issue. *Industrial and Engineering Chemistry Research.* 47, 7721–7732.

Floudas, C.A., Lin, X., 2004. Continuous-time versus discrete-time approaches for scheduling of chemical processes: A review. *Computers and Chemical Engineering.* 28, 2109–2129.

Giannelos, N.F., Georgiadis, M.C., 2002. A simple new continuous-time formulation for short-term scheduling of multipurpose batch processes. *Industrial and Engineering Chemistry Research.* 41, 2178–2184.

Gupta, S., Karimi, I.A., 2003a. Scheduling a two-stage multiproduct process with limited product shelf life in intermediate storage. *Industrial and Engineering Chemistry Research.* 42, 490–508.

Gupta, S., Karimi, I.A., 2003b. An improved MILP formulation for scheduling multiproduct, multistage batch plants. *Industrial and Engineering Chemistry Research.* 42, 2365–2380.

Hui, C.W., Gupta, A., 2000. A novel MILP formulation for short-term scheduling of multistage multi-product batch plants. *Computers and Chemical Engineering.* 24, 2705–2717.

Ierapetritou, M.G., Floudas, C.A., 1998. Effective continuous-time formulation for short-term scheduling: 1. Multipurpose batch processes. *Industrial and Engineering Chemistry Research.* 37, 4341–4359.

Ierapetritou, M.G., Floudas, C.A., 1999. Effective continuous time formulation for short-term scheduling. 1. Multiple intermediate due dates. *Industrial and Engineering Chemistry Research.* 38, 3446–3461.

Janak, S.L., Floudas, C.A., 2008. Improving unit-specific event based continuous time approaches for batch processes: Integrality gap and task splitting. *Computers and Chemical Engineering.* 32, 913–955.

Janak, S.L., Lin, X., Floudas, C.A., 2004. Enhanced continuous-time unit-specific event-based formulation for short-term scheduling of multipurpose batch processes: Resource constraints and mixed storage policies. *Industrial and Engineering Chemistry Research.* 43, 2516–2533.

Janak, S.L., Lin, X., Floudas, C.A., 2007. A new robust optimization approach for scheduling under uncertainty: II. Uncertainty with known probability distribution. *Computers and Chemical Engineering.* 31, 171–195.

Karimi, I.A., McDonald, C.M., 1997. Planning and scheduling of parallel semicontinuous processes. II. Short-term scheduling. *Industrial and Engineering Chemistry Research.* 36, 2701–2714.

Kondili, E., Pantelides, C.C., Sargent, R.W.H., 1993. A general algorithm for short-term scheduling of batch operations. I. MILP formulation. *Computers and Chemical Engineering.* 17, 211–227.

Lamba, N., Karimi, I.A., 2002a. Scheduling parallel production lines with resource constraints. I. Model formulation. *Industrial and Engineering Chemistry Research.* 41, 779–789.

Lamba, N., Karimi, I.A., 2002b. Scheduling parallel production lines with resource constraints. II. Decomposition algorithm. *Industrial and Engineering Chemistry Research.* 41, 790–800.

Lim, M.F., Karimi, I.A., 2003. Resource-constrained scheduling of parallel production lines using asynchronous slots. *Industrial and Engineering Chemistry Research.* 42, 6832–6842.

Lin, X., Floudas, C.A., 2001. Design, synthesis and scheduling of multipurpose batch plants via an effective continuous-time formulation. *Computers and Chemical Engineering.* 25, 665–674.

Lin, X., Floudas, C.A., Modi, S., Juhasz, N.M., 2002. Continuous-time optimization approach for medium-range production scheduling of a multiproduct batch plant. *Industrial and Engineering Chemistry Research.* 41, 3884–3906.

Lin, X., Janak, S.L., Floudas, C.A., 2004. A new robust optimization approach for scheduling under uncertainty: I. Bounded uncertainty. *Computers and Chemical Engineering.* 28, 1069–1085.

Liu, Y., 2006. Scheduling of multi-stage multi-product batch plants with parallel units. PhD thesis. Singapore: Department of Chemical and Biomolecular Engineering, National University of Singapore.

Liu, Y., Karimi, I.A., 2007a. Scheduling multistage, multiproduct batch plants with nonidentical parallel units and unlimited intermediate storage. *Chemical Engineering Science.* 62, 1549–1566.

Liu, Y., Karimi, I.A., 2007b. Novel continuous-time formulations for scheduling multi-stage batch plants with identical parallel units. *Computers and Chemical Engineering.* 31, 1671–1693.

Liu, Y., Karimi, I.A., 2008. Scheduling multistage batch plants with parallel units and no interstage storage. *Computers and Chemical Engineering.* 32, 671–693.

Majozi, T., Friedler, F., 2006. Maximization of throughput in a multipurpose batch plant under a fixed time horizon: S-graph approach. *Industrial and Engineering Chemistry Research.* 45, 6713–6720.

Majozi, T., Zhu, X.X., 2001. A novel continuous-time MILP Formulation for multipurpose batch plants. *Industrial and Engineering Chemistry Research.* 40, 5935–5949.

Maravelias, C.T., Grossmann, I.E., 2003. New general continuous-time state-task network formulation for short-term scheduling of multipurpose batch plants. *Industrial and Engineering Chemistry Research.* 42, 3056–3074.

McDonald, C.M., Karimi, I.A., 1997. Planning and scheduling of parallel semicontinuous processes. I. Production planning. *Industrial and Engineering Chemistry Research.* 36, 2691–2700.

Méndez, C.A., Cerdá, J., 2000. Optima scheduling of a resource-constrained multiproduct batch plant supplying intermediates to nearby end-product facilities. *Computers and Chemical Engineering.* 24, 369–376.

Méndez, C.A., Cerdá, J., Grossmann, I.E., Harjunkoski, I., Fahl, M., 2006. State of-the-art review of optimization methods for short-term scheduling of batch processes. *Computers and Chemical Engineering.* 30, 913–946.

Méndez, C.A., Henning, G.P., Cerdá, J., 2000. Optimal scheduling of batch plants satisfying multiple product orders with different due-dates. *Computers and Chemical Engineering.* 24, 2223–2245.

Méndez, C.A., Henning, G.P., Cerdá, J., 2001. An MILP continuous-time approach to short-term scheduling of resource-constrained multistage flowshop batch facilities. *Computers and Chemical Engineering.* 25, 701–711.

Pantelides, C.C., 1993. Unified frameworks for optimal process planning and scheduling. In D.W.T. Rip-pin, J.C. Hale, J. Davis (Eds.), *Proceedings of the Second International Conference on Foundations of Computer-Aided Process Operations* (pp. 253–274). Crested Butte, CO.

Pekny, J.F., Zentner, M.G., 1993. Learning to solve process scheduling problems: The role of rigorous knowledge acquisition frameworks. In D.W.T. Rippin, J.C. Hale, J. Davis (Eds.), *Proceedings of the Second International Conference on Foundations of Computer-Aided Process Operations* (pp. 275–309). Crested Butte, CO.

Pinto, J., Grossmann, I.E., 1997. A logic based approach to scheduling problems with resource constraints. *Computers and Chemical Engineering.* 21, 801–818.

Pinto, J.M., Grossmann, I.E., 1994. Optimal cyclic scheduling of multistage continuous multiproduct plants. *Computers and Chemical Engineering.* 18, 797–816.

Pinto, J.M., Grossmann, I.E., 1995. A continuous-time mixed-integer linear programming model for short-term scheduling of multistage batch plants. *Industrial and Engineering Chemistry Research.* 34, 3037–3051.

Reddy, P.C.P., Karimi, I.A., Srinivasan R., 2004. A new continuous-time formulation for scheduling crude oil operations. *Chemical Engineering Science.* 59, 1325–1341.

Reklaitis, G.V., 1991. Perspectives of scheduling and planning of process operations. *Process Systems and Engineering PSE'91.* Montebello, Quebec, Canada.

Reklaitis, G.V., 1982. Review of scheduling of process operations. *AIChE Symposium.* 78, 119–133.

Sahinidis, N.V., Grossmann, I.E., 1991. MINLP model for cyclic multiproduct scheduling on continuous parallel lines. *Computers and Chemical Engineering.* 15, 85–103.

Sanmarti, E., Friedler, F., Puigjaner, L., 1998. Combinatorial technique for short term scheduling of multipurpose batch plants based on schedule-graph representation. *Computers and Chemical Engineering.* 22, 847–850.

Sanmarti, E., Holczinger, T., Puigjaner, L., Friedler, F., 2002. Combinatorial framework for effective scheduling of multipurpose batch plants. *AIChE Journal.* 48, 2557–2570.

Schilling, G., Pantelides, C., 1996. A simple continuous-time process scheduling formulation and a novel solution algorithm. *Computers and Chemical Engineering.* 20, 1221–1226.

Shah, N., Pantelides, C.C., Sargent, R.W.H., 1993. A general algorithm for short-term scheduling of batch operations II. Computational issues. *Computers and Chemical Engineering.* 17, 229–244.

Shaik, M.A., Floudas, C.A., 2007. Improved unit-specific event-based model continuous-time model for short-term scheduling of continuous processes: Rigorous treatment of storage requirements. *Industrial and Engineering Chemistry Research.* 46, 1764–1779.

Shaik, M.A., Floudas, C.A., 2008. Unit-specific event-based continuous time approach for short-term scheduling of batch plants using RTN framework. *Computers and Chemical Engineering.* 32, 260–274.

Shaik, M.A., Floudas, C.A., 2009. Novel unified modeling approach for short term scheduling. *Industrial and Engineering Chemistry Research.* 48, 2947–2964.

Shaik, M.A., Janak, S.L., Floudas, C.A., 2006. Continuous-time models for short-term scheduling of multipurpose batch plants: A comparative study. *Industrial and Engineering Chemistry Research.* 45, 6190–4209.

Sherali, H.D., Adams, W.P., 1994. A hierarchy of relaxations and convexhull characterizations for mixed-integer zero–one programming problems. *Discrete and Applied Mathematics.* 52, 83–106.

Sundaramoorthy, A., Karimi, I.A., 2005. A simpler better slot-based continuous-time formulation for short-term scheduling in multipurpose batch plants. *Chemical Engineering Science.* 60, 2679–2702.

Susarla, N., Li, J., Karimi, I.A., 2010. A novel approach to scheduling of multipurpose batch plants using unit slots. *AICHE Journal.* 56, 1859–1879.

Wang, S., Guignard, M., 2002. Redefining event variables for efficient modelling of continuous-time batch processing. *Annals of the Operation Research.* 116, 113–126.

Xia, Q.S., Macchietto, S., 1997. Design and synthesis of batch plants MINLP solution based on a stochastic method. *Computers and Chemical Engineering.* 21, S697–S702.

Zhang, X., Sargent, R.W.H., 1996. The optimal operation of mixed production facilities-a general formulation and some approaches for the solution. *Computers and Chemical Engineering.* 20, 897–904.

2 Modelling for Effective Solutions
Reduction of Binary Variables

2.1 INTRODUCTION

Scheduling of multipurpose batch plants has gotten considerable attention in the last two decades where considerable scheduling models have been proposed. The research direction has been to improve the existing scheduling models in terms of model size, computational time and optimal objective value. This chapter also has the same objective of presenting a better scheduling model that is compact and computationally efficient. The model is rigorous in handling the different intermediate storage policies and unit wait time. The binary variables, continuous variables, constraints and computational time required are reduced by almost 41%, 15%, 43% and 42%, respectively, when compared to the recent rigorous unit-specific event-based model for short-term scheduling of batch plants.

2.2 NECESSARY BACKGROUND

The use of optimization techniques for scheduling of multipurpose batch plants has been receiving considerable attention by researchers from academics and industries due to the significant economic benefit they bring to industries. The features of multipurpose batch plants allow flexibility in the production of small-volume, high value–added products by sharing common resources like processing equipment, storage tank, utilities, manpower, etc. However, the same flexibility of multipurpose batch plants result production scheduling to be a challenging task, which leads to the existence of different scheduling models. All these developed models can be generally grouped into three major categories (Sundaramoorthy and Maravelias, 2011). In the first category, every product follows a sequential production pattern where every task consumes and produces one material and batch mixing and splitting are not permissible. Most of the models considered a simplified approach of allocating batches to processing units to optimize different objectives (minimizing makespan, earliness and tardiness of due date, inventory, etc.) without considering storage and utility constraints. The sequence-based or precedence-based representation uses either direct precedence or indirect precedence sequencing of pairs of tasks on units. In the direct precedence models (Pinto and Grossmann, 1997; Méndez and Cerdá, 2000; Gupta and Karimi, 2003; Liu and Karimi, 2007), batch i' has to follow immediately after batch i is processed in the same unit. The number of binary variables required in these models increases quadratically with the number of batches. Having this draw-back, indirect precedence models (Méndez et al., 2001; Méndez and Cerdá, 2004; Ferrer-Nadal et al., 2008) have been developed to reduce the binary variables required by allowing batch i' to be processed not necessarily immediately after batch i iharjun the same unit. However, it is not trivial to identify subsequent tasks, sequence-dependent costs and prevent certain processing sequences in indirect precedence models. Kopanos et al. (2009, 2010) have overcome this issue efficiently by combining direct and indirect precedence decision variables in the same model. Models developed in the second category are based on network processing approach where batch mixing and splitting is allowed and storage and utility constraints are also catered for. This chapter focuses and gives a comprehensive review of models developed under this category, since it is the interest and motivation of this work. Lastly, it is the hybrid processing environment which constitutes both sequential and network processing system (Sundaramoorthy and Maravelias, 2011; Velez and Maravelias, 2013, 2014).

A considerable amount of research work exists for addressing scheduling problem for network-based batch processing system. In order to capture the essence of time in batch processing, researchers adopted discrete- and continuous-time representation. In the discrete-time representation, the time horizon is divided into equal intervals, and the starting and finishing time of a task should be aligned at the interval (Kondili et al., 1993; Shah et al., 1993; Dedopoulos and Shah, 1995). This time representation allows easy and straightforward modelling of due dates, equipment unavailability, time-dependent utility pricing and holding and backlog costs. However, models based on discrete-time representation have the following draw-backs: (1) model accuracy; (2) large discrete-time models due to sequence-dependent changeover or non-instantaneous material transfer times between units and (3) cannot handle variable processing times. Continuous-time models are accurate, handle different magnitude of task processing times without increasing the model size and can easily be integrated with lower-level control layer. The major drawback is that they turn to be more complex when handling sequencing of tasks than their discrete-time counterparts. The different continuous-time models proposed in the literature for scheduling of multipurpose batch plants can be broadly classified into two distinct categories: single-grid and multi-grid time representations (Floudas and Lin, 2004; Shaik et al., 2006; Shaik and Floudas, 2008, Harjunkoskia et al., 2014). The term time-grid designates all time representations in scheduling models that employ time slots/periods/points/events. The starting and finishing time of tasks are mapped onto one or more time reference grids (Harjunkoskia et al., 2014).

In the single-grid-based models, the starting time associated with different tasks at the same grid point is identical (Schilling and Pantelides, 1996; Zhang and Sargent, 1996; Pinto and Grossmann, 1997; Maravelias and Grossmann, 2003; Sundaramoorthy and Karimi, 2005). On the other hand, in multiple-grid time representation the time horizon is divided into unknown lengths where the starting times of tasks in different units at the same grid point can take different values. The sequencing of different tasks is achieved through special big-M constraints (Ierapetritou and Floudas, 1998; Majozi and Zhu, 2001; Janak et al., 2004, 2007; Shaik et al. 2006; Janak and Floudas, 2008; Shaik and Floudas, 2009; Susarla et al., 2010; Vooradi and Shaik, 2012). These models perform better in terms of objective value and computational time due to the non-uniform time grid representation as a result of reducing the number of grid points required for solving the scheduling problem. In all the previous models that are based on multiple-grid time representation, the sequence constraints that are core in the formulation result in unconditional sequencing, which implies that different tasks in different units are always aligned without the actual material transfer between these units. The drawback was first addressed by Seid and Majozi (2012) by formulating sequence constraints between different tasks in different units to be conditional, thereby leading to improved objective value and reduced event points to solve the scheduling problem compared to the previously developed models. Recently, Vooradi and Shaik (2013) identified that the model by Seid and Majozi (2012) is actually partially conditional, improved the concept further and developed a rigorous conditional sequencing that results in better optimal objective value and reduced event points required in some of the case studies they investigated. However, the model size and complexity increased considerably, hence this contribution. The reader can get a more comprehensive and detailed review on production scheduling in processing industry from Harjunkoskia et al. (2014).

The main objective of this chapter is to present a rigorous conditional sequencing formulation based on STN representation with significantly reduced size and complexity compared to published work in literature. The chapter is organized as follows. In Section 2.3, the general problem statement for scheduling of network-based batch facilities is given. The developed mathematical model is presented in Section 2.4 followed by case studies from literature in Section 2.6 in order to compare the efficiency of the developed model in this chapter. Finally, in Section 2.7, conclusions are drawn from this work.

2.3 PROBLEM STATEMENT

A typical scheduling problem of multipurpose batch plants can be stated as follows. Given: (1) the production recipe that shows the steps and sequences the raw materials should follow to become products, (2) the size of the units and their suitability to perform the different tasks, (3) storage capacity and their suitability to store different material and (4) the time horizon, using these data it is required to determine (1) the maximum achievable profit of the plant or the minimum makespan if throughput is fixed a priori and (2) a production schedule related to the optimal resource utilization.

2.4 MODEL FORMULATION

The mathematical formulation presented in this work is based on STN representation and contains the following constraints.

2.4.1 ALLOCATION CONSTRAINTS

Constraint (2.1) allocates one task at any given event point p in a unit. This constraint is used when a unit is multipurpose and capable of performing more than one task.

$$\sum_{i \in I_j} y(i, j, p) \le 1, \quad \forall i \in I, \quad j \in J, \quad p \in P \tag{2.1}$$

2.4.2 Capacity Constraints

Constraint (2.2) enforces the amount of material inside the unit for process should be between the lower and upper capacity of the unit.

$$V_{i,j}^{L} y(i,j,p) \le mu(i,j,p) \le V_{i,j}^{U} y(i,j,p), \quad \forall i \in I, \quad j \in J, \quad p \in P \qquad (2.2)$$

2.4.3 Material Balance around a Storage Unit

Constraint (2.3) determines the amount of intermediate material in the storage at event point p.

$$q(s,p) = q(s,p-1) + \sum_{i \in I_s^p} \sum_{j \in J_i} mu(i,j,p-1) - \sum_{i' \in I_s^c} \sum_{j' \in J_{i'}} mu(i',j',p),$$

$$\forall s \in S^i, \quad p \in P, \quad p \ge 1 \qquad (2.3)$$

Constraint (2.4) is used for material balance around the intermediate storage at the first event point, that is, $p = 1$.

$$q(s,p) = Q_0(s) - \sum_{i' \in I_s^c} \sum_{j' \in J_{i'}} mu(i',j',p), \quad \forall s \in S^i, \quad p \in P, \quad p = 1 \qquad (2.4)$$

Constraint (2.5) states that the amount of product material stored at each event point p is the amount stored at the previous event point $p-1$ and product state s produced by tasks at event point p.

$$q(s,p) = q(s,p-1) + \sum_{i \in I_s^p} \sum_{j \in J_i} mu(i,j,p), \quad s \in S^p, \quad p \in P, \quad p \ge 1 \qquad (2.5)$$

Constraint (2.6) is used for the amount of product material stored at event point $p = 1$.

$$q(s,p) = Q_0(s) + \sum_{i \in I_s^p} \sum_{j \in J_i} mu(i,j,p), \quad s \in S^p, \quad p \in P, \quad p = 1 \qquad (2.6)$$

Constraint (2.7) states that the amount of raw material in storage at each event point p is the difference between the amount stored at the previous event point $p-1$ and material state s consumed by the consuming tasks at event point p.

$$q(s,p) = q(s,p-1) - \sum_{i' \in I_s^c} \sum_{j' \in J_{i'}} mu(i',j',p), \quad s \in S^r, \quad p \in P, \quad p \ge 1 \qquad (2.7)$$

Constraint (2.8) is used for the amount of raw material stored at event point $p = 1$.

$$q(s,p) = QO(s) - \sum_{i \in I_s^c} \sum_{j \in J_i} mu(i,j,p), \quad s \in S^r, \quad p \in P, \quad p = 1 \qquad (2.8)$$

2.4.4 DURATION CONSTRAINTS

Constraints (2.9) and (2.10) describe the duration constraints. In this mathematical formulation, the duration constraint is based on processing unit which allows for the reduction of variables and constraints, hence the compactness of the model. Constraints (2.9) and (2.10) are only used for tasks that produce stable intermediate states, where the states can wait for a while before they can be consumed by the subsequent tasks.

$$tu(j,p) \geq tp(j,p) + \sum_{i \in I_j} \alpha(i,j) y(i,j,p) + \sum_{i \in I_j} \beta(i,j) mu(i,j,p), \quad \forall j \in J, \quad p \in P \tag{2.9}$$

$$tu(j,p) \leq tp(j,p) + \sum_{i \in I_j} \alpha(i,j) y(i,j,p) + \sum_{i \in I_j} \beta(i,j) mu(i,j,p)$$

$$+ \sum_{i \in I_j} UW(i,j), \quad \forall j \in J, \quad p \in P \tag{2.10}$$

For tasks that produce unstable states, which require states to be consumed immediately by subsequent tasks, as encountered in zero-wait condition, Constraints (2.9) and (2.11) are used.

$$tp(j,p) \leq tu(j,p) + \sum_{i \in I_j} \alpha(i,j) y(i,j,p) + \sum_{i \in I_j} \beta(i,j) mu(i,j,p) + H\left(1 - \sum_{i \in I_j} y(i,j,p)\right),$$

$$\forall i \in I_{ZW}, \quad j \in J, \quad p \in P \tag{2.11}$$

2.4.5 SEQUENCE CONSTRAINTS FOR PROCESSING UNITS

Constraint (2.12) states that the starting time of the processing unit at event point p is after the completion time of the same unit at previous event point $p-1$. Basing duration and sequence constraints on units rather than tasks allows the avoidance of sequencing constraints for different tasks in different units, as encountered in other formulations for unit-specific event representation. For example, in the previous literature models based on unit-specific event-point time representation if the unit performs four different tasks, we would need 8 variables for defining the starting and finishing times for the tasks and 16 constraints relating to sequence constraints for the same task, as well as different tasks, in the same unit. However, in the proposed model we need only two variables relating to the starting and finishing time for a unit and one constraint to sequence this unit, thereby leading to a compact model.

$$tu(j,p) \geq tp(j,p-1), \quad \forall j \in J, \quad p \in P \tag{2.12}$$

2.4.6   Sequence Constraints for Different Tasks in Different Units

The constraints under this section are used to conditionally sequence the consuming and producing units. In the previous unit-specific event-point-based formulations, the producing and consuming units are aligned even if there is no material transfer and this shortcoming was first observed and improved to be conditional by Seid and Majozi (2012) and later improved to make it rigorously conditional by Vooradi and Shaik (2013). However, the method proposed by Vooradi and Shaik (2013) results in increased number of binary variables and big-M constraints. The model proposed in this manuscript is significantly reduced the number of binary variables and big-M constraints.

2.4.6.1   Sequence Constraints for Different Tasks in Different Units for Stable State That Can Wait after It Is Produced

Constraints (2.13) through (2.16) are used to conditionally align the consuming units with the producing units for stable intermediate states. They trace whether the amount of material that is consumed in a unit comes from storage or producing units or from both producing units and storage.

$$\sum_{i' \in I_s^c}\sum_{j' \in J_{i'}} \theta(s,i') mu(i',j',p) \le q(s,p-1) + \sum_{j' \in J_s^c}\sum_{j \in J_s^p} mex(s,j,j',p-1), \quad \forall s \in S^{cw}, \ p \in P$$

$$(2.13)$$

Constraint (2.14) ensures that the amount of material that is consumed in the unit is greater than the amount of same material coming from all the producing units, since some of the material could be taken from storage.

$$\theta(s,i') mu(i,j',p) \ge \sum_{j \in J_s^p} mes(s,j,j',p), \quad \forall s \in S^{cw}, \ j' \in J_s^c, \ p \in P \quad (2.14)$$

Constraint (2.15) ensures that the amount of state s sent from producing unit to consuming units is limited by the amount of state produced by the producing unit.

$$\delta(s,i) mu(i,j,p) \ge \sum_{j' \in J_s^c} mes(s,j,j',p), \quad \forall s \in S^{cw}, \ j \in J_s^p, \ p \in P \quad (2.15)$$

Constraint (2.16) is used to trigger the binary variable to take a value of one if indeed material, that is, state s is transferred between the consuming and producing units, otherwise they take a value of zero.

$$\sum_s mes(s,j,j',p) \le V_j^U t(j,j',p), \quad \forall s \in S^{cw}, \ j \in J_s^p, \ j' \in J_s^c, \ p \in P \quad (2.16)$$

2.4.6.2   Sequence Constraints for Different Tasks in Different Units for Unstable States with Zero Wait

Constraints (2.17) through (2.19) have the same concept as Constraints (2.14) through (2.16), but apply to zero-wait states.

$$\theta(s,i')mu(i',j',p) = \sum_{j \in J_s^p} mes(s,j,j',p), \quad \forall s \in S^{zw}, \quad j' \in J_s^c, \quad p \in P \quad (2.17)$$

$$\delta(s,i)mu(i,j,p) = \sum_{j' \in J_s^c} mes(s,j,j',p), \quad \forall s \in S^{zw}, \quad j \in J_s^p, \quad p \in P \quad (2.18)$$

$$\sum_{s} mes(s,j,j',p) \le V_j^U t(j,j',p), \quad \forall s \in S^{zw}, \quad j \in J_s^p, \quad j' \in J_s^c, \quad p \in P \quad (2.19)$$

2.4.6.3   Sequence Constraint That Links the Starting Time of the Consuming Units with the Finishing Time of Producing Units

Constraint (2.20) ensures that the starting time of the consuming unit is later than the finishing time of the producing unit if material transfer occurs between the units.

$$tu(j',p) \ge tp(j,p-1) - H\big(1 - t(j,j',p-1)\big), \quad \forall s \in S^{cw}, \quad j \in J_s^p, \quad j' \in J_s^c, \quad p \in P \quad (2.20)$$

Constraint (2.21) is used to prevent real-time storage violation if state s is used by consuming units from storage after it was produced and stored.

$$tu(j',p) \ge tp(j,p-2) - H\left(1 - \sum_{i \in I_j} y(i,j,p-2)\right),$$
$$\forall s \in S^{cw}, \quad j \in J_s^p, \quad j' \in J_s^c, \quad p \in P \quad (2.21)$$

Constraints (2.21) and (2.22) are used for zero-wait condition which forces the consuming unit j to start immediately after the producing unit j', if consuming unit j receives materials from producing unit j'. Otherwise, these two constraints are relaxed.

$$tu(j',p) \le tp(j,p-1) - H\left(2 - \sum_{i} y(i,j,p) - t(j,j',p-1)\right),$$
$$\forall s \in S^{zw}, \quad j' \in J_s^c, \quad j \in J_s^p, \quad p \in P \quad (2.22)$$

2.4.7   Sequence Constraints for FIS Policy

Constraint (2.23) states that if the amount of material produced from the producing units exceeds the capacity of the storage, then the variable $ma(s,j,p)$ should take a value indicating that there is not enough storage for the material produced from the producing unit j. Consequently, Constraint (2.24) enforces the binary variable to take a value of one. However, if there is enough storage capacity to store material that is produced from the producing unit j, then $ma(s, j, p)$ takes a value of zero, which ultimately forces the binary variable $x(j, p)$ to take the value of zero.

$$\sum_{i \in I_s^p} \sum_{j \in J_s^p} \delta(s,i) mu(i, j, p) + q(s, p) \leq Q^U(s) + \sum_{j \in J_s^p} ma(s, j, p) \qquad (2.23)$$

$$\sum_{s} ma(s, j, p) \leq V_j^U x(j, p), \quad \forall s \in S^{cw}, \quad j \in J_s^p, \quad p \in P \qquad (2.24)$$

Constraint (2.25) is used to ensure that the consuming unit finishes earlier compared to the finishing time of the producing unit so that the producing unit can send its material immediately to the available consuming unit if this is necessary at a particular event point for an optimum overall schedule. In this instance, both the transfer binary variable, $t(j, j', p)$, and the binary variable signifying absence of storage, $x(j, p)$, take a value of one. Otherwise, the constraint is relaxed.

$$tp(j', p) \leq tp(j, p) + H(2 - t(j, j', p) - x(j, p)), \quad \forall s \in S^{cw}, \quad j' \in J_s^c, \quad j \in J_s^p \quad (2.25)$$

2.4.8 TIME HORIZON CONSTRAINTS

Constraints (2.26) and (2.27) ensure that the starting and finishing times of the processing units are within the time horizon of interest.

$$tu(j, p) \leq H, \quad \forall p \in P, \quad j \in J \qquad (2.26)$$

$$tp(j, p) \leq H, \quad \forall p \in P, \quad j \in J \qquad (2.27)$$

2.4.9 OBJECTIVE FUNCTION

Equation 2.28 is the objective function for maximization of profit.

$$\text{maximize } pro = \sum_{s \in S^P} price(s) q(s, p), \quad \forall p = P \qquad (2.28)$$

In the event that the demand for product is stipulated a priori, Constraint (2.29) is necessary and the objective function becomes minimization of makespan as shown in Constraint (2.30). Worthy of note is the fact that for makespan minimization problem, H is a variable instead of a parameter.

$$\sum_{s \in S^P} q(s, p) \geq demand(s), \quad \forall p = P \qquad (2.29)$$

$$\text{minimize } makespan = H \qquad (2.30)$$

2.5 ILLUSTRATIVE EXAMPLE

This literature example is taken from Shaik and Floudas (2009). Two products are produced from four raw materials. The facility has a heater, three reactors, a separator and two mixers. The heater and reactors are suitable for more than one task. Figure 2.1 shows a unit performing multiple tasks, multiple units suitable for a task, dedicated units for specific tasks and recycle of material which are common characteristics of multipurpose batch plants. The corresponding scheduling data for this example are shown in Tables 2.1 and 2.2. The price for product 1 and 2 is $5/m.u.

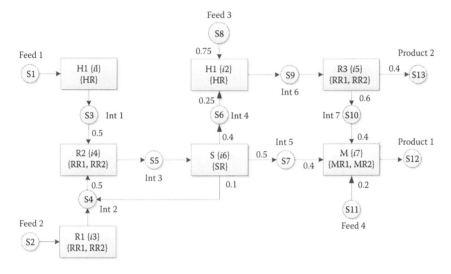

FIGURE 2.1   STN representation for the Example.

TABLE 2.1
Batch Size and Processing Time Data

Task	Lable i	Unit	Lable j	$\alpha(i, j)$	$\beta(i, j)$	$V_{i,j}^{L} - V_{i,j}^{U}$
			Example 1			
Heating-1	$i1$	Heater	HR	0.667	0.00667	0–100
Heating-2	$i2$	Heater	HR	1	0.01	0–100
Reaction-1	$i3$	Reactor-1	RR-1	1.333	0.01333	0–100
	$i3$	Reactor-2	RR-2	1.333	0.00889	0–150
Reaction-2	$i4$	Reactor-1	RR-1	0.667	0.00667	0–100
	$i4$	Reactor-2	RR-2	0.667	0.00445	0–150
Reaction-3	$i5$	Reactor-1	RR-1	1.333	0.0133	0–100
	$i5$	Reactor-2	RR-2	1.333	0.00889	0–150
Separation	$i6$	Separator	SR	2	0.00667	0–300
Mixing	$i7$	Mixer-1	MR-1	1.333	0.00667	20–200
	$i7$	Mixer-2	MR-2	1.333	0.00667	20–200

TABLE 2.2

Initial Inventory and Storage Size Data

States	Storage Capacity (m.u.)	Initial Inventory (m.u.)
1	UL	AA
2	UL	AA
3	100	AA
4	100	0
5	300	0
6	150	50
7	150	50
8	UL	0
9	150	0
10	150	0
11	UL	AA
12	UL	0
13	UL	0

2.5.1 RESULTS AND DISCUSSION

The computational statistics for the different models are given in Table 2.3. In maximization of profit for the time horizons of 8 and 12 h, all models get the same objective value. However, for the time horizons of 10 and 16 h the proposed model and the model by V.S (2013) give better objective values compared to other models. The improvement in objective values by the proposed model and V.S (2013) is attributed to the fact that these models rigorously align sequence constraints between consuming and producing tasks. So far, the recent model by V.S (2013) provides better objective values compared to other previous literature models based on continuous-time representation. However, the model size in terms of binary variables, continuous variables and constraints has increased considerably in order to rigorously align sequence constraints. V.S (2013) has pointed out that future work should address reducing the model size so that it can be easy to solve. With this motivation, this work reduced the model size without compromising the objective values. For example, for the time horizon of 16 h the proposed model reduced the binary variables required by 49.4%, continuous variables by 16.9% and constraints by 41.8% when compared to V.S (2013). It is difficult to compare the model statistics of this work with models by SLK2 (2010), V.S (2012) and S&M (2012) since these models do not rigorously align sequence constraints which lead to suboptimal results. As a consequence, for the rest of the examples in this paper the results obtained by this work are compared to the recent model by V.S (2013). The Gantt chart for the time horizon of 16 h for this example is given in Figure 2.2.

TABLE 2.3

Computational Results for Example 1

Model	Events	Binary Variables	Continuous Variables	Constraints	MILP	CPU Time (s)
			H = 8			
SLK2 (2010)	7	102	655	1017	1583.44	—
V.S (2012)	6	41	256	608	1583.44	—
S&M (2012)	5	107	418	889	1583.44	—
V.S (2013)	5	234	408	1209	1583.44	4
This work	5	125	354	726	1583.44	1.9
			H = 10			
SLK2 (2010)	9	136	853	1428	2337.36	—
V.S (2012)	9	137	457	1403	2345.3	—
S&M (2012)	6	131	516	1093	2345.3	—
V.S (2013)	7	358	576	1785	2358.2	1211.7
This work	7	185	486	1052	2358.2	894.9
			H = 12			
SLK2 (2010)	8	119	754	1249	3041.26	—
V.S (2012)	7	52	302	302	3041.26	—
S&M (2012)	7	155	604	1297	3041.26	—
V.S (2013)	7	358	576	1785	3041.26	0.9
This work	7	185	486	1052	3041.26	0.8
			H = 16			
SLK2 (2010)	11	170	1051	1786	4241.5	—
V.S (2012)	11	181	571	1755	4262.8	—
S&M (2012)	10	227	868	1909	4261.9	—
V.S (2013)	10	544	828	2649	4262.8	104.1
This work	10	275	684	1541	4262.8	95.9

(*Continued*)

TABLE 2.3 (*Continued*)
Computational Results for Example 1

Model	Events	Binary Variables	Continuous Variables	Constraints	MILP	CPU Time (s)
			$d(s12) = 100$, $d(s13) = 200$ m.u. and $M = 50$			
SLK2 (2010)	9	136	859	1448	13.366	—
V.S (2012)	9	74	394	977	13.366	—
S&M (2012)	7	155	605	1299	13.366	—
V.S (2013)	7	358	577	1644	13.366	0.8
This work	7	185	487	976	13.366	0.6
			$d(s12) = 250$, $d(s13) = 250$ m.u. and $M = 50$			
SLK2 (2010)	11	820	170	1057	17.025	—
V.S (2012)	10	159	514	1591	17.025	—
V.S (2013)	10	544	829	2442	17.025	1.8
This work	10	275	685	1429	17.025	0.9

Note: SLK2, Susarla et al.; V.S, Vooradi and Shaik; S&M, Seid and Majozi.

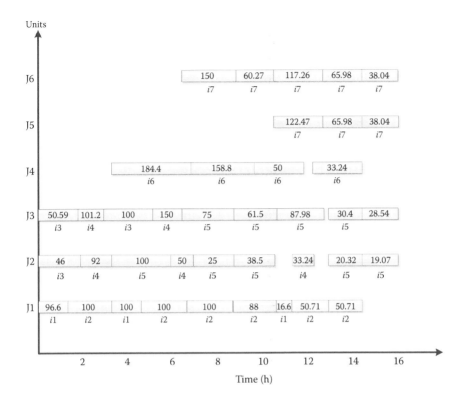

FIGURE 2.2 Gantt chart for the time horizon of 16 h.

2.6 CASE STUDIES

This section discusses case studies on scheduling and prediction of time points for multipurpose batch plants for the readers to have an in-depth understanding of the application of scheduling models.

2.6.1 CASE STUDY I

This case was first studied by Kondili et al. (1993), and it is one of the most common examples that appear in literature. The STN representation which illustrates the steps and sequences the raw materials should follow to produce two products is given in Figure 2.3. The scheduling data required to solve this problem is given in Tables 2.4 and 2.5. The price of products 1 and 2 is $10/m.u.

2.6.1.1 Results and Discussions

The computational results for this case study are given in Table 2.6. The proposed model considerably reduced the binary variables required by 38.9%–44% when it is compared to the literature model by V.S (2013) for the different scenarios considered. The continuous variables and constraints required by this work are also lesser

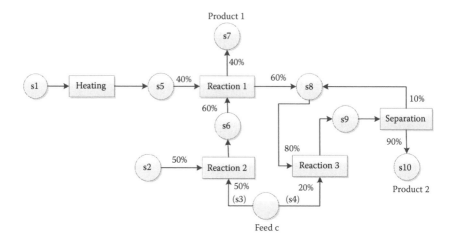

FIGURE 2.3   STN representation for Case Study I.

TABLE 2.4

Batch Size and Processing Time Data for Case Studies I and II

Task	Lable i	Unit	Lable j	$\alpha(i, j)$	$\beta(i, j)$	$V_{i,j}^L - V_{i,j}^U$
			Case Study I			
Heating	$i1$	Heater	HR	0.667	0.00667	0–100
Reaction-1	$i2$	Reactor-1	RR-1	1.334	0.02664	0–50
	$i2$	Reactor-2	RR-2	1.334	0.01665	0–80
Reaction-2	$i3$	Reactor-1	RR-1	1.334	0.02664	0–50
	$i3$	Reactor-2	RR-2	1.334	0.01665	0–80
Reaction-3	$i4$	Reactor-1	RR-1	0.667	0.01332	0–50
	$i4$	Reactor-2	RR-2	0.667	0.00833	0–80
Separation	$i5$	Separator	SR	1.3342	0.00666	0–200
			Case Study II			
Task-1	$i1$	Unit-1	U-1	2	0	0–260
Task-2	$i2$	Unit-2	U-2	3	0	0–140
	$i2$	Unit-3	U-3	2	0	0–120
Task-3	$i3$	Unit-4	U-4	3	0	0–120
	$i3$	Unit-5	U-5	2	0	0–140
Task-4	$i4$	Unit-1	U-1	1	0	0–1
	$i4$	Unit-2	U-2	3	0	0–1

TABLE 2.5

Initial Inventory and Storage Size Data for Case Studies I and II

States	Case Study I		Case Study II	
	Storage Capacity (m.u.)	Initial Inventory (m.u.)	Storage Capacity (m.u.)	Initial Inventory (m.u.)
1	UL	AA	UL	AA
2	UL	AA	10	0
3	UL	AA	10	0
4	100	0	UL	0
5	200	0	UL	0
6	150	0	—	—
7	200	0	—	—
8	UL	0	—	—
9	UL	0	—	—
10	—	—	—	—
11	—	—	—	—
12	—	—	—	—
13	—	—	—	—

by 12%–17% and 35%–70%, respectively, when compared to the literature model of V.S (2013) for the different scenarios considered. The proposed model solved the case study in much shorter CPU time. For the last scenario, the model is solved by allowing the processing unit to store its produced material in the next time points when the unit is idle. By doing so, a better optimal value is obtained.

2.6.2   CASE STUDY II

This example is taken from Vooradi and Shaik (2013) and presented in this work for comparison. The STN representation is depicted in Figure 2.4. Two products are produced from one raw material using different production paths. The necessary scheduling data for this example are given in Tables 2.1 and 2.2. The price of products 1 and 2 is $1/m.u.

2.6.2.1   Results and Discussion

The computational results for Case Study II are shown in Table 2.7. For the time horizon of 9 and 16 h, both models obtained the same optimal objective value. Even for this small problem considerable reduction in model size is obtained by the proposed model. For instance, for 16 h time horizon the binary variables, continuous variables and constraints required are reduced by 32.4%, 15.6%, 26.5%, respectively, using the proposed formulation, when compared to the literature model. The computation time required for both instances is lower when using the proposed model. Gantt chart that shows the type of task performed in each unit, the starting and finishing time for a task and batch size for the time horizon of 16 h, is depicted in Figure 2.5. It is worth

TABLE 2.6
Computational Results for Case Study I

Model	Events	Binary Variables	Continuous Variables	Constraints	MILP	CPU Time (s)
			H = 8			
V.S (2013)	4	108	231	615	1498.57	0.23
This work	4	66	203	374	1498.58	0.15
			H = 10			
V.S (2013)	5	146	292	806	1962.69	0.44
This work	5	86	251	485	1962.69	0.3
			H = 12			
V.S (2013)	6	184	353	997	2658.52	1.3
This work	6	106	299	596	2658.53	0.5
			H = 16			
V.S (2013)	8	260	475	1379	3738.38	42.1
This work	8	146	395	822	3738.39	21.8
			$d(s7) = d(s10) = 200$ m.u. and $M = 50$			
V.S (2013)	9	298	537	1436	19.34	5.6
This work	9	166	444	931	19.34	1.4
			$d(s7) = 500, d(s10) = 400$ m.u. and $M = 100$			
V.S (2013) ($\Delta n = 1$)	21	876	1387	4459	47.687	40,000
This work	21	490	1356	2493	47.6835	40,000

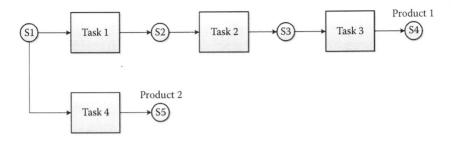

FIGURE 2.4 STN representation for Case Study II.

TABLE 2.7
Computational Results for Case Study II

Model	Events	Binary Variables	Continuous Variables	Constraints	MILP	CPU Time (s)
			$H = 9$			
V.S (2013)	7	147	265	691	386	14.4
This work	7	102	228	514	386	5.7
			$H = 16$			
V.S (2013)	10	222	379	1018	1115	292
This work	10	150	321	784	1115	211

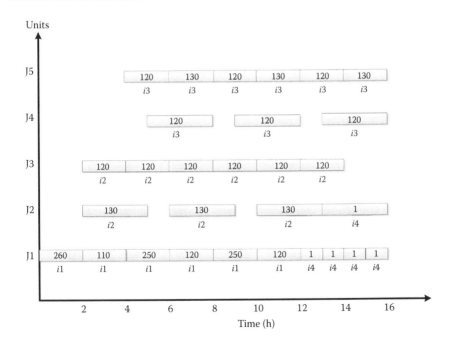

FIGURE 2.5 Gantt chart for the time horizon of 16 h for Case Study II.

to note that industrial scale scheduling problems require large variables and constraints; the reduction of the model size by the proposed technique will significantly improve the computational efficiency.

2.7 CONCLUSIONS

Compact MILP scheduling formulation for scheduling of multipurpose batch plants based on multi-grid time-point representation is presented in this work. The model is based on rigorous conditional sequencing between consuming and producing tasks while achieving smaller model size. Through illustrative examples, this work demonstrated that the proposed model leads to considerable reduction in terms of binary, continuous variables and constraints required, when compared to the recent rigorous conditional sequencing model that is based on unit-specific event points as published in literature.

NOMENCLATURE

SET

I	Tasks
I_{zw}	Tasks with zero wait
I_j	Tasks that can be performed in unit j
I_s^p	All tasks which produce material state s
I_s^c	All tasks which consume material state s
S	Any material state s
S^i	Intermediate state s
S^p	Product state s
S^r	Raw material state s
S^{cw}	Any state s which can wait in a unit after it is produced
S^{zw}	Zero wait state s
P	Event points
J	Processing units
J_i	Units suitable to execute task i
J_s^c	State s consuming units
J_s^p	State s producing units

PARAMETERS

$V_{i,j}^L$	Minimum batch size for task i in unit j
$V_{i,j}^U$	Maximum batch size for task i in unit j
V_j^U	Maximum batch size capacity for unit j
H	Time horizon
$Q_0(s)$	Initial amount available for material state s

$Q^U(s)$	Maximum storage capacity for material state s
$\alpha(i,j)$	Coefficient for fixed processing time for task i in unit j
$\beta(i,j)$	Coefficient for variable processing time for task i in unit j
$UW(i,j)$	Waiting time for task i in unit j after it is produced
$\theta(s,i)$	Stoichiometric coefficient of reactant
$\delta(s,i)$	Stoichiometric coefficient of product
$price(s)$	Selling price for product state s
$demand(s)$	Market demand for product state s

BINARY VARIABLES

$y(i,j,p)$ — Binary variable to denote task i performed in unit j at event point p

$t(j,j',p)$ — Binary variable to denote if material transfer between the consuming unit j' and producing unit j occurs

$x(j,p)$ — Binary variable to denote the amount of material produced by unit j is beyond the storage capacity and needs to be consumed immediately by the consuming units

POSITIVE CONTINUOUS VARIABLES

$mu(i,j,p)$ — Amount of batch size for task i performed in unit j at event point p

$q(s,p)$ — Amount of material state s stored at event point p

$tu(j,p)$ — Starting time of unit j at time event p

$tp(j,p)$ — Finishing time of unit j at time event p

$ma(s,j,p)$ — The amount of material state s produced by unit j is beyond the storage capacity and needs to be consumed by the consuming units immediately

$mex(s,j,j',p)$ — Amount of material state s sent from the producing unit j' to the consuming unit j

$Profit$ — Profit obtained form selling products

$mekspan$ — The total length of the schedule, i.e. when all the jobs have finished processing

REFERENCES

Dedopoulos, I.T., Shah, N., 1995. Optimal short-term scheduling of maintenance and production for multipurpose plants. *Industrial Engineering and Chemical Research.* 34, 192–201.

Ferrer-Nadal, S., Capón-Garćia, E., Méndez, C.A., Puigjaner, L., 2008. Material transfer operations in batch scheduling. A critical modelling issue. *Industrial Engineering and Chemical Research.* 47, 7721–7732.

Floudas, C.A., Lin, X., 2004. Continuous-time versus discrete-time approaches for scheduling of chemical processes: A review. *Computer and Chemical Engineering.* 28, 2109–2129.

Gupta, S., Karimi, I.A., 2003. Scheduling a two-stage multiproduct process with limited product shelf life in intermediate storage. *Industrial Engineering and Chemical Research.* 42, 490–508.

Harjunkoskia, I., Maravelias, C.T., Bongers, P., Castro, P.M., Engell, S., Grossmann, I.E., Hooker, J., 2014. Scope for industrial applications of production scheduling models and solution methods. *Computer and Chemical Engineering.* 62, 161–193.

Ierapetritou, M.G., Floudas, C.A., 1998. Effective continuous-time formulation for short-term scheduling: 1 Multipurpose batch processes. *Industrial Engineering and Chemical Research.* 37, 4341–4359.

Janak, S.L., Floudas, C.A., 2008. Improving unit-specific event based continuous-time approaches for batch processes: Integrality gap and task splitting. *Computer and Chemical Engineering.* 32, 913–955.

Janak, S.L., Lin, X., Floudas, C.A., 2004. Enhanced continuous-time unit-specific event-based formulation for short-term scheduling of multipurpose batch processes: Resource constraints and mixed storage policies. *Industrial Engineering and Chemical Research.* 43, 2516–2533.

Janak, S.L., Lin, X., Floudas, C.A., 2007. A new robust optimization approach for scheduling under uncertainty: II Uncertainty with known probability distribution. *Computer and Chemical Engineering.* 31, 171–195.

Kondili, E., Pantelides, C.C., Sargent, R.W.H., 1993. A general algorithm for short-term scheduling of batch operations I. MILP formulation. *Computer and Chemical Engineering.* 17, 211–227.

Kopanos, G.M., Laínez, J.M., Puigjaner, L., 2009. An efficient mixed-integer linear programming scheduling framework for addressing sequence-dependent setup issues in batch plants. *Industrial Engineering and Chemical Research.* 48, 6346–6357.

Kopanos, G.M., Méndez, C.A., Puigjaner, L., 2010. MIP-based decomposition strategies for large-scale scheduling problems in multiproduct multistage batch plants: A benchmark scheduling problem of the pharmaceutical industry. *European Journal of Operation Research.* 207, 644–655.

Liu, Y., Karimi, I.A., 2007. Novel continuous-time formulations for scheduling multi-stage batch plants with identical parallel units. *Computer and Chemical Engineering.* 31, 1671–1693.

Majozi, T., Zhu, X.X., 2001. A novel continuous-time MILP Formulation for multipurpose bach plants. *Industrial Engineering and Chemical Research.* 40, 5935–5949.

Maravelias, C.T., Grossmann, I.E., 2003. New general continuous-time state-task network formulation for short-term scheduling of multipurpose batch plants. *Industrial Engineering and Chemical Research.* 42, 3056–3074.

Méndez, C.A., Cerdá, J., 2000. Optimal scheduling of a resource-constrained multiproduct batch plant supplying intermediates to nearby end product facilities. *Computer and Chemical Engineering.* 2, 369–376.

Méndez, C.A., Cerdá, J., 2004. Short-term scheduling of multistage batch processes subject to limited finite resources. *Computer Aided Chemical Engineering.* 15B, 984–989.

Méndez, C.A., Henning, G.P., Cerdá, J., 2001. An MILP continuous-time approach to short-term scheduling of resource-constrained multistage flowshop batch facilities. *Computer and Chemical Engineering.* 25, 701–711.

Pinto, J., Grossmann, I.E., 1997. A logic based approach to scheduling problems with resource constraints. *Computer and Chemical Engineering.* 21, 801–818.

Schilling, G., Pantelides, C., 1996. A simple continuous-time process scheduling formulation and a novel solution algorithm. *Computer and Chemical Engineering.* 20, 1221–1226.

Seid, R., Majozi, T., 2012. A robust mathematical formulation for multipurpose batch plants. *Chemical Engineering Science.* 68, 36–53.

Shah, N., Pantelides, C.C., Sargent, R.W.H., 1993. A general algorithm for short-term scheduling of batch operations II Computational issues. *Computer and Chemical Engineering.* 17, 229–244.

Shaik, M.A., Floudas, C.A., 2008. Unit-specific event-based continuous-time approach for short-term scheduling of batch plants using RTN framework. *Computer and Chemical Engineering.* 32, 260–274.

Shaik, M.A., Floudas, C.A., 2009. Novel unified modeling approach for short term scheduling. *Industrial Engineering and Chemical Research.* 48, 2947–2964.

Shaik, M.A., Janak, S.L., Floudas, C.A., 2006. Continuous-time models for short-term scheduling of multipurpose batch plants: A comparative study. *Industrial Engineering and Chemical Research.* 45, 6190–4209.

Sundaramoorthy, A., Karimi, I.A., 2005. A simpler better slot-based continuous-time formulation for short-term scheduling in multipurpose batch plants. *Chemical Engineering Science.* 60, 2679–2702.

Sundaramoorthy, A., Maravelias, C.T., 2011. A general framework for process scheduling. *AICHE Journal.* 57, 695–710.

Susarla, N., Li, J., Karimi, I., 2010. A novel approach to scheduling of multipurpose batch plants using unit slots. *AICHE Journal.* 56, 1859–1879.

Velez, S., Maravelias, C.T., 2013. Mixed-integer programming model and tightening methods for scheduling in general chemical production environments. *Industrial Engineering and Chemical Research.* 52, 3407–3423.

Velez, S., Maravelias, C.T., 2014. Advances in mxed-integer programming methods for chemical production scheduling. *Annual Review of Chemical and Biomolecular Engineering.* 5, 97–121.

Vooradi, R., Shaik, M.A., 2012. Improved three-index unit-specific event-based model for short-term scheduling of batch plants. *Computer and Chemical Engineering.* 43, 148.

Vooradi, R., Shaik, M.A., 2013. Rigorous unit-specific event-based model for short-term scheduling of batch plants using conditional sequencing and unit-wait times. *Industrial Engineering and Chemical Research.* 52, 12950–12972.

Zhang, X., Sargent, R.W.H., 1996. The optimal operation of mixed production facilities-a general formulation and some approaches for the solution. *Computer and Chemical Engineering.* 20, 897–904.

3 Methods to Reduce Computational Time
Prediction of Time Points

3.1 INTRODUCTION

This chapter presents a mathematical technique for the prediction of the optimal number of time points in short-term scheduling of multipurpose batch plants. The mathematical formulation is based on the state sequence network (SSN) representation. The developed method is based on the principle that the optimal number of time points depends on how frequently the critical unit is used throughout the time horizon. In the context of this work, a critical unit refers to a unit that is most frequently used and it is active for most of the time points when it is compared to other units. A linear model is used to predict how many times the critical unit is used. In conjunction with knowledge of recipe, this information is used to determine the optimal number of time points. The statistical R-squared value obtained between the predicted and actual number of optimal time points in all the problems considered was 0.998, which suggests that the developed method is accurate in determining the optimal number of time points. Consequently this avoids costly computational times due to iterations. In the model by Majozi and Zhu (2001), the sequence constraint that pertains to tasks that consume and produce the same state, the starting time of the consuming task at time point p must be later than the finishing time of the producing task at the previous time point $p - 1$. This constraint is relaxed by the proposed models if the state is not used at the current time point p. This relaxation gives a better objective value as compared to previous models. An added feature of the proposed models is their ability to exactly handle fixed intermediate storage (FIS) operational philosophy, which has proven to be a subtle drawback in published scheduling techniques.

3.2 MOTIVATION

In the continuous-time representation, the optimal number of time points which gives the optimal objective value is found through iteration. This is done by increasing the number of time points at each iteration by one until the objective value converges. The objective value may not change with an increment of one additional time point, but may change with an increment of two or more. For example in Case Study I in this chapter, where duration constraints depend on batch size, for a time horizon of 36 h, the objective value for both time points 11 and 12 is 445.5. The iteration is stopped at this point giving an optimal objective value of 445.5. However, if the time points are increased by one, a better objective value of 447 is obtained, so that the optimal number of time points is not 11 but 13. This indicates that the iteration method of obtaining the optimal number of time points, where the criterion is to stop when the solution does not improve by adding one time point to the previous one, is subject to a suboptimal solutions and needs to be verified by further addition of time points.

Again, for Case Study I where duration is fixed, which is not dependent on batch size, for a time horizon of 168 h the objective value is 3,525 at time points 73 and requires more than 40,000 s of CPU time. At time points 74, the objective value is 3,550 and requires a CPU time of more than 40,000 s. At time points 75, the objective value is 3550. Increasing the number of time points beyond 74 does not improve the objective value. As a result, the optimal number of time points is 74. Consequently, a time horizon of 168 h needs 74 time points. In the iteration method of getting the optimal objective value, the CPU time required is the sum of the CPU times of each iteration. This becomes computationally costly as the time horizon increases. For complicated problems where each iteration takes a day, a number of days will be required to obtain the objective value which is not desirable for batch plants where it is usually a norm to schedule on a daily or weekly basis. Moreover, process shifts might necessitate a schedule revisit in the order of hours, thereby militating against this iterative procedure.

3.3   PROBLEM STATEMENT

In the scheduling of multipurpose batch plants, the following are given: (1) the production recipe that indicates the sequence of unit processes whereby the raw materials are changed into products, (2) the capacity of a unit and the type of tasks the unit can perform, (3) the maximum storage capacity for each material and (4) the time horizon of interest.

Using the given data, it is required to determine (1) the maximum achievable profit of the plant, (2) the minimum makespan if throughput is given and (3) a production schedule related to the optimal resource utilization.

3.4   MATHEMATICAL MODEL USING STATE
SEQUENCE NETWORK

It is important to explain the effective state in the SSN representation because it renders the opportunity to reduce the number of binary variables. If a process requires multiple raw materials to make a particular product, then it is a fact that if one of the raw materials is fed, then all the other feed materials also exist to make the product. By noting this, it is easy to see that only one of the states needs to be defined as an effective state. The choice of effective state is not unique; however, once the choice of effective state has been made, it should remain consistent throughout the formulation. Effective states are considered in defining the binary variables. The mathematical model presented in this chapter entails the following constraints. For more explanation on effective states and SSN representation, the reader can refer to the paper by Majozi and Zhu (2001). Presented here is the condensed mathematical formulation. Detailed formulation on the subject can be found in Chapter 2.

3.4.1 ALLOCATION CONSTRAINTS

$$\sum_{s_{in,j} \in S^*_{in,j}} y\left(s_{in,j}, p\right) \le 1, \quad \forall j \in J, \quad p \in P \tag{3.1}$$

3.4.2 CAPACITY CONSTRAINTS

$$V^L_{s_{in,j}} y\left(s_{in,j}, p\right) \le mu\left(s_{in,j}, p\right) \le V^U_{s_{in,j}} y\left(s_{in,j}, p\right), \quad \forall p \in P, \quad j \in J, \quad s_{in,j} \in S_{in,J} \tag{3.2}$$

3.4.3 MATERIAL BALANCE FOR STORAGE

$$q_s\left(s, p\right) = q_s\left(s, p-1\right) - \sum_{s_{in,j} \in S^{sc}_{in,J}} \rho^{sc}_{s_{in,j}} mu\left(s_{in,j}, p\right) + \sum_{s_{in,j} \in S^{sp}_{in,J}} \rho^{sp}_{s_{in,j}} mu\left(s_{in,j}, p-1\right),$$

$$\forall p \in P, \quad s \in S \tag{3.3}$$

$$q_s\left(s^p, p\right) = q_s\left(s^p, p-1\right) + \sum_{s_{in,j} \in s^{s^p p}_{in,J}} \rho^{sp}_{s_{in,j}} mu\left(s_{in,j}, p\right), \quad \forall p \in P, \quad s^p \in S^p \tag{3.4}$$

3.4.4 DURATION CONSTRAINTS

$$t_p\left(s_{in,j'}, p\right) \ge t_u\left(s_{in,j'}, p\right) + \tau\left(s_{in,j'}\right) y\left(s_{in,j'}, p\right),$$

$$\forall j \in J, \quad p \in P, \quad s_{in,j} \in S^{sp}_{in,J}, \quad s_{in,j'} \in S^{sc}_{in,J} \tag{3.5}$$

3.4.5 SEQUENCE CONSTRAINTS

The two subsections address the proper allocation of tasks in a given unit that ensures the starting time of a new task to be later than the finishing time of the previous task.

3.4.5.1 Same Task in Same Unit

$$t_u\left(s_{in,j}, p\right) \ge t_p\left(s_{in,j}, p-1\right), \quad \forall j \in J, \quad p \in P, \quad s_{in,j} \in S^*_{in,j} \tag{3.6}$$

3.4.5.2 Different Tasks in Same Unit

$$t_u\left(s_{in,j}, p\right) \ge t_p\left(s'_{in,j}, p-1\right), \quad \forall j \in J, \quad p \in P, \quad s_{in,j} \ne s'_{in,j}, \quad s_{in,j}, s'_{in,j}, \in S^*_{in,j} \tag{3.7}$$

$$t_p\left(s^{usp}_{in,j}, p-1\right) \ge t_u\left(s^{usc}_{in,j}, p\right) - H\left(1 - y\left(s^{usp}_{in,j}, p-1\right)\right),$$

$$\forall j \in J, \quad p \in P, \quad s^{usc}_{in,j} \in S^{usc}_{in,j}, \quad s^{usp}_{in,j} \in S^{usp}_{in,j} \tag{3.8}$$

3.4.6 DIFFERENT TASKS IN DIFFERENT UNITS

3.4.6.1 If an Intermediate State Is Produced from One Unit

$$\rho\left(s_{in,j}^{sp}\right)mu\left(s_{in,j},p-1\right)\leq qs\left(s,p\right)+V_{j}^{U}t\left(j,p\right) \quad \forall j\in J, \quad p\in P, \quad s_{in,j}\in S_{in,j}^{sp} \quad (3.9)$$

$$tu\left(s_{in,j'},p\right)\geq tp\left(s_{in,j},p\right)-H\left(\left(2-y\left(s_{in,j},p-1\right)-t\left(j,p\right)\right)\right),$$
$$\forall j\in J, \quad p\in P, \quad s_{in,j}\in S_{in,J}^{sp}, \quad s_{in,j'}\in S_{in,J}^{sc} \quad (3.10)$$

3.4.6.2 If an Intermediate State Is Produced from More than One Unit

$$\sum_{s_{in,j}\in S_{in,J}^{sc}}\rho_{s_{in,j}}^{sc}mu\left(s_{in,j},p\right)$$

$$\leq qs\left(s,p-1\right)+\sum_{s_{in,j}\in S_{in,J}^{sp}}\rho_{s_{in,j}}^{sp}mu\left(s_{in,j},p-1\right)t\left(j,p\right), \quad \forall j\in J, \quad p\in P \quad (3.11)$$

$$t_{u}\left(s_{in,j'},p\right)\geq t_{p}\left(s_{in,j},p-2\right)-H\left(1-y\left(s_{in,j},p-2\right)\right),$$
$$\forall j\in J, \quad p\in P, \quad s_{in,j}\in S_{in,J}^{sp}, \quad s_{in,j'}\in S_{in,J}^{sc} \quad (3.12)$$

3.4.7 SEQUENCE CONSTRAINT FOR FIXED INTERMEDIATE STORAGE POLICY

$$\sum_{s_{in,j}\in S_{in,J}^{sp}}\rho_{s_{in,j}}^{ps}mu\left(s_{in,j},p-1\right)+qs\left(s,p-1\right)\leq QS^{U}+\sum_{j\in J_{s}}V_{j}^{U}\left(1-x\left(s,p\right)\right)$$

$$\forall j\in J, \quad p\in P, \quad s\in S \quad (3.13)$$

$$tp\left(s_{in,j'},p-1\right)\leq tp\left(s_{in,j},p-1\right)+H\left(2-y\left(s_{in,j'},p\right)-y\left(s_{in,j},p-1\right)\right)+H\left(x\left(s,p\right)\right)$$
$$\forall j\in J, \quad p\in P, \quad s_{in,j}\in S_{in,J}^{sp}, \quad s_{in,j'}\in S_{in,J}^{sc} \quad (3.14)$$

3.4.8 STORAGE CONSTRAINTS

$$q_{s}\left(s,p\right)\leq QS^{U}+\sum_{s_{in,j}\in S_{in,J}^{sp}}u\left(s_{in,j},p\right) \quad \forall s\in S, \quad p\in P, \quad j\in J \quad (3.15)$$

$$u\left(s_{in,j},p\right)\leq\rho_{s_{in,j}}^{sp}mu\left(s_{in,j},p-1\right)+u\left(s_{in,j},p-1\right) \quad \forall p\in P, \quad j\in J \quad (3.16)$$

$$u\left(s_{in,j},p\right) \le V_j^U \left(1 - \sum_{s_{in,j} \in S_{in,j}^*} y\left(s_{in,j},p\right)\right) \quad \forall p \in P, \quad j \in J \tag{3.17}$$

3.4.9 TIME HORIZON CONSTRAINTS

$$t_u\left(s_{in,j},p\right) \le H, \quad \forall s_{in,j} \in S_{in,J}, \quad p \in P, \quad j \in J \tag{3.18}$$

$$t_p\left(s_{in,j},p\right) \le H, \quad \forall s_{in,j} \in S_{in,J}, \quad p \in P, \quad j \in J \tag{3.19}$$

3.4.10 CALCULATION OF THE TOTAL DURATION A UNIT IS ACTIVE WITHIN THE TIME HORIZON

The knowledge of this value, that is the total duration a unit is active over a time horizon, is fundamental in the prediction of the optimal number of time points. The following two sections show the calculation of this value in the case of fixed and variable duration constraints, respectively.

3.4.10.1 Duration Constraints Are Fixed

Constraint (3.20) is an important constraint which is used to predict optimal number of time points. The equation is used in the model when the model is solved as a linear programming maximization (LP MAX) and a linear programming minimization (LP MIN). In solving LP MAX and LP MIN, the binary variable $y(s_{in,j}, p)$ is considered as a continuous variable with a value between 0 and 1. Constraint (3.20) expresses that the total duration a unit is active is limited by the time horizon, the sum of mean processing times of previous tasks and the sum of mean processing times of all subsequent tasks leading to a product as prescribed by the recipe.

$$\sum_p \sum_{s_{in,j}} \tau\left(s_{in,j}\right) y\left(s_{in,j},p\right) \le H - \sum_{s_{in,j'} \in S_{in,J}^{PT}} \tau\left(s_{in,j'}\right) - \sum_{s_{in,j'} \in S_{in,J}^{ST}} \tau\left(s_{in,j'}\right),$$

$$\forall j \in J, \quad p \in P, \quad S_{in,j} \ne s_{in,j'} \tag{3.20}$$

In the beginning, the parameter $\sum_{s_{in,j'} \in S_{in,J}^{PT}} \tau(s_{in,j'})$ caters for the situation where the unit j must wait for all the previous tasks to finish before it can start processing. If there is an intermediate at the beginning of the time horizon so that unit j can start its operation, then the parameter becomes 0. The parameter $\sum_{s_{in,j'} \in S_{in,J}^{ST}} \tau(s_{in,j'})$ states that at the end of its operation unit j gives its intermediate to the subsequent tasks which lead to a product (the shortest path of subsequent tasks lead to a product), so that the unit is not active at this time. If the unit is *the one producing a product*, then the term becomes 0; this assumption is used not to underestimate time points because of underutilization of a unit. Constraint (3.20) tightens the model better than one simply stating that the sum of the durations of all tasks suitable in each unit should be less than the scheduling time horizon without loss of any generality.

3.4.10.2   Duration Constraints Dependent on Batch Size

Constraint (3.21) is similar to (3.20), but for the situation where duration is a function of batch size.

$$\sum_{p}\sum_{s_{in,j}}\tau\left(s_{in,j}\right)y\left(s_{in,j},p\right)+\beta\left(s_{in,j}\right)mu\left(s_{in,j},p\right)\le H-\sum_{s_{in,j'}\in S_{in,J}^{PT}}\tau\left(s_{in,j'}\right)-\sum_{s_{in,j''}\in S_{in,J}^{ST}}\tau\left(s_{in,j'}\right),$$

$$\forall j \in J, \quad p \in P, \quad s_{in,j} \ne s_{in,j'} \tag{3.21}$$

3.4.11   Objective Function

Constraints (3.1) through (3.8), (3.10), (3.13) through (3.15), (3.18) through (3.21) together with the objective function given in Constraint (3.22) constitute LP MAX, and LP MIN comprises of Constraints (3.1) through (3.8), (3.13) through (3.15), (3.18) through (3.21), (3.20) through (3.24). Sequence Constraint (3.10) and (3.15) for LP MAX and LP MIN can be expressed as

$$tu\left(s_{in,j'},p\right)\ge tp\left(s_{in,j},p-1\right)-H\left(1-y\left(s_{in,j},p-1\right)\right)$$

$$\forall j \in J, \quad p \in P, \quad s_{in,j} \in S_{in,J}^{sp}, \quad s_{in,j'} \in S_{in,J}^{sc} \tag{3.10'}$$

$$q_s\left(s,p\right)\le QS^{\max} \quad \forall s \in S, \quad p \in P \tag{3.15'}$$

Model ML2 constitutes Constraints (3.1) through (3.22). For the model ML2, Constraints (3.20) and (3.21) are expressed as

$$\sum_{p}\sum_{s_{in,j}\in S_{in,j}^{*}}\tau\left(s_{in,j}\right)y\left(s_{in,j},p\right)\le H, \quad \forall j \in J, \quad p \in P \tag{3.20'}$$

$$\sum_{p}\sum_{s_{in,j}\in S_{in,j}^{*}}\tau\left(s_{in,j}\right)y\left(s_{in,j},p\right)+\beta\left(s_{in,j}\right)mu\left(s_{in,j},p\right)\le H, \quad \forall j \in J, \quad p \in P \tag{3.21'}$$

The objective of the scheduling problem is to maximize the product throughput or to minimize the makespan, as given in Constraint (3.22).

$$\text{maximize}\sum_{s}\text{price}\left(s^{p}\right)qs\left(s^{p},p\right), \quad \forall p = P, \quad s^{p} \in S^{p}$$

or

$$\text{minimize } H \tag{3.22}$$

Worthy of mention is that the number of time points is set at $\left\lceil \dfrac{H}{\tau^L} \right\rceil$, where τ^L is the minimum constant term of processing time among the given tasks, when solving LP MAX and LP MIN. Also, the choice of either Constraint (3.20) or Constraint (3.21) is dependent on whether task durations are fixed or allowed to vary with batch size. Nevertheless, to emphasize, this formulation will entail a large number of variables due to the excessive number of time points than required when solving the model as mixed integer linear programming (MILP). However, this need not place any computational burden on the solution since the formulation is LP.

3.4.12 Prediction of Optimal Number of Time Points

Figure 3.1 shows a procedure used to solve the sequence of scheduling problems necessary for the prediction of time points. The first step is to solve the model as a LP MAX problem using the maximum number of time points calculated with $\left\lceil \dfrac{H}{\tau^L} \right\rceil$. In this step, the maximum achievable throughput that is very close to the dual objective value at the optimal number of time points is determined. The number of time points each unit can be active in the time horizon is also determined. Equation 3.23 is used to determine the total time points required using the frequency of utilization of unit j obtained from the LP MAX problem. The unit which gives the maximum number of time points is the unit most frequently used in the time horizon. This becomes the *critical unit*; therefore, the optimal number of time points is dependent on the frequency of utilization of this unit. In essence, this particular observation is the cornerstone of this contribution.

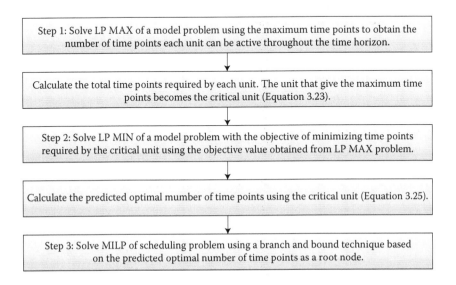

FIGURE 3.1   Procedure to solve scheduling problem.

The first term on the right of Equation 3.23 $\sum_{s_{in,j} \in S^*_{in,j}} \sum_p y\left(s_{in,j}, p\right)$ is used to determine how many times a unit j starts processing throughout the time horizon of interest.

$$p_j = \left\lceil \sum_{s_{in,j} \in S^*_{in,j}} \sum_p y\left(s_{in,j}, p\right) + \sum_{i \in I^{PT}} i + \sum_{i \in I^{ST}} i \right\rceil, \quad j \in J, \quad p \in P \tag{3.23}$$

The term $\sum_{i \in I^{PT}} i$ is to cater for those time points not shared by a unit because at the beginning, a unit must wait until all the previous tasks have been processed. If there is an intermediate initially in storage to start a task at the beginning of the time horizon, then $\sum_{i \in I^{PT}} i = 0$. Unit j at the end of its operation passes its intermediate to the subsequent units (that lead to a product). The time points associated with the subsequent tasks can be calculated with $\sum_{i \in I^{ST}} i$; these time points are not shared by unit j. If the unit j is a unit producing a product, then the term $\sum_{i \in I^{ST}} i$ becomes 0; this assumption prevents the overestimation of the predicted optimal number of time points.

The second step is to minimize the number of time points the critical unit is active, as given in Constraint (3.24). This is done by fixing the objective value and number of time points for each unit other than the critical unit to the values obtained from LP MAX. The model used in this step, that is, LP MIN, comprises of Constraints (3.1) through (3.8), (3.13) through (3.15), (3.18) through (3.24). The term j^c represents the critical unit. The predicted optimal number of time points which is very close to the actual optimal number of time points is then given in Equation 3.25. Equation 3.23 for LP MIN can be expressed as

$$z_j = \left\lceil \sum_{s_{in,j} \in S^*_{in,j}} \sum_p y\left(s_{in,j}, p\right) \right\rceil, \quad p \in P, \quad j \in J \tag{3.23'}$$

$$\text{Minimize} \sum_{s_{in,j^c}} \sum_p y\left(s_{in,j^c}, p\right) \tag{3.24}$$

$$p^{pre} = \left\lceil \sum_{s_{in,j^c}} \sum_p y(s_{in,j^c}, p) + \sum_{i \in I^{PT}} i + \sum_{i \in I^{ST}} i \right\rceil \tag{3.25}$$

The third step is to solve the exact MILP problem using the branch and bound strategy around the predicted optimal number of time points. Solving the MILP problem starts at the root node p^{pre}, the left node is a node that is smaller than p^{pre} and the right node is a node that is larger than p^{pre}. After solving the MILP problem at the predicted optimal number of time points, the iteration starts at the immediate node $p^{pre} - 1$. If the optimal objective value at $p^{pre} - 1$ is worse than that of p^{pre}, it is not necessary to continue any time point lower than $p^{pre} - 1$. Otherwise, continue solving until the objective value becomes worse. If the objective value at $p^{pre} - 1$ is worse than at p^{pre}, then start solving the right node at $p^{pre} + 1$. If the objective value at $p^{pre} + 1$ is better continue solving $p^{pre} + 2...$ until the solution converges. Otherwise, stop the iteration.

3.5 CASE STUDIES

In order to illustrate this method for predicting the optimal number of time points, four case studies found in published literature are presented. The results are given under each case study and discussed. All the results were obtained using GAMS CPLEX 9.1.2 in a 2.4 GHz, 4GB of RAM, Acer TravelMate 5740G computer. The prediction method is applied to the model ML2 which gives the lowest time points and better objective values when compared to the other model.

3.5.1 CASE STUDY I: SIMPLE BATCH PLANT INVOLVING THREE TASKS

This case study was conducted by Ierapetritou and Floudas (1998a) and Majozi and Zhu (2001). The flowsheet of this well-published literature example is given in Figure 3.2. The processed material passes through mixing, reaction and purification as it is converted to product. The STN and SSN representations are shown in Figure 3.3. The parameters for Case Study I are given in Table 3.1.

3.5.1.1 Results and Discussions

The computational statistics that show the optimal objective value, number of constraints required and the total number of continuous and binary variables required by the model for different time horizons are given in Table 3.2. The duration constraints in the model are fixed. In the iteration method, a lower time point is chosen first and then increased by one until no change in the objective value is observed. For a time horizon of 12 h, the iteration was started at time points 2. The objective value was 0. The time points were increased by one to time points 3 and a better objective value of 50 was recorded. This process of increasing the number of time points by one was continued and the objective value of 100 did not change after time points 4. Hence, the optimal objective value was 100. The CPU time required to solve the problem

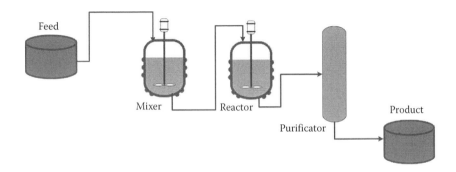

FIGURE 3.2   Flowsheet for Case Study I.

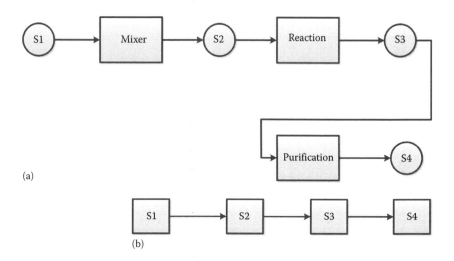

(a)

(b)

FIGURE 3.3 STN (a) and SSN (b) representations of Case Study I.

TABLE 3.1

Given Parameters for Case Study I

Unit	Label j	Capacity	Suitability	$\tau\left(s_{in,j}\right)$	$\beta\left(s_{in,j}\right)$
Heater		100	Mixing	3 for batch dependent	0.03
				4.5 for fixed duration	
Reactor		75	Reaction	2 for batch dependent	0.0266
				3 for fixed duration	
Separator		50	Purification	1 for batch dependent	0.02
				1.5 for fixed duration	

State	Storage Capacity	Initial Amount	Price
1	Unlimited	Unlimited	0
2	100	0	0
3	100	0	0
4	Unlimited	0	1

---

consists not only of the CPU time required at optimal number of time points 4, but is the sum of the CPU times of each iteration from time points 2 to 5, which is (0.07 + 0.036 + 0.086 + 0.051) s.

For a time horizon of 24 h, a similar procedure was followed until the solution converged. The optimal objective value of 350 was obtained at time points 9 (after which there was no improvement in value). For the time horizon of 24 h, the model consisted of 351 constraints, 215 continuous variables and 61 binary variables at the optimal number of time points. The model was solved to 0% relative

TABLE 3.2

Results Found for Case Study I under Maximization of Profit (Fixed Duration)

Model	p	CPU Time (s)	Nodes	RMILP	MILP	B.V	C.V	Constraints	Non-Zeros	Relative Gap (%)
					$(H = 12)$					
ML2	2	0.07	0	0	0	12	47	78	163	
ML2	3	0.036	0	50	50	19	71	117	267	—
ML2	4	0.086	2	100	100	26	95	156	365	—
ML2	5	0.051	10	150	100	33	119	195	467	
					$(H = 24)$					
ML2	8	0.01	0	300	300	54	191	312	787	
ML2	9	0.08	10	350	350	61	215	351	875	—
ML2	10	0.5	1301	400	350	68	239	390	977	—
					$(H = 36)$					
ML2	14	0.12	25	600	600	96	335	546	1385	—
ML2	15	42.5	92,785	650	625	103	359	585	1487	—
ML2	16	246.7	454,229	700	625	110	383	624	1589	—
					$(H = 168)$					
ML2	73	40,000	2,054,460	3550	3525	509	1751	2847	7403	0.7
ML2	74	40,000	2,436,601	3600	3550	516	1775	2886	7607	1.4
ML2	75	40,000	1,903,820	3650	3550	523	1799	2925	7705	2.8

gap and required a CPU time of 0.08 s. The model was also solved for a time horizon of 168 h to see how the computational time increased with this time horizon. The problem required 75 iterations to solve. The optimal number of time points was 74 with an objective value of 3550. The relative gap does not close in the specified CPU time of 40,000 s. It can be observed, as is expected, that the CPU time required in solving the problem increases exponentially as the number of time points increases. This is due to an increasing number of binary variables. Evidently, the iteration method for obtaining the optimal objective value is computationally costly, which renders it impractical as the time horizon for scheduling problems increases. This necessitates prediction of the optimal number of time points beforehand.

The results for this literature example obtained from the prediction method are explained here in detail. For the first scenario, which is a time horizon of 12 h, the first step is to solve a LP MAX problem to maximize the throughput. The time points for the LP model can be calculated with $\left\lceil \dfrac{H}{\tau^L} \right\rceil$ (12 h/1.5 h) to be 8. From Table 3.3, it can be seen that the LP MAX problem for the time horizon of 12 h resulted in an objective value of 150. The value of 1.5 for heater implies that the number of times the heater can be utilized is two (rounded up to the nearest integer). The number of times the reactor can be used is two and the number of times the separator can be used is three. Equation 3.23 was used to determine the time points required by using the frequency of utilization of each unit obtained from the LP MAX problem. For the mixer it becomes $p_1 = \left\lceil \sum_p y(s_1, p) + \sum_{i \in I^{PT}} i + \sum_{i \in I^{ST}} i = 1.5 + 0 + 2 \right\rceil = 4$. The parameter $\sum_{i \in I^{PT}} i$ in this equation is zero because there is no task before the mixer that is producing an intermediate. The parameter $\sum_{i \in I^{ST}} i$ in this equation is two because at the end of the process mixing, the mixer passes its intermediate to the reactor and separator to be converted into product. There are two time points associated with the reactor and separator after the end of the process by the mixer. Applying the same thought as that of p_1, the time points required by the reactor and separator were calculated.

TABLE 3.3

Computational Results for LP MAX and LP MIN for Case Study I under Maximization of Profit (Duration Constraints Are Fixed)

H	Obj. Value	Heater	Reactor	Separator	CPU Time (s) for LP MAX	Separator after LP MIN	CPU Time for LP MIN	p^{pre}
12	150	1.5	2	3	0.015	3	0.015	5
24	433.3	4.33	5.8	8.7	0.015	8.7	0.031	11
36	700	7	9.33	14	0.031	14	0.046	16
168	3633.3	36.3	48.4	72.7	0.33	72.7	0.34	75

$$p_2 = \left\lceil \sum_p y(s_2, p) + \sum_{i \in I^{PT}} i + \sum_{i \in I^{ST}} i = 2 + 1 + 1 \right\rceil = 4 \quad \text{and}$$

$$p_3 = \left\lceil \sum_p y(s_3, p) + \sum_{i \in I^{PT}} i + \sum_{i \in I^{ST}} i = 3 + 2 + 0 \right\rceil = 5$$

The maximum number of time points obtained from the LP MAX problem is, therefore, 5. Consequently, the separator becomes the critical unit.

The second step is to solve the LP MIN problem by setting the objective value to 150, $\sum_p y(s_1, p) = 1.5$, $\sum_p y(s_2, p) = 2.0$ and minimizing $\sum_p y(s_3, p)$, to obtain a tighter solution for the number of time points the critical unit is used throughout the time horizon. From the minimization problem, it was found that the critical unit is used three times. Equation 3.25 was used to predict the optimal number of time points.

$$p^{pre} = \left\lceil \sum_p y(s_3, p) + \sum_{i \in I^{PT}} i + \sum_{i \in I^{ST}} i + 1 = 3 + 2 + 0 = 5, \right.$$

which is close to the actual optimal number of time points, that is, 4.

For the time horizon of 36 h, the first step of the LP MAX problem gave an objective value of 700. Using Equation 3.23, $\sum_p y(s_1, p) = 7$, $\sum_p y(s_2, p) = 9.33$ and $\sum_p y(s_3, p) = 14$, the value of p_1 was calculated to be 9 (7 + 0 + 2), p_2 was calculated to be $12 = \lceil 9.33 + 1 + 1 \rceil$ and p_3 was calculated to be 16 (14 + 2 + 0). From these values, the maximum of p_j (highest integer value) is 16; hence, the separator is still the critical unit.

The second step is to solve the LP MIN using the objective value (700) and the number of time points each unit can be active, obtained from the LP MAX problem to minimize $\sum_p y(s_3, p)$. $\sum_p y(s_3, p)$ was found to be 14 implying that the critical unit is used 14 times. Using Equation 3.25, the predicted optimal number of time p^{pre} was 16 (14 + 2 + 0), which is very close to the actual optimal number of time points, 15.

For the time horizon of 168 h, the first step of LP MAX problem gave an objective value of 3633.3. Using Equation 3.23, $\sum_p y(s_1, p) = 36.3$, $\sum_p y(s_2, p) = 48.4$ and $\sum_p y(s_3, p) = 72.7$, the value of p_1 was calculated to be 39 $\lceil 36.3 + 0 + 2 \rceil$, p_2 was calculated to be 51 $\lceil 48.4 + 1 + 1 \rceil$ and p_3 was calculated to be 75 $\lceil 72.7 + 2 + 0 \rceil$. From these values, the maximum of p_j (highest integer value) is 75; hence the separator is the critical unit.

The second step is to solve the LP MIN using the objective value (3633.3) and the number of time points each unit can be active, obtained from the LP MAX problem to minimize $\sum_p y(s_3, p)$. $\sum_p y(s_3, p)$ was found to be 72.7 implying that the critical unit is used 73 times (rounded up to the nearest integer). Using Equation 3.25, the predicted optimal number of time points p^{pre} was $\lceil 72.7 + 2 + 0 \rceil = 75$, which is very

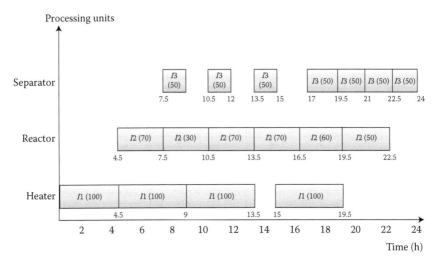

FIGURE 3.4   Gantt chart for time horizon of 24 h.

close to the actual optimal number of time points, 74. It is also interesting to see that the CPU time in solving both LP MAX and LP MIN was 0.67 (0.33 + 0.34), which is less than 1 s. This implies that the method is also accurate in predicting over long time horizon in few seconds of processing time, which makes the prediction method more valuable in avoiding computational costly iterations, especially as the time horizon increases. The Gantt chart which shows the amount of material used at each time by a task as well as the starting and finishing time of a task for a time horizon of 24 h is given in Figure 3.4.

The computational results obtained for Case Study I by modelling duration constraints as a function of batch size are given in Table 3.4. The problem was solved for time horizons between 12 and 168 h. As shown in Table 3.4, for a 12 h time horizon, an optimal objective value of 71.52 was obtained at time points 4. For the time horizon of 36 h, the iteration was started at time points 11. It is interesting to see that if the iteration principle is followed, the iteration for this time horizon was stopped at time points 12, since the objective value at time points 11 and 12 is 445.5. However, for this time horizon, the optimal objective value was found at time points 13 with an optimal objective value of 447. This indicates that the criteria of stopping the iteration when the objective value are the same as the previous time points is subjected to suboptimal results. For the time horizon of 168 h, the model gave an optimal objective value of 2645 using optimal number of time points 56. For this time horizon, the problem is solved to 2% relative gap in a specified CPU time of 40,000 s.

The proposed model used to predict the optimal number of time points was also tested for this case study. Table 3.5 shows the computational results for both the LP MAX and LP MIN problems for different time horizons. For example, for a time horizon of 36 h, the model reveals that the critical unit was the separator and is used 11 times throughout the time horizon, requiring 11 time points. At the beginning of the time horizon, the separator accepts its intermediate after it is processed from

TABLE 3.4

Computational Results for Case Study I under Maximization of Profit (Duration Constraints Dependent on Batch Size)

Model	p	CPU Time (s)	Nodes	RMILP	MILP	B.V	C.V	Constraints	Non-Zeros	Relative Gap (%)
					(H = 12)					
ML2	2	0	0	0	0	12	47	78	173	—
ML2	3	0.062	0	50	50	19	71	117	281	—
ML2	4	0.093	10	100	71.52	26	95	156	389	—
ML2	5	0.23	283	150	71.52	33	119	195	497	—
					(H = 24)					
ML2	7	0.328	252	250	245.5	47	167	273	713	
ML2	8	1.5	1,932	300	250.2	54	191	312	821	—
ML2	9	33.7	48,438	350	250.2	61	215	351	929	—
					(H = 36)					
ML2	11	15	13,938	450	445.5	75	263	429	1145	—
ML2	12	303	303,937	500	445.5	82	287	468	1253	—
ML2	13	1,549	1,366,014	550	447	89	311	507	1361	—
ML2	14	40,000	3,145,705	600	447	96	335	546	1469	—
					(H = 168)					
ML2	56	40,000	6,764,693	2700	2645.5	390	1343	2184	6005	2
ML2	57	40,000	5,637,699	2750	2645.5	397	1367	2223	6113	3.9

TABLE 3.5

Computational Results for LP MAX and LP MIN for Case Study I (Duration Constraints Dependent on Batch Size)

H	Obj. Value	Heater	Reactor	Separator	CPU Time (s) for LP MAX	Separator after LP MIN	CPU Time (s) for LP MIN	p^{pre}
12	150	1.5	2	3	0.093	3	0.015	5
24	350	3.5	4.7	7	0.098	7	0.14	9
36	550	5.5	7	11	0.12	11	0.14	13
168	2750	27.5	36.7	55	0.72	55	0.73	57

the mixer and reactor, which require two time points not shared by the separator. The predicted optimal number of time points then becomes 13, which is the same as the actual value 13. For the time horizon of 168 h, the model predicts 57 time points, which is again close to the actual optimal number of time points 56 in a CPU time of 1.45 s. The proposed method predicts the optimal number of time points, which is almost the same as the actual optimal number of time points, even for long time horizon, proving that it is very useful in avoiding a large number of iterations, which is encountered as the time horizon increases. It is difficult to compare the CPU time required by the proposed method with that of Li and Floudas (2010) since they employed iteration method to determine the maximum number of event points.

3.5.2 CASE STUDY II: BATCH PLANT WITH A RECYCLE STREAM

At first this case study was conducted by Kondili et al. (1993), and it becomes one of the most common examples that appeared in published literature. The flowsheet of this well-published literature example is given in Figure 3.5. This batch plant produces two different products sharing the same processing units. The unit operations consist of preheating, three different reactions and separation. The STN and SSN representations of the flowsheet are shown in Figure 3.6. The given parameters for this case study are given in Table 3.6.

3.5.2.1 Results and Discussions

Table 3.7 shows the results for Case Study II. For a time horizon of 8 h, the problem was solved with a 0% relative gap, giving an optimal objective value of 1498.6 with the corresponding optimal number of time points 4. The proposed method also predicts the same time points, which is 4. The time required to solve the problem was less than 1 s for each iteration. As the time horizon increased, the binary variables also increased resulting in an increase in CPU time. For the time horizon of 16 h, the model requires 8 time points to give an optimal objective value of 3738.4. As it is seen from Table 3.7, a time horizon 16 h requires CPU time of 79.7 s, which is high when compared to time horizon of 8 h with a CPU time of 0.156 s. This indicates that

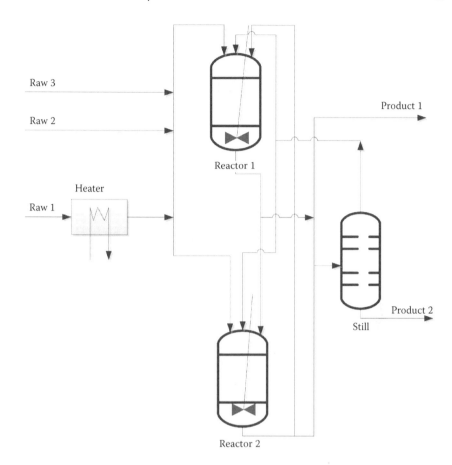

FIGURE 3.5 Flowsheet for Case Study II.

as the time horizon increases, CPU time required increases drastically, even worse
if iteration method is followed.

Table 3.8 shows the computational results of the method used to predict the optimal number of time points. For a time horizon of 8 h, the first step was to solve the
LP MAX problem using the determined maximum number of time points, which was
$\left\lceil \dfrac{8}{0.667} \right\rceil = 12$. The problem was solved in less than 1 s giving an objective value of
2252.6. From Table 3.8, the number of times reactor 1 can be used was 4 throughout
the 8 h time horizon of interest. Similarly, the number of times reactor 2 can be
used was 4. Using Equation 3.23, the time points required using reactor 2 was 4. The
term $\sum_{i \in I^{PT}} i$, in Equation 3.23, for reactor 2 becomes 0 since there is no task before
reactor 2. Since reactor 2 is a unit producing a product and also a unit producing an
intermediate used by the separator, the term $\sum_{i \in I^{ST}} i$ becomes 0; this assumption is
used not to overestimate the number of time points required even if there is a subsequent task separator.

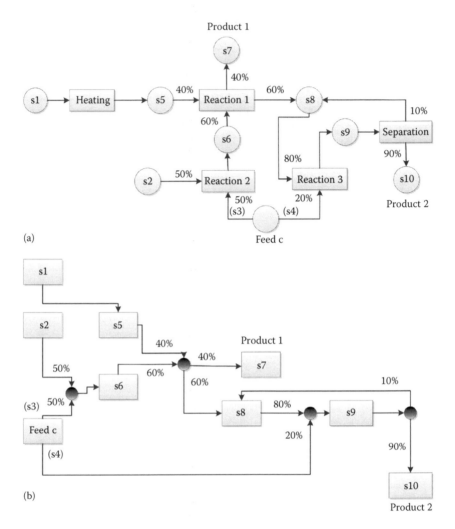

(a)

(b)

FIGURE 3.6   (a) STN and (b) SSN representations of Case Study II.

The second step of LP MIN also gave the same result for the number of times the critical unit (R2) was used as that of LP MAX, which is 4. Using Equation 3.25, the predicted optimal number of time points then becomes 4, which is the same as the actual optimal number of time points, i.e. 4. For the time horizon of 10 and 12 h, the model predicts 5 and 6 time points, respectively, which is close to the actual optimal number of time points 6 for 10 h and 7 for 12 h. From Table 3.8, the results obtained using the proposed method to predict the optimal number of time points, the predicted optimal number of time points obtained almost matched the actual optimal number of time points, indicating that the method is rigorous. Figure 3.7 shows the resultant Gantt chart for a time horizon of 16 h.

TABLE 3.6

Given Data for Case Study II

Task	$V_{s_{in,j}}^L$ $V_{s_{in,j}}^U$	Unit	$\tau\left(s_{in,j}\right)$	$\beta\left(s_{in,j}\right)$
Heating (i1)	0–100	Heater (HR)	0.667	0.00667
Reaction 1 (i2)	0–50	Reactor 1 (RR-1)	1.334	0.02664
Reaction 2 (i3)	0–50	Reactor 1 (RR-1)	1.334	0.02664
Reaction 3 (i4)	0–50	Reactor 1 (RR-1)	0.667	0.01332
Reaction 1 (i2)	0–80	Reactor 2 (RR-2)	1.334	0.01665
Reaction 2 (i3)	0–80	Reactor 2 (RR-2)	1.334	0.01665
Reaction 3 (i4)	0–80	Reactor 2 (RR-2)	0.667	0.00833
Separation (i5)	0–200	Separator (SR)	1.3342	0.00666
State	**Storage Capacity**	**Initial Amount**	**Price**	
s2	Unlimited	Unlimited	0	
s3,s4	Unlimited	Unlimited	0	
s5	100	0	0	
s8	200	0	0	
s6	150	0	0	
s9	200	0	0	
s7	Unlimited	0	10	
s10	Unlimited	0	10	

3.5.3 CASE STUDY III: MULTIPURPOSE BATCH PLANT WITH SIX UNITS AND EIGHT TASKS

At first Sundaramoorthy and Karimi (2005) conducted this case study, which is relatively a more complex problem, and later often used in literature to check the efficiency of models in terms of optimal objective value and CPU time required to get optimal objective value. The plant accommodates many common features of multipurpose batch plants such as unit performing multiple tasks, multiple units suitable for a task and dedicated unit for specific tasks. The STN and SSN representations of this case study are depicted in Figure 3.8. The data required for the case study are given in Table 3.9. The problem is solved for 8, 10, 12 and 16 h.

3.5.3.1 Results and Discussions

The proposed model for the time horizon of 8 h requires 5 time points to get the optimal objective value of 1583.4 in a CPU time of 1.25 s, as shown in Table 3.10. The proposed method used to predict optimal number of time points predicted 5 time points which is exactly the same as the actual optimal number of time points. For the time horizon of 10 h requires 6 time points to get the optimal objective value of 2345.3 in a CPU time of 8 s. For this case also the prediction method predicted the same number of time points as that of the actual number of time points 6. For the time horizon of 16 h, an optimal objective value of 4261.9 was obtained in a specified CPU time of 10,000 s using the optimal number of time points 10. The prediction

TABLE 3.7

Computational Results of Case Study II under Maximization of Profit

Model	P	RMIP	MIP	Constraints	C.V	B.V	Non-Zeros	CPU Time (s)
				(H = 8)				
ML2	3	975	866.67	382	206	48	1075	0.093
ML2	4	1730.6	1498.6	528	275	64	1518	0.156
ML2	5	2123.3	1498.6	676	344	80	1991	0.608
				(H = 10)				
ML2	5	2436.7	1917	674	344	80	1961	0.327
ML2	6	2730.7	1962.7	908	403	96	2548	11.3
ML2	7	2780.2	1962.7	1054	470	112	2985	816
				(H = 12)				
ML2	6	3076.6	2646.8	908	403	96	2548	1.36
ML2	7	3301	2658.5	1054	470	112	2985	35.4
ML2	8	3350.5	2658.5	1200	537	128	3422	1255.3
				(H = 16)				
ML2	7	3799.6	3638.7	1054	470	112	2985	0.84
ML2	8	4291.7	3738.38	1200	537	128	3422	79.7
ML2	9	4439	3738.38	1258	620	144	3733	887

TABLE 3.8

Computational Results of LP MAX and LP MIN for Case Study II

H	Obj. Value	Reactor 1 (RR-1)	Reactor 2 (RR-2)	Separator (SR)	Heater (HR)	CPU Time (s) for LP MAX	CPU Time (s) for LP MIN	Reactor 2 after LP MIN	P^{pre}
8	2252.6	3.52	3.8	0.82	0.76	0.031	0.05	3.8	4
10	2815.8	4.4	4.8	1.03	0.96	0.031	0.05	4.8	5
12	3379	5.33	5.64	1.23	1.15	0.062	0.08	5.64	6
16	4505.4	7.03	7.42	1.64	2.5	0.11	0.09	7.42	8

method only requires a CPU time of 0.12 s to predict exactly the same number of time points as that of the actual number of time points 10. This indicates that the method is powerful in predicting optimal number of time points in low CPU time.

The method used to predict optimal number of time points is explained here for the time horizon of 16 h. The computational results of LP MAX and LP MIN are given in Table 3.11.

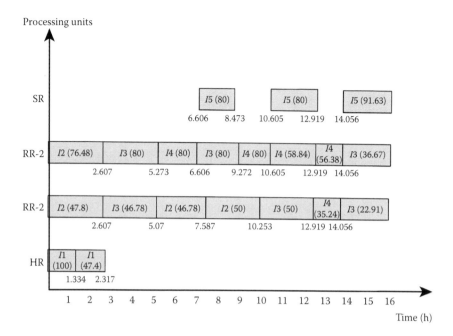

FIGURE 3.7 Gantt chart for a time horizon of 16 h.

The first step was to solve the LP MAX problem using the determined maximum number of time points, which was $\left\lceil \dfrac{16}{0.667} = 24 \right\rceil$. The problem was solved in less than 1 s giving an objective value of 5201.9. From Table 3.11, the number of times reactor 1 can be used was 4 (rounded up to the nearest integer) throughout the 16 h time horizon of interest. For reactor 2, also implying that the number of times that reactor 2 can be used was 7. The number of times the heater, separator, mixer 1 and mixer 2 can be used throughout the 16 h time horizon of interest are 9, 2, 3 and 2, respectively. Equation 3.23 was used to determine the time points required by using the frequency of utilization of each unit obtained from the LP MAX problem. For reactor 1, it becomes $\left\lceil \sum_{p} \sum_{s_{in,j} \in S_{in,j}^{*}} y\left(s_{in,j}, p\right) + \sum_{i \in I^{PT}} i + \sum_{i \in I^{ST}} i = 3.5 + 0 + 0 \right\rceil = 4$, for reactor 2 it becomes $\left\lceil \sum_{p} \sum_{s_{in,j} \in S_{in,j}^{*}} y\left(s_{in,j}, p\right) + \sum_{i \in I^{PT}} i + \sum_{i \in I^{ST}} i = 6.72 + 0 + 0 \right\rceil = 7$.

The term $\sum_{i \in I^{PT}} i$ for reactor 1 and reactor 2 equals to zero because both reactors can start at the beginning of the time horizon by processing reaction 1. The term $\sum_{i \in I^{ST}} i$ for both reactors equals to zero since most of the time the units produce a product at the last time point rather than producing an intermediate and make the unit ideal. This assumption can be justified when looking at the Gantt chart for

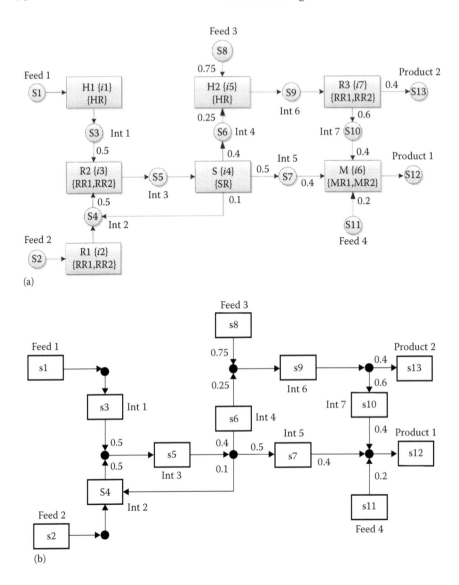

FIGURE 3.8 (a) STN and (b) SSN representations for Case Study III.

the time horizon of 10 h in Figure 3.9; the reactors produce a product at the last time point. Consequently, the assumption helps to predict a fair number of time points instead of overestimating the time points which result to excessive computational time.

For the heater, Equation 3.23 gives $\left[\sum_p \sum_{s_{in,j} \in S^*_{in,j}} y\left(s_{in,j}, p\right) + \sum_{i \in I^{PT}} i + \sum_{i \in I^{ST}} i = 9 + 0 + 1 = 10\right]$. The term $\sum_{i \in I^{PT}} i$ for the heater becomes zero

TABLE 3.9
Data for Case Study III

Task	Unit	$\tau\left(s_{in,j}\right)$	$\beta\left(s_{in,j}\right)$	$V^L_{in,j} - V^U_{in,j}$
Heating 1 (i1)	Heater (HR)	0.667	0.00667	0–100
Heating 2 (i5)	Heater (HR)	1	0.01	0–100
Reaction 1 (i2)	Reactor 1 (RR-1)	1.333	0.01333	0–100
Reaction 1 (i2)	Reactor 2 (RR-2)	1.333	0.00889	0–150
Reaction 2 (i3)	Reactor 1 (RR-1)	0.667	0.00667	0–100
Reaction 2 (i3)	Reactor 2 (RR-2)	0.667	0.00445	0–150
Reaction 3 (i7)	Reactor 1 (RR-1)	1.333	0.0133	0–100
Reaction 3 (i7)	Reactor 2 (RR-2)	1.333	0.00889	0–150
Separation (i4)	Separator (SR)	2	0.00667	0–300
Mixing (i6)	Mixer 1 (MR-1)	1.333	0.00667	20–200
Mixing (i6)	Mixer 2 (MR-2)	1.333	0.0067	20–200

State	Initial Amount	Storage Capacity	Price
s1	Available as required	Unlimited	0
s2	Available as required	Unlimited	0
s3	0	100	0
s4	0	100	0
s5	0	300	0
s6	50	150	0
s7	50	150	0
s8	Available as required	Unlimited	0
s9	0	150	0
s10	0	150	0
s11	Available as required	Unlimited	0
s12	0	Unlimited	5
s13	0	Unlimited	5

since the heater can start processing at the beginning of the time horizon. The term $\sum_{i \in I^{ST}} i$ for the heater becomes one since the heater at the last time point gives its intermediate to the reactors to be converted into product. This time point is not shared by the heater. For the separator, Equation 3.23 gives

$$\left[\sum_p \sum_{s_{in,j} \in S^*_{in,j}} y\left(s_{in,j}, p\right) + \sum_{i \in I^{PT}} i + \sum_{i \in I^{ST}} i = 1.9 + 2 + 1\right] = 5. \text{ The term } \sum_{i \in I^{PT}} i$$

for the separator becomes 2 because at the beginning the separator gets its intermediate after reaction 1 and reaction 2 are processed which requires two time points associated to each task not shared by the separator. The term $\sum_{i \in I^{ST}} i$ for the separa-

tor becomes 1 since at the end of its operation the separator passes its intermediate to the subsequent task, mixing (the shortest path to form product 1) instead of heating (the longest path which is the path to form product 2). This assumption is also a fair

TABLE 3.10
Computational Results for Case Study III under Maximization of Profit

Model	Time Horizon (H)	P	CPU Time (s)	Nodes	RMILP	MILP	B.V	C.V	Constraints	Non-Zeros	Relative Gap (%)
ML2	8	5	1.25	3,662	2100	1583.4	107	428	889	2680	—
ML2	10	6	8	9,250	2871.9	2345.3	131	516	1093	3308	—
ML2	12	7	128	103,886	3465.6	3041.3	155	604	1297	3936	—
ML2	16	10	10,000	4,455,151	4653.1	4262.8	227	868	1909	5820	2

TABLE 3.11

Computational Results of LP MAX and LP MIN for Case Study III

H	Obj. Value	RR-1	RR-2	SR	HR	MR-1	MR-2	CPU Time (s) for LP max	CPU Time (s) for LP min	HR after LP min	P^{pre}
8	2488	1.95	3	0.71	3.7	1	1	0.02	0.02	3.7	5
10	3166.4	1.47	4.7	1	4.8	0	2.5	0.026	0.026	4.8	6
12	3844.9	1	5.7	1.3	6	0.73	2.3	0.032	0.032	6	7
16	5201.9	3.5	6.72	1.9	8.3	3	1.1	0.06	0.06	8.3	10

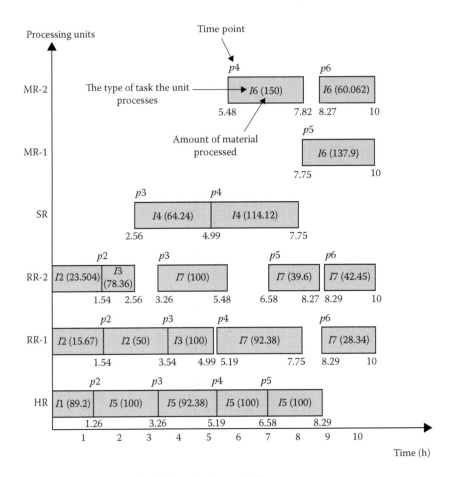

FIGURE 3.9 Gantt chart for the time horizon of 10 h.

assumption which maximizes the utilization of the separator and prevents overestimating the time points. If the separator is a critical unit, the last intermediates either state $s6$ is consumed by the heating followed by reaction 3 or state $s7$ is consumed by the mixing task. Now let us examine each case, finally produced state $s6$ will take a minimum of 1 h for heating and 1.333 h for reaction 3, which forces the separator unit to finish 2.333 h earlier than the time horizon. For the case of state $s7$, it will take a minimum of 1.333 h which will force the separator to finish 1.333 h earlier than the time horizon. Using the principle that the critical unit is the unit that is frequently used and active most of the time horizon, the shortest path to form product 2 allows this principle for the separator to be active for longer than the longest path to form product 1.

For mixer 1, Equation 3.23 gives $\left\lceil \sum_p \sum_{s_{in,j} \in S^*_{in,j}} y\left(s_{in,j}, p\right) + \sum_{i \in I^{PT}} i + \sum_{i \in I^{ST}} i = 3 + 2 + 0 = 4 \right\rceil$. The term $\sum_{i \in I^{PT}} i$, for mixer 1, becomes 2 since at the beginning the mixer can only start its operation after heating two and reaction three are processed, which requires two time points not shared by the mixer. The term $\sum_{i \in I^{ST}} i$ becomes zero since there is no subsequent task after mixing operation, and it is the unit producing a product and can be active until the last time point.

For mixer 2, Equation 3.23 gives $\left[\sum_p \sum_{s_{in,j} \in S^*_{in,j}} y(s_{in,j}, p) + \sum_{i \in I^{PT}} i + \sum_{i \in I^{ST}} i = 1.1 + 2 + 0 = 4 \right]$. The unit that gives the highest time points is the heater; as a result it becomes the critical unit.

The second step of LP MIN also gave the same result for the number of times the critical unit heater was used as that of LP MAX, that is, 9. Equation 3.25 is used to predict the optimal number of time points. The term $\sum_{s_{in,j^c}} \sum_p y(s_{in,j^c}, p)$ in Equation 3.25 for this case becomes 9. The term $\sum_{i \in I^{PT}} i$ for this case is zero because the heater can start processing at the beginning of the time horizon. The parameter $\sum_{i \in I^{ST}} i$ for this case is one because at the end of the process heating, the heater passes its intermediate to the reactor to be converted into product. There is one time point associated with the reactor after the end of the process by the heater. The predicted optimal number of time points then becomes 10, which is the same as the actual optimal number of time points. The predicted optimal number of time points obtained matched the actual optimal number of time points, indicating that the method is rigorous.

3.5.4 Case Study IV: Industrial Case Study

This case study was taken from the book by Majozi and Zhu (2001). Figure 3.10 is the flowsheet for the case study. The STN and SSN representations are shown in Figure 3.11. The process consists of five consecutive steps. The first step involves

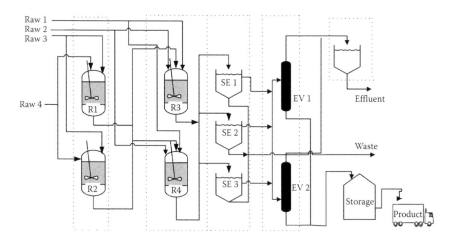

FIGURE 3.10 Flowsheet for the industrial case study.

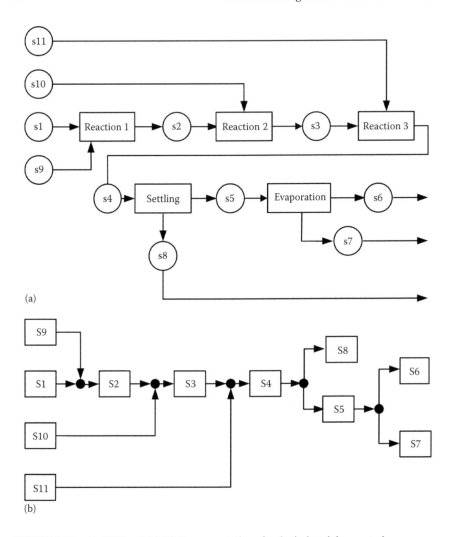

FIGURE 3.11   (a) STN and (b) SSN representations for the industrial case study.

a reaction which forms an arsenate salt. This reaction requires two raw materials, raw 3 and raw 4, and can be conducted in either reactor R1 or R2. The arsenate salt from the first step is then transferred to either reactor R3 or R4, wherein two consecutive reactions take place. The first of these reactions is aimed at converting the arsenate salt to a disodium salt using raw material 1 (raw 1). The disodium salt is then reacted further to form the monosodium salt using raw material 2 (raw 2). The monosodium salt solution is then transferred to the settling step in order to remove the solid by-product. Settling can be conducted in any of the three settlers, that is SE1, SE2 or SE3. The solid by-product is dispensed with as waste and the remaining monosodium salt solution is transferred to the final step. This step consists of two evaporators, EV1 and EV2, which remove the excess amount of water from the monosodium solution. Evaporated water is removed as effluent and monosodium salt (product) is taken to storage. States $s1$ and $s9$ in the SSN represent raw 3 and raw 4, respectively. States $s10$ and $s2$ represent raw 1 and the arsenate salt, while states $s11$ and $s3$ represent raw 2 and the disodium salt, respectively. State $s4$ is the monosodium solution that is transferred to the settlers to form states $s8$ (solid by-product) and $s5$ (remaining monosodium solution). State $s5$ is separated into states $s7$ (water) and $s6$ (product). Table 3.12 shows the data for the case study. The stoichiometric data are included in order to perform material balances in each

TABLE 3.12
Data for Industrial Case Study

Unit	$v^L_{sin,j} - v^U_{sin,j}$	Task	$\tau(s_{in,j})$
RR1	0–10	Reaction 1 (R1)	1.6
RR2	0–10	Reaction 1 (R1)	1.6
RR3	0–10	Reaction 2 (R2)	2.4
RR3	0–10	Reaction 3 (R3)	0.8
RR4	0–10	Reaction 2 (R2)	2.4
RR4	0–10	Reaction 3 (R3)	0.8
SE1	0–10	Settling (SR)	0.8
SE2	0–10	Settling (SR)	0.8
SE3	0–10	Settling (SR)	0.8
EV1	0–10	Evaporation (EV)	2.4
EV2	0–10	Evaporation (EV)	2.4

State	Storage Capacity	Initial Amount
$s1$	Unlimited	Unlimited
$s2$	100	0
$s3$	100	0
$s4$	100	0
$s5$	100	0
$s6$	100	0
$s7$	100	0
$s8$	100	0
$s9$	Unlimited	Unlimited
$s10$	Unlimited	Unlimited
$s11$	Unlimited	Unlimited

Stoichiometric Data

State	Ton/Ton Output	Ton/Ton Product
$s1$	0.2	
$s9$	0.25	
$s10$	0.35	
$s11$	0.2	
$s7$		0.7
$s8$		1

unit operation. The second column of the stoichiometric data shows the amount of raw material required (tons) per unit mass (tons) of the overall output, that is, $s6 + s7 + s8$. The third column shows the ratio of each by-product ($s7$ and $s8$) to product ($s6$) in ton/ton product. The objective function is the maximization of product ($s6$) output.

3.5.4.1 Results and Discussions

This case study is solved using fixed processing time where duration of a task is independent of batch size. The computational results for this industrial case study are given in Table 3.13. The model was solved for a time horizon of 8, 10, 12 and 16 h. For a time horizon of 12 h, an optimal objective value of 18.5 was found using 8 time points. For a time horizon of 16 h, an optimal objective value was obtained using 12 time points. The formulation resulted in 10,515 constraints, 9,397 continuous variables and 264 binary variables. The optimal objective value of 30.6 is obtained in a specified CPU time of 10,000 s.

The proposed method for predicting the optimal number of time points was also tested for this case study. Table 3.14 shows the computational results obtained from the LP MAX and LP MIN of the problem. For a time horizon of 16 h, using Equation 3.25, the critical unit is RR4 and it was used nine times. Considering RR4

TABLE 3.13

Computational Results for Industrial Case Study under Maximization of Profit

Model	H	P	Nodes	RMILP	MILP	B.V	C.V	Cons-traints	CPU Time (s)	Non-Zeros
ML2	8	6	494	12.34	9.26	132	4699	5,211	0.7	7,156
ML2	10	8	3,852,196	16.5	13.9	176	6265	6,979	5,271	7,156
ML2	12	8	900	20.6	18.5	176	6265	6,979	2.5	9,678
ML2	16	12	4,330,203	37	30.6	264	9397	10,515	10,000[a]	14,722

[a] Relative gap, 17%.

TABLE 3.14

Computational Results Obtained from LP MAX and LP MIN for Industrial Case Study

H	Obj. Value	RR1	RR2	RR3	RR4	SE1	SE2	SE3	EV1	EV2	P^{pre}
8	8.7	0.74	0.31	1.57	2.7	1	1	0.35	1	0.48	6
10	14.16	0.5	1.23	2.93	3.96	3.14	0	0.69	1.7	0.57	7
12	19.6	0	2.38	4.74	4.78	3.27	0	2	2.7	0.7	8
16	30.5	0.15	2.2	6.12	8.7	3	2.3	2.3	0.85	4.3	12

as the critical unit and using Equation 3.25, the predicted optimal number of time point becomes 12. The term $\sum_{i \in I^{PT}} i$ for this case is 1 since there is one task before reaction two (R2), which is reaction 1 (R1). At the beginning of the time horizon, RR4 starts processing after reaction R1 is conducted and this task (reaction 1) has one time point associated with it. This time point is not shared by RR4. The term $\sum_{i \in I^{PT}} i$ becomes 2 since RR4, at the end of its operation, passes its intermediate to the next processes which are settling and evaporation with one time point associated with each of these tasks. As a result, the predicted optimal number of time points becomes 12, which is the same as the actual optimal number of time points 12. As shown in Table 3.14, for all time horizons the proposed model predicts the optimal number of time points close to the actual optimal number of time points.

Table 3.15 summarizes the results obtained in all the cases considered in terms of actual and predicting optimal number of time points.

Figure 3.12 shows the comparison between the actual optimal number of time points and the predicted optimal number of time points using the data from Table 3.15. The R^2 value is 0.998, which suggests that the proposed model gives

---

TABLE 3.15

Results of Predicted and Actual Optimal Number of Time Points for Each Case Study

H	Case Studies	Actual Optimal Number of Time Points	Predicted Optimal Number of Time Points
12	I fixed duration	4	5
24	I fixed duration	9	11
36	I fixed duration	15	16
168	I fixed duration	74	75
12	I duration batch dependent	4	5
24	I duration batch dependent	8	9
36	I duration batch dependent	13	13
168	I duration batch dependent	56	57
8	II	4	4
10	II	6	5
12	II	7	6
16	II	8	8
8	III	5	5
10	III	6	6
12	III	7	7
16	III	10	10
8	IV	6	6
10	IV	8	7
12	IV	8	8
16	IV	12	12

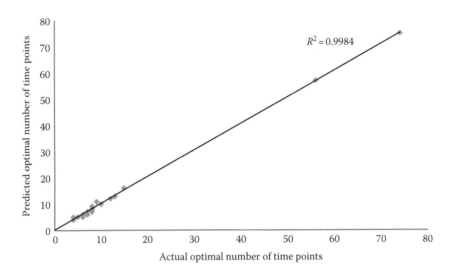

FIGURE 3.12   Comparison between predicted and actual optimal number of time points.

accurate predictions. Application of this proposed method is vital to avoid unnecessary costly iteration, especially as the time horizon increases.

3.6   CONCLUSIONS

A method for predicting the optimal number of time points has been presented in this chapter. Linear programming maximization and minimization models were used to identify the critical unit and the number of times this critical unit was used. Based on the principle that the optimal number of time points depends on how many times this critical unit is used, an equation was developed to predict the optimal number of time points. The model was tested for a number of case studies taken from published literature. A variance of 0.998 was obtained when comparing predicted optimal number of time points with the actual optimal number of time points. This indicates that the method is accurate. Application of the proposed method will avoid costly computational iteration as the time horizon increases.

NOTATION

SETS

I^{PT} {Previous tasks that are performed before task i in the recipe}
S {$s \mid s$ any state s}
I {$i \mid i$ is a task}
I^{ST} {Subsequent tasks that are performed after task i in the recipe}
$S_{in,J}^{PT}$ {$s_{in,j} \mid s_{in,j}$ previous task before task $s_{in,j}$}
$S_{in,J}^{ST}$ {$s_{in,j} \mid s_{in,j}$ subsequent task before task $s_{in,j}$}

J $\{j \mid j$ is a unit$\}$

$S^c_{in,j}$ $\{s^c_{in,j} \mid s^c_{in,j}$ task which consume state $s\}$

$S^*_{in,j}$ $\{s^*_{in,j} \mid s^*_{in,j}$ state used by a task$\}$

$S_{in,j}$ $\{s_{in,j} \mid s_{in,j}$ effective state$\}$

$S^{uc}_{in,j}$ $\{s^{uc}_{in,j} \mid s^{uc}_{in,j}$ task which consume unstable state $s\}$

$S^p_{in,j}$ $\{s^p_{in,j} \mid s^p_{in,j}$ task which produce state $s\}$

$S^{up}_{in,j}$ $\{s^{up}_{in,j} \mid s^{up}_{in,j}$ task which produce unstable state $s\}$

S^P $\{s^P \mid s^P$ a state which is a product$\}$

P $\{p \mid p$ is a time point$\}$

J_s $\{j_s \mid j_s$ is a unit producing state $s\}$

VARIABLES

$t_p\left(s_{in,j}, p\right)$ Time at which task ends at time point p, $s_{in,j} \in S_{in,j}$

$t_u\left(s_{in,j}, p\right)$ Time at which task starts at time point p, $s_{in,j} \in S_{in,j}$

$mu\left(s_{in,j}, p\right)$ Amount of material processed by a task at time point p

$q_s\left(s, p\right)$ Amount of state s stored at time point p

$y\left(s_{in,j}, p\right)$ Binary variable for assignment of task at time point p

$t\left(j, p\right)$ Binary variable associated with usage of state produced by unit j at time point p

$t\left(j, s, p\right)$ Binary variable associated with usage of state s produced by unit j at time point p if the unit produces more than one intermediate at time point p, $s \in S^p_{in,j}$

$x\left(s, p\right)$ Binary variable associated with availability of storage for state s at time point p

$u\left(s_{in,j}, p\right)$ Amount of material stored in unit j at time point p

Z_j The number of time points a unit is active obtained from LP MAX problem

PARAMETERS

$V^U_{s_{in,j}}$ Maximum capacity of unit j to process a particular task

$V^L_{s_{in,j}}$ Minimum capacity of unit j to process a particular task

H Time horizon of interest

QS^o Initial amount of state s stored

QS^U Maximum capacity of storage to store a state s

$\tau\left(s_{in,j}\right)$ Coefficient of constant term of processing time of state s, $s \in S_{in,j}$

$\beta\left(s_{in,j}\right)$ Coefficient of variable term of processing time of state s, $s \in S_{in,j}$

$\rho\left(s^{sp}_{in,j}\right)$ Portion of state s produced by a task

$\rho\left(s^{sc}_{in,j}\right)$ Portion of state s consumed by a task

REFERENCES

Ierapetritou, M.G., Floudas, C.A., 1998a. Effective continuous-time formulation for short-term scheduling: 1. Multipurpose batch processes. *Industrial and Engineering Chemistry Research.* 37, 4341–4359.

Kondili, E., Pantelides, C.C., Sargent, R.W.H., 1993. A general algorithm for short-term scheduling of batch operations. I. MILP formulation. *Computer and Chemical Engineering.* 17, 211–227.

Li, J., Floudas, C.A., 2010. Optimal event point determination for short-term scheduling of multipurpose batch plants via unit-specific event-based continuous-time approaches. *Industrial and Engineering Chemistry Research.* 49, 7446–7469.

Majozi, T., Zhu, X.X., 2001. A novel continuous-time MILP Formulation for multipurpose bach plants. *Industrial and Engineering Chemistry Research.* 40, 5935–5949.

Sundaramoorthy, A., Karimi, I.A., 2005. A simpler better slot-based continuous-time formulation for short-term scheduling in multipurpose batch plants. *Chemical Engineering Science.* 60, 2679–2702.

4 Integration of Scheduling and Heat Integration
Minimization of Energy Requirements

4.1 INTRODUCTION

Presented in this chapter is a mathematical technique for simultaneous heat integration and process scheduling in multipurpose batch plants. Taking advantage of the intermittent continuous behaviour of process streams during transfer from one processing unit to another, as determined by the recipe, the presented formulation aims to maximize the coincidence of availability of hot and cold stream pairs with feasible temperature driving forces, while taking into consideration process scheduling constraints. Distinct from similar contributions in the published literature, time is treated as one of the key optimization variables instead of a parameter fixed a priori. Heat integration during stream transfer has the added benefit of shortened residence time in processing units, which invariably improves the throughput, as more batches are likely to be processed within a given time horizon, compared to conventional cases with heating and cooling in situ.

4.2 NECESSARY BACKGROUND

The trend towards batch processing calls for the development of effective techniques for production planning and scheduling. Extensive research has been conducted on developing mathematical models to improve batch plant efficiency (Harjunkoski et al., 2014), with recent advances in computer technology allowing large-scale and more complex problems to be handled by using optimization techniques. In addition to process scheduling, heat integration may be considered for batch plants to reduce external utility (e.g. steam and cooling water) requirements for tasks involving heating or cooling, such as endothermic and exothermic reactions. This is driven by not only cost considerations, but also increased environmental awareness and the pursuit of energy efficiency. Consequently, the past two decades or so have been characterized by a significant body of research on heat integration of batch plants (Fernández et al., 2012).

Heat integration in batch processes can be carried out directly or indirectly (Majozi, 2010). Direct integration requires the hot and cold tasks to coincide in order for process–process heat exchange to take place, whereas indirect integration uses thermal storage enabling heat exchange between tasks performed in different time intervals, thus providing further and more flexible heat recovery. The methodologies developed for batch heat integration can be categorized into pinch-based and mathematical techniques. Early studies for minimizing energy consumption in batch plants were mostly based on pinch analysis and conducted by extending methods developed for continuous processes. Mathematical optimization, on the other hand, has been increasingly adopted for cases with multiple design criteria and/or practical constraints. More importantly, the use of mathematical techniques allows time to be treated as a variable, in which case the production schedule is allowed to change. This implies the incorporation of heat integration into batch process scheduling and a simultaneous solution for more optimal results.

4.3 MOTIVATING EXAMPLE

While there have been many works addressing scheduling and heat integration of batch plants, most only consider direct and/or indirect integration between processing units/vessels. The opportunity for energy savings through heat exchange between process streams during material transfer was overlooked and is exploited in this chapter.

The motivation and rationale behind the contribution is illustrated in Figures 4.1 and 4.2. As shown in Figure 4.1, the material transferred from processing unit 1 to 4

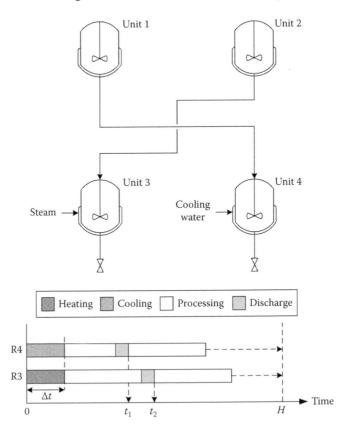

FIGURE 4.1 Impact of overlooking process–process heat integration on time.

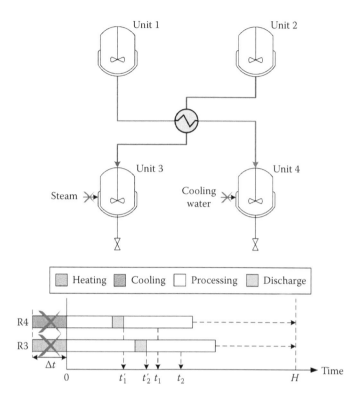

FIGURE 4.2 Impact of exploiting process–process heat integration on time.

requires cooling prior to the commencement of a subsequent processing task in unit 4. On the other hand, the material transferred from processing units 2 to 3 requires heating before the next processing task in unit 3. The times required for both heating and cooling in units 3 and 4 are denoted as Δt for illustration purposes, given that these need not necessarily be of equal duration.

It is evident that if the thermal driving forces permit and the transfer times from units 1 and 2 to units 4 and 3 coincide, an opportunity for stream–stream heat integration could arise and be exploited with significant benefits. Figure 4.2 summarizes this particular observation. Heat integration of streams from units 1 to 2 is effected through an adequately sized heat exchanger, during the transfer of materials to subsequent processing units (i.e. units 3 and 4). Consequently, the need for external hot and cold utilities, as well as dedicated time for heating and cooling inside the processing units (in situ) prior to the commencement of processing tasks, is eliminated, with potentially substantial positive cost implications. In particular, more batches are likely to be produced in units 3 and 4 with the elimination of dedicated in situ heating and cooling times. The implication of this time benefit is depicted in Figure 4.2, where it is clearly shown that $t_1 - t_1' = t_2 - t_2' = \Delta t$.

The overall idea, therefore, is to maximize the coincidence of availability of hot and cold stream pairs with feasible thermal driving forces within a comprehensive process scheduling framework. This would require a mathematical formulation to adequately address simultaneous heat integration and process scheduling where time is a key optimization variable. This is the essence of the contribution.

4.4 PROBLEM STATEMENT

The problem addressed in this chapter can be briefly stated as follows. Given: (1) production scheduling data, including the recipe for each product, task durations, equipment capacities, the time horizon of interest, raw material costs and product selling prices; (2) operating temperatures for processing tasks, supply temperatures of raw materials and storage temperatures for final products; (3) specific heat capacities of states and (4) costs of external hot and cold utilities, determine an optimal production schedule that maximizes the profit, which is defined as product revenue minus raw material and utility costs. For simplicity, the following assumptions are made in the problem:

- Heat capacities are constant.
- Heat losses for temporary storage are negligible.
- Countercurrent heat exchangers are used for heat integration.

4.5 MODEL FORMULATION

Figure 4.3 shows the superstructure representations on which the mathematical modelling is based. It can be seen that each processing unit can receive material from and send material to other units or storage. The input material may need to be heated or cooled in the processing unit before the task commences in order to meet the operating temperature; similarly, the output material may be heated or cooled in situ before leaving for further processing or storage. To reduce the use of external hot and cold utilities, heat recovery between output (i.e. intermediate and product) streams (Figure 4.3a) may be considered, as well as heat exchange for raw material pre-heating (Figure 4.3b). Note that output streams are assumed to have heat exchange before entering downstream processing units or storage, while raw material streams are, after leaving storage. If there is more than one state produced in a processing unit, these states are assumed to exit the unit at the same temperature and treated as separate streams for heat integration. It is also assumed that external utilities can only be used to unit jackets – no utility heaters or coolers are available for use. In addition to the necessary scheduling constraints (see Appendix), the mathematical model consists mainly of mass and energy balances and heat integration constraints, as presented here. Notation used is given in the Nomenclature.

4.5.1 Mass and Energy Balances for Processing Units

Constraints (4.1) and (4.2) describe the inlet mass balance for unit j at any time point p. The input state, either a raw material or an intermediate, may come from storage or upstream units (j') where it is produced. Note that there is nothing transferred from producing units at the first time point ($p = 1$) as no intermediates have been produced yet. The only possible input at this point in time is raw material from storage. It is

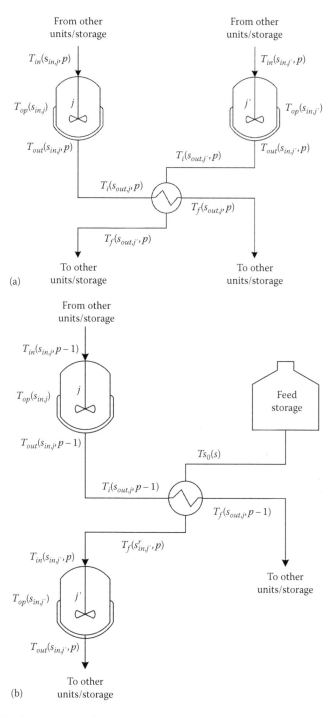

FIGURE 4.3 Superstructure for heat integration between (a) outlet streams and (b) outlet and inlet (raw material) streams.

also worth noting that this set of constraints also cater for stoichiometric require-ments of inputs into a unit, hence the coefficient $\rho_s^c\left(s_{in,j}\right)$.

$$\rho_s^c\left(s_{in,j}\right)m_u\left(s_{in,j},p\right)=ms_{out}\left(s,j,p\right)+\sum_{j'\in J_s^p}mt\left(s,j',j,p-1\right)$$

$$\forall j\in J_s^c,\quad p\in P,\quad p>1,\quad s\in S,\quad s_{in,j}\in S_{in,j}^{eff} \tag{4.1}$$

$$\rho_s^c\left(s_{in,j}\right)m_u\left(s_{in,j},p\right)=ms_{out}\left(s,j,p\right)\quad\forall j\in J_s^c,\quad p\in P,\quad p=1,\quad s\in S,\quad s_{in,j}\in S_{in,j}^{eff} \tag{4.2}$$

Constraint (4.3) describes the outlet mass balance for unit j at time point p. The output state, an intermediate or a final product, may be sent to storage or downstream consuming units j'. Again, this constraint takes into account the implications of stoi-chiometry through $\rho_s^p\left(s_{in,j}\right)$.

$$\rho_s^p\left(s_{in,j}\right)m_u\left(s_{in,j},p\right)=ms_{in}\left(s,j,p\right)+\sum_{j'\in J_s^c}mt\left(s,j,j',p\right)$$

$$\forall j\in J_s^p,\quad p\in P,\quad s\in S,\quad s_{in,j}\in S_{in,j}^{eff} \tag{4.3}$$

Constraint (4.4) stipulates that the total amount of states consumed in the task at time point p cannot exceed the capacity of the processing unit. This constraint is only active if a particular task is active in unit j. In the absence of a task in a unit, this constraint ensures that no material is processed. It is also worth noting that the minimum amount of material to be processed for a particular task in a unit need not be zero. These last two conditions explain the presence of a binary variable in the constraint.

$$V_j^L y\left(s_{in,j},p\right)\leq m_u\left(s_{in,j},p\right)\leq V_j^U y\left(s_{in,j},p\right)\quad\forall j\in J,\quad p\in P,\quad s_{in,j}\in S_{in,j}^{eff} \tag{4.4}$$

Constraints (4.5) and (4.6) describe the inlet energy balance for processing units. It is important to note that if unit j uses a raw material ($s\in S^r$), the correspond-ing Ts_{out} (s, p) term on the right-hand side of both constraints will be replaced by $T_f\left(s_{in,j}^r,p\right)$ or $Ts_0(s)$ for $p=1$, since the raw material stream may have heat exchange before entering unit j from the storage. On the other hand, the outlet energy balance need not be performed because all the output states are assumed to exit the unit at the same temperature, T_{out} ($s_{in,j}$,p), as mentioned earlier.

$$m_u\left(s_{in,j},p\right)Cp_{in}\left(s_{in,j}\right)T_{in}\left(s_{in,j},p\right)=\sum_{s\in S_j^c}ms_{out}\left(s,j,p\right)Cp\left(s\right)Ts_{out}\left(s,p\right)$$

$$+\sum_{s\in S_j^c}\sum_{j'\in J_s^p}mt\left(s,j',j,p-1\right)Cp\left(s\right)T_f\left(s_{out,j'},p-1\right)$$

$$\forall j\in J,\quad p\in P,\quad p>1,\quad s_{in,j}\in S_{in,j}^{eff},\quad s_{out,j'}\in S_{out,j} \tag{4.5}$$

$$m_u\left(s_{in,j},p\right)Cp_{in}\left(s_{in,j}\right)T_{in}\left(s_{in,j},p\right)=\sum_{s\in S_j^c}ms_{out}\left(s,j,p\right)Cp\left(s\right)Ts_{out}\left(s,p\right)$$

$$\forall j\in J,\quad p\in P,\quad p=1,\quad s_{in,j}\in S_{in,j}^{eff} \tag{4.6}$$

where

$$Cp_{in}\left(s_{in,j}\right)=\sum_{s\in S}\rho_s^c\left(s_{in,j}\right)Cp\left(s\right).$$

4.5.2 Mass and Energy Balances for Storage

Constraints (4.7) through (4.12) describe the mass balances for the storage of raw material, intermediate and product states. All these constraints are formulated such that the amount of material stored at time point p is the amount stored at the previous time point $(p-1)$ adjusted by the amounts consumed and/or produced by the tasks. Note that these constraints assume different forms depending on whether they apply to a feed, intermediate or final product stream/state. The main distinction between intermediates and products is that the latter are only stored after production, whereas the former may be stored or consumed after production.

$$qs\left(s,p\right)=qs\left(s,p-1\right)-\sum_{j\in J_s^c}ms_{out}\left(s,j,p\right)\quad\forall p\in P,\quad p>1,\quad s\in S^r \tag{4.7}$$

$$qs\left(s,p\right)=Q_0\left(s\right)-\sum_{j\in J_s^c}ms_{out}\left(s,j,p\right)\quad\forall p\in P,\quad p=1,\quad s\in S^r \tag{4.8}$$

$$qs\left(s,p\right)=qs\left(s,p-1\right)+\sum_{j\in J_s^p}ms_{in}\left(s,j,p-1\right)$$

$$-\sum_{j\in J_s^c}ms_{out}\left(s,j,p\right)\quad\forall p\in P,\quad p>1,\quad s\in S^i \tag{4.9}$$

$$qs\left(s,p\right)=Q_0\left(s\right)-\sum_{j\in J_s^c}ms_{out}\left(s,j,p\right)\quad\forall p\in P,\quad p=1,\quad s\in S^i \tag{4.10}$$

$$qs\left(s,p\right)=qs\left(s,p-1\right)+\sum_{j\in J_s^p}ms_{in}\left(s,j,p\right)\quad\forall p\in P,\quad p>1,\quad s\in S^P \tag{4.11}$$

$$qs\left(s,p\right)=Q_0\left(s\right)+\sum_{j\in J_s^p}ms_{in}\left(s,j,p\right)\quad\forall p\in P,\quad p=1,\quad s\in S^P \tag{4.12}$$

Constraints (4.13) and (4.14) state that the amount of state s stored at any time point cannot exceed the maximum capacity of the storage. It is imperative that constraint (4.14) takes into account the amount of material produced at time point p.

$$qs(s,p) \leq Q_s^U \quad \forall p \in P, \quad s \in S^r \cup S^p \tag{4.13}$$

$$qs(s,p) + \sum_{j \in J_s^p} ms_{in}(s,j,p) \leq Q_s^U \quad \forall p \in P, \quad s \in S^i \tag{4.14}$$

Constraints (4.15) and (4.16) are analogous to previous Constraints (4.5) and (4.6), but only apply to intermediate states in storage. Similar constraints for raw material and product storage are not needed, assuming that raw materials are only sent out for processing at their supply temperatures and final products only sent in at their storage temperatures.

$$qs(s,p)Ts_{out}(s,p) = qs(s,p-1)Ts_{out}(s,p-1) + \sum_{j \in J_s^p} ms_{in}(s,j,p-1)T_f(s_{out,j},p-1)$$

$$- \sum_{j \in J_s^c} ms_{out}(s,j,p)Ts_{out}(s,p) \quad \forall p \in P, \quad p > 1, \quad s \in S^i \tag{4.15}$$

$$qs(s,p)Ts_{out}(s,p) = Q_0(s)Ts_0(s) - \sum_{j \in J_s^c} ms_{out}(s,j,p)Ts_{out}(s,p)$$

$$\forall p \in P, \quad p = 1, \quad s \in S^i \tag{4.16}$$

4.5.3   Heating and Cooling in Processing Units

As mentioned earlier, the input and output materials may be heated or cooled in processing units before and after tasks. Such potential heating and cooling requirements can be identified from the temperature data of the production. For example, if a processing unit ($j = 1$) operating at a temperature of T_1 produces an intermediate to be further processed in a downstream unit ($j = 2$) where the operating temperature is higher ($T_2 > T_1$), the intermediate produced at T_1 needs to be heated to T_2. This heating can take place either in the first or the second unit using external hot utility, or during material transfer between the units through heat exchange.

For processing units that may have heating for their input materials, Constraint (4.17) is used to calculate the heat load, which can be zero for an active task, indicating that the heating has been finished elsewhere. In this case, the temperature of the input material is the same as the operating temperature.

$$q_{in}^h(s_{in,j},p) = m_u(s_{in,j},p)Cp_{in}(s_{in,j})\left(T_{op}(s_{in,j}) - T_{in}(s_{in,j},p)\right)$$

$$\forall j \in J_{in}^h, \quad p \in P, \quad s_{in,j} \in S_{in,j}^{eff} \tag{4.17}$$

For units that may need cooling for their input materials, Constraint (4.18) is used:

$$q_{in}^c\left(s_{in,j},p\right)=m_u\left(s_{in,j},p\right)Cp_{in}\left(s_{in,j}\right)\left(T_{in}\left(s_{in,j},p\right)-T_{op}\left(s_{in,j}\right)\right)$$

$$\forall j\in J_{in}^c,\quad p\in P,\quad s_{in,j}\in S_{in,j}^{eff}\tag{4.18}$$

If either heating or cooling would be needed, for example, in the case of multiple input states with varying temperatures, constraints (4.19) through (4.23) can be used. Constraint (4.19) states that unit j may have either heating or cooling, but not both, for its input material, whenever the task is active ($y(s_{in,j},p)=1$). When heating is to take place $\left(y_{in}^h\left(s_{in,j},p\right)=1;\ y_{in}^c\left(s_{in,j},p\right)=0\right)$, Constraint (4.20) quantifies the heat load, while Constraints (4.21) and (4.22) become redundant and Constraint (4.23) forces the cooling load to be zero. Overall, this situation reduces Constraint (4.20) to Constraint (4.17). Similarly, when cooling is to take place ($y_{in}^h\left(s_{in,j},p\right)=0$; $y_{in}^c\left(s_{in,j},p\right)=1$), Constraints (4.20) and (4.23) become redundant, Constraint (4.21) forces the heating load to be zero and Constraint (4.22) reduces to Constraint (4.18).

$$y_{in}^h\left(s_{in,j},p\right)+y_{in}^c\left(s_{in,j},p\right)=y\left(s_{in,j},p\right)\quad\forall j\in J_{in}^*,\quad p\in P,\quad s_{in,j}\in S_{in,j}^{eff}\tag{4.19}$$

$$Q^U\left(y_{in}^h\left(s_{in,j},p\right)-1\right)\le q_{in}^h\left(s_{in,j},p\right)-m_u\left(s_{in,j},p\right)Cp_{in}\left(s_{in,j}\right)\left(T_{op}\left(s_{in,j}\right)-T_{in}\left(s_{in,j},p\right)\right)$$

$$\le Q^U\left(1-y_{in}^h\left(s_{in,j},p\right)\right)\quad\forall j\in J_{in}^*,\quad p\in P,\quad s_{in,j}\in S_{in,j}^{eff}\tag{4.20}$$

$$q_{in}^h\left(s_{in,j},p\right)\le Q^U y_{in}^h\left(s_{in,j},p\right)\quad\forall j\in J_{in}^*,\quad p\in P,\quad s_{in,j}\in S_{in,j}^{eff}\tag{4.21}$$

$$Q^U\left(y_{in}^c\left(s_{in,j},p\right)-1\right)\le q_{in}^c\left(s_{in,j},p\right)-m_u\left(s_{in,j},p\right)Cp_{in}\left(s_{in,j}\right)\left(T_{in}\left(s_{in,j},p\right)-T_{op}\left(s_{in,j}\right)\right)$$

$$\le Q^U\left(1-y_{in}^c\left(s_{in,j},p\right)\right)\quad\forall j\in J_{in}^*,\quad p\in P,\quad s_{in,j}\in S_{in,j}^{eff}\tag{4.22}$$

$$q_{in}^c\left(s_{in,j},p\right)\le Q^U y_{in}^c\left(s_{in,j},p\right)\quad\forall j\in J_{in}^*,\quad p\in P,\quad s_{in,j}\in S_{in,j}^{eff}\tag{4.23}$$

Constraints (4.24) and (4.25) are used for processing units that may have heating and those that may have cooling for their output materials, respectively.

$$q_{out}^h\left(s_{in,j},p\right)=m_u\left(s_{in,j},p\right)Cp_{out}\left(s_{in,j}\right)\left(T_{out}\left(s_{in,j},p\right)-T_{op}\left(s_{in,j}\right)\right)$$

$$\forall j\in J_{out}^h,\quad p\in P,\quad s_{in,j}\in S_{in,j}^{eff}\tag{4.24}$$

$$q_{out}^c\left(s_{in,j},p\right)=m_u\left(s_{in,j},p\right)Cp_{out}\left(s_{in,j}\right)\left(T_{op}\left(s_{in,j}\right)-T_{out}\left(s_{in,j},p\right)\right)$$

$$\forall j\in J_{out}^c,\quad p\in P,\quad s_{in,j}\in S_{in,j}^{eff}\tag{4.25}$$

where $Cp_{out}\left(s_{in,j}\right)=\sum_{s\in S}\rho_s^p\left(s_{in,j}\right)Cp\left(s\right)$. If either heating or cooling would be needed, for example, in the case of multiple output states proceeding to different destinations with varying target temperatures, Constraints (4.26) through (4.30) can be used. Worth the emphasis is the fact that Constraints (4.26) through (4.30) are directly analogous to Constraints (4.19) through (4.23), with the latter applicable to input, and the former to output states.

$$y_{out}^h\left(s_{in,j},p\right)+y_{out}^c\left(s_{in,j},p\right)\le y\left(s_{in,j},p\right) \quad \forall j\in J_{out}^*, \quad p\in P, \quad s_{in,j}\in S_{in,j}^{eff} \quad (4.26)$$

$$Q^U\left(y_{out}^h\left(s_{in,j},p\right)-1\right)\le q_{out}^h\left(s_{in,j},p\right)-m_u\left(s_{in,j},p\right)Cp_{out}\left(s_{in,j}\right)\left(T_{out}\left(s_{in,j},p\right)-T_{op}\left(s_{in,j}\right)\right)$$
$$\le Q^U\left(1-y_{out}^h\left(s_{in,j},p\right)\right) \quad \forall j\in J_{out}^*, \quad p\in P, \quad s_{in,j}\in S_{in,j}^{eff} \quad (4.27)$$

$$q_{out}^h\left(s_{in,j},p\right)\le Q^U y_{out}^h\left(s_{in,j},p\right) \quad \forall j\in J_{out}^*, \quad p\in P, \quad s_{in,j}\in S_{in,j}^{eff} \quad (4.28)$$

$$Q^U\left(y_{out}^c\left(s_{in,j},p\right)-1\right)\le q_{out}^c\left(s_{in,j},p\right)-m_u\left(s_{in,j},p\right)Cp_{out}\left(s_{in,j}\right)\left(T_{op}\left(s_{in,j}\right)-T_{out}\left(s_{in,j},p\right)\right)$$
$$\le Q^U\left(1-y_{out}^c\left(s_{in,j},p\right)\right) \quad \forall j\in J_{out}^*, \quad p\in P, \quad s_{in,j}\in S_{in,j}^{eff} \quad (4.29)$$

$$q_{out}^c\left(s_{in,j},p\right)\le Q^U y_{out}^c\left(s_{in,j},p\right) \quad \forall j\in J_{out}^*, \quad p\in P, \quad s_{in,j}\in S_{in,j}^{eff} \quad (4.30)$$

Constraint (4.31) gives the time required for heating, while Constraint (4.32) gives the time required for cooling of input materials in processing units dedicated to either heating or cooling. Constraint (4.33), on the other hand, applies to those units that are capable of both heating and cooling, depending on the nature of input materials.

$$tr_{in}\left(s_{in,j},p\right)=q_{in}^h\left(s_{in,j},p\right)/\left(\dot{M}^{st}\lambda^{st}\right) \quad \forall j\in J_{in}^h, \quad p\in P, \quad s_{in,j}\in S_{in,j}^{eff} \quad (4.31)$$

$$tr_{in}\left(s_{in,j},p\right)=q_{in}^c\left(s_{in,j},p\right)/\left(\dot{M}^{cw}Cp^{cw}\left(T_{out}^{cw}-T_{in}^{cw}\right)\right) \quad \forall j\in J_{in}^c, \quad p\in P, \quad s_{in,j}\in S_{in,j}^{eff}$$
$$(4.32)$$

$$tr_{in}\left(s_{in,j},p\right)=\left(q_{in}^h\left(s_{in,j},p\right)/\left(\dot{M}^{st}\lambda^{st}\right)\right)+\left(q_{in}^c\left(s_{in,j},p\right)/\left(\dot{M}^{cw}Cp^{cw}\left(T_{out}^{cw}-T_{in}^{cw}\right)\right)\right)$$
$$\forall j\in J_{in}^*, \quad p\in P, \quad s_{in,j}\in S_{in,j}^{eff} \quad (4.33)$$

Constraints (4.34) through (4.36) are directly analogous to Constraints (4.31) through (4.33), but apply to output states.

$$tr_{out}\left(s_{in,j},p\right)=q_{out}^h\left(s_{in,j},p\right)/\left(\dot{M}^{st}\lambda^{st}\right) \quad \forall j\in J_{out}^h, \quad p\in P, \quad s_{in,j}\in S_{in,j}^{eff} \quad (4.34)$$

$$tr_{out}\left(s_{in,j},p\right)=q_{out}^c\left(s_{in,j},p\right)/\left(\dot{M}^{cw}Cp^{cw}\left(T_{out}^{cw}-T_{in}^{cw}\right)\right) \quad \forall j\in J_{out}^c, \quad p\in P, \quad s_{in,j}\in S_{in,j}^{eff}$$
$$(4.35)$$

$$tr_{out}\left(s_{in,j}, p\right) = \left(q_{out}^{h}\left(s_{in,j}, p\right) / \left(\dot{M}^{st}\lambda^{st}\right)\right) + \left(q_{out}^{c}\left(s_{in,j}, p\right) / \left(\dot{M}^{cw}Cp^{cw}\left(T_{out}^{cw} - T_{in}^{cw}\right)\right)\right)$$

$$\forall j \in J_{out}^{*}, \quad p \in P, \quad s_{in,j} \in S_{in,j}^{eff} \tag{4.36}$$

Taking into account the heating and cooling times in processing units, the duration constraint is formulated as in Constraint (4.37). Note that the processing time is assumed to be constant and independent of the batch size. The inequality caters for further residence time in the processing unit in the case of unavailable intermediate storage. The latter could be due to deliberate omission of intermediate storage in design or unavailability due to finite intermediate storage.

$$t_{p}\left(s_{in,j}, p\right) \geq t_{u}\left(s_{in,j}, p\right) + tr_{in}\left(s_{in,j}, p\right) + \tau\left(s_{in,j}\right)y\left(s_{in,j}, p\right) + tr_{out}\left(s_{in,j}, p\right)$$

$$\forall j \in J, \quad p \in P, \quad s_{in,j} \in S_{in,j}^{eff} \tag{4.37}$$

4.5.4 Heat Integration Constraints

In order to reduce utility consumption, heat exchange among output and raw material streams is considered. As in the case of heating and cooling in processing units, the potential hot and cold streams can be identified from the temperature data.

For temperature assignment, Constraint (4.38) states that the initial temperature of an outlet (intermediate or product) stream for heat exchange is the temperature at which it leaves the processing unit.

$$T_{i}\left(s_{out,j}, p\right) = T_{out}\left(s_{in,j}, p\right) \quad \forall j \in J, \quad p \in P, \quad s_{in,j} \in S_{in,j}^{eff}, \quad s_{out,j} \in S_{out,j} \tag{4.38}$$

Constraints (4.39) through (4.41) state that an outlet stream can be heat integrated with another outlet stream or a raw material stream after leaving the producing unit, and that a raw material stream may exchange heat with an outlet stream for preheating before entering the consuming unit. Additionally, there will be no heat exchanged if the corresponding unit is not active ($y(s_{in,j}, p) = 0$). It should be noted that raw material pre-cooling is not considered in this work, not common in practice. Thus, raw material streams will only act as cold streams in the context of heat integration. Note also that the one-to-one heat integration arrangement imposed by these constraints is designed to simplify process operation based on practical considerations.

$$\sum_{s_{out,j'} \in S_{*}^{c}} x\left(s_{out,j}, s_{out,j'}, p\right) + \sum_{s_{in,j'}^{r} \in S_{*}^{c}} x\left(s_{out,j}, s_{in,j'}^{r}, p\right) \leq y\left(s_{in,j}, p\right)$$

$$\forall j \in J, \quad p \in P, \quad s_{in,j} \in S_{in,j}^{eff}, \quad s_{out,j} \in S_{*}^{h} \tag{4.39}$$

$$\sum_{s_{out,j'} \in S_{*}^{h}} x\left(s_{out,j'}, s_{out,j}, p\right) \leq y\left(s_{in,j}, p\right) \quad \forall j \in J, \quad p \in P, \quad s_{in,j} \in S_{in,j}^{eff}, \quad s_{out,j} \in S_{*}^{c} \tag{4.40}$$

$$\sum_{s_{out,j'} \in S_*^h} x\left(s_{out,j'}, s_{in,j}^r, p-1\right) \leq y\left(s_{in,j}, p\right) \quad \forall j \in J, \quad p \in P, \quad p > 1, \quad s_{in,j} \in S_{in,j}^{eff}, \quad s_{in,j}^r \in S_*^c$$

$$(4.41)$$

The amounts of heat exchanged between hot and cold streams are given by Constraints (4.42) through (4.44). For non-existing matches, Constraint (4.45) ensures zero heat loads.

$$\sum_{s_*^c \in S_*^c} q_{ex}\left(s_{out,j}, s_*^c, p\right) = \rho_s^p\left(s_{in,j}\right) m_u\left(s_{in,j}, p\right) Cp(s)\left(T_i\left(s_{out,j}, p\right) - T_f\left(s_{out,j}, p\right)\right)$$

$$\forall j \in J_s^p, \quad p \in P, \quad s \in S, \quad s_{in,j} \in S_{in,j}^{eff}, \quad s_{out,j} \in S_*^h \qquad (4.42)$$

$$\sum_{s_{out,j'} \in S_*^h} q_{ex}\left(s_{out,j'}, s_{out,j}, p\right) = \rho_s^p\left(s_{in,j}\right) m_u\left(s_{in,j}, p\right) Cp(s)\left(T_f\left(s_{out,j}, p\right) - T_i\left(s_{out,j}, p\right)\right)$$

$$\forall j \in J_s^p, \quad p \in P, \quad s \in S, \quad s_{in,j} \in S_{in,j}^{eff}, \quad s_{out,j} \in S_*^c \qquad (4.43)$$

$$\sum_{s_{out,j'} \in S_*^h} q_{ex}\left(s_{out,j'}, s_{in,j}^r, p-1\right) = \rho_s^c\left(s_{in,j}\right) m_u\left(s_{in,j}, p\right) Cp(s)\left(T_f\left(s_{in,j}^r, p\right) - Ts_0(s)\right)$$

$$\forall j \in J_s^c, \quad p \in P, \quad p > 1, \quad s \in S^r, \quad s_{in,j} \in S_{in,j}^{eff}, \quad s_{in,j}^r \in S_*^c \qquad (4.44)$$

$$q_{ex}\left(s_*^h, s_*^c, p\right) \leq Q^U x\left(s_*^h, s_*^c, p\right) \quad \forall p \in P, \quad s_*^h \in S_*^h, \quad s_*^c \in S_*^c \qquad (4.45)$$

where s_*^h only represents $s_{out,j}$, while s_*^c can be $s_{out,j}$ or $s_{in,j}^r$. Constraints (4.46) through (4.49) ensure feasible temperature driving forces for heat exchange between a pair of outlet streams and between an outlet and a raw material stream, respectively.

$$T_i\left(s_{out,j}, p\right) - T_f\left(s_{out,j'}, p\right) \geq \Delta T_{min} - \Gamma\left(1 - x\left(s_{out,j}, s_{out,j'}, p\right)\right)$$

$$\forall j, j' \in J, \quad p \in P, \quad s_{out,j} \in S_*^h, \quad s_{out,j'} \in S_*^c \qquad (4.46)$$

$$T_f\left(s_{out,j}, p\right) - T_i\left(s_{out,j'}, p\right) \geq \Delta T_{min} - \Gamma\left(1 - x\left(s_{out,j}, s_{out,j'}, p\right)\right)$$

$$\forall j, j' \in J, \quad p \in P, \quad s_{out,j} \in S_*^h, \quad s_{out,j'} \in S_*^c \qquad (4.47)$$

$$T_i\left(s_{out,j}, p-1\right) - T_f\left(s_{in,j'}^r, p\right) \geq \Delta T_{min} - \Gamma\left(1 - x\left(s_{out,j}, 0, s_{in,j'}^r, p-1\right)\right)$$

$$\forall j, j' \in J, \quad p \in P, \quad p > 1, \quad s_{in,j'}^r \in S_*^c, \quad s_{out,j} \in S_*^h \qquad (4.48)$$

$$T_f\left(s_{out,j}, p\right) - Ts_0(s) \geq \Delta T_{min} - \Gamma\left(1 - x\left(s_{out,j}, s_{in,j'}^r, p\right)\right)$$

$$\forall j \in J, \quad j' \in J_s^c, \quad p \in P, \quad s \in S^r, \quad s_{in,j'}^r \in S_*^c, \quad s_{out,j} \in S_*^h \qquad (4.49)$$

It is assumed that the temperatures of available hot and cold utilities are adequate to achieve the desired heating and cooling. While the temperature of the input material is bounded by the operating temperature, the highest and lowest achievable temperatures of output materials are limited by utility temperatures. This is given in Constraints (4.50) and (4.51).

$$T_{sat}^{st} - T_{out}\left(s_{in,j}, p\right) \geq \Delta T_{min} \quad \forall j \in J_{out}^{h}, \quad p \in P, \quad s_{in,j} \in S_{in,j}^{eff} \tag{4.50}$$

$$T_{out}\left(s_{in,j}, p\right) - T_{in}^{cw} \geq \Delta T_{min} - \Gamma\left(1 - y\left(s_{in,j}, p\right)\right) \quad \forall j \in J_{out}^{c}, \quad p \in P, \quad s_{in,j} \in S_{in,j}^{eff} \tag{4.51}$$

To ensure that the product achieves its storage temperature, Constraints (4.52) and (4.53) are used for product streams to have and not to have heat exchange, respectively:

$$T_f\left(s_{out,j}, p\right) \leq T_{stor}\left(s\right) \quad \forall j \in J_s^p, \quad p \in P, \quad s \in S^p, \quad s_{out,j} \in S_{out,j} \tag{4.52}$$

$$T_{out}\left(s_{in,j}, p\right) \leq T_{stor}\left(s\right) \quad \forall j \in J_s^p, \quad p \in P, \quad s \in S^p, \quad s_{in,j} \in S_{in,j}^{eff} \tag{4.53}$$

The following constraints are necessary for heat integration ensuring that the two process streams appear at the same time if there is a heat exchange between them. Constraints (4.54) and (4.55) are applicable to outlet stream matches and Constraints (4.56) and (4.57) to output/raw material stream matches.

$$t_p\left(s_{in,j}, p\right) \geq t_p\left(s_{in,j'}, p\right) - H\left(1 - x\left(s_{out,j}, s_{out,j'}, p\right)\right)$$
$$\forall j, j' \in J, \quad p \in P, \quad s_{in,j}, s_{in,j'} \in S_{in,j}^{eff}, \quad s_{out,j} \in S_*^h, \quad s_{out,j'} \in S_*^c \tag{4.54}$$

$$t_p\left(s_{in,j}, p\right) \leq t_p\left(s_{in,j'}, p\right) + H\left(1 - x\left(s_{out,j}, s_{out,j'}, p\right)\right)$$
$$\forall j, j' \in J, \quad p \in P, \quad s_{in,j}, s_{in,j'} \in S_{in,j}^{eff}, \quad s_{out,j} \in S_*^h, \quad s_{out,j'} \in S_*^c \tag{4.55}$$

$$t_p\left(s_{in,j}, p-1\right) \geq t_u\left(s_{in,j'}, p\right) - tt - H\left(1 - x\left(s_{out,j}, s_{in,j'}^r, p-1\right)\right)$$
$$\forall j, j' \in J, \quad p \in P, \quad p > 1, \quad s_{in,j}, s_{in,j'} \in S_{in,j}^{eff}, \quad s_{out,j} \in S_*^h, \quad s_{in,j'}^r \in S_*^c \tag{4.56}$$

$$t_p\left(s_{in,j}, p-1\right) \leq t_u\left(s_{in,j'}, p\right) - tt + H\left(1 - x\left(s_{out,j}, s_{in,j'}^r, p-1\right)\right)$$
$$\forall j, j' \in J, \quad p \in P, \quad p > 1, \quad s_{in,j}, s_{in,j'} \in S_{in,j}^{eff}, \quad s_{out,j} \in S_*^h, \quad s_{in,j'}^r \in S_*^c \tag{4.57}$$

4.5.5 OBJECTIVE FUNCTION

For profit maximization, the objective function may be formulated as the difference between the product revenue (*PR*) and the costs of raw materials (*RC*) and external utilities (*UC*), as shown in Equation 4.58. It is evident that Equation 4.58 is aimed at reducing the utility costs, which is concomitant with maximizing process–process heat integration.

$$\max Profit = PR - RC - UC \tag{4.58}$$

where

$$PR = \sum_{s \in S^P} Price(s) qs(s,p)\Big|_{p=|P|} \tag{4.59}$$

$$RC = \sum_{s \in S^r} \sum_{j \in J_s^c} \sum_{p \in P} Cost(s) ms_{out}(s,j,p) \tag{4.60}$$

$$UC = \sum_{p \in P} CHU \left(\sum_{s_{in,j} \in S_{in,j}^{eff}} \left(q_{in}^H(s_{in,j},p) + q_{out}^H(s_{in,j},p) \right) \right)$$

$$+ \sum_{p \in P} CCU \left(\sum_{s_{in,j} \in S_{in,j}^{eff}} \left(q_{in}^C(s_{in,j},p) + q_{out}^C(s_{in,j},p) \right) \right) \tag{4.61}$$

Due to the presence of bilinear terms in Constraints (4.5), (4.6), (4.15) through (4.18), (4.20), (4.22), (4.24), (4.25), (4.27), (4.29) and (4.42) through (4.44) and the use of binary variables, the overall model is a mixed integer nonlinear program (MINLP), for which global optimality may not be guaranteed. In the next section, the application of the presented model is demonstrated through a modified literature case study. The model is implemented in the GAMS environment (Rosenthal, 2015) on a Core i7–4790, 3.60 GHz processor with BARON (Tawarmalani and Sahinidis, 2005) as the MINLP solver.

4.6   ILLUSTRATIVE CASE STUDY

A multipurpose batch plant is to be scheduled to produce three products ($s9$–$s11$) from three raw materials ($s1$, $s2$ and $s7$) over a 16-h time horizon. The STN (state-task network) process representation is shown in Figure 4.4, where states $s3$–$s6$ and $s8$ are all intermediates. This case study is adapted from Pinto et al. (2008), with added heating and cooling requirements of materials to explore heat integration opportunities. Tables 4.1 and 4.2 present the pertinent data for the case study. For simplicity, it is assumed that each state has a dedicated storage vessel and each processing task is performed in a dedicated reactor. Note that the supply temperature in Table 4.2 refers to the initial temperature of a state in the storage vessel. Therefore, the supply temperatures of intermediates and products are taken as zero.

It can be seen in Figure 4.4 that unit R1 is likely to have heating for its input $s1$ and cooling for its output $s4$, as the supply temperature of $s1$ is lower than the

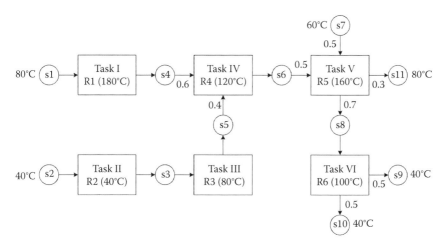

FIGURE 4.4   STN representation for the case study.

TABLE 4.1
Processing Data for the Case Study

Task	Unit	Duration (h)	Operating Temperature (°C)	Capacity Limits (kg)
I	R1	2	180	30–120
II	R2	2	40	20–80
III	R3	4	80	20–80
IV	R4	2	120	50–200
V	R5	1	160	50–200
VI	R6	1	100	35–140

operating temperature of R1, and the latter is higher than the operating temperature of R4. Similarly, potential heating requirements for the inputs of R3 and R5 and the outputs of R2–R4 can be identified from Figure 4.4, together with potential cooling requirements for the outputs of R5 and R6. However, some uncertainties remain for R4 and R6. The input of R4 consists of $s4$ and $s5$, which are produced at different temperatures and can give a mixture temperature higher or lower than its operating temperature. In the case of R6, if the product storage temperature of $s11$ (80°C) is achieved by utility cooling inside R5, then $s8$ will leave R5 at 80°C, which is lower than the operating temperature of R6 (100°C). However, if product cooling is achieved by heat exchange during the transfer, the final temperature of $s8$ may be higher than the operating temperature of R6. Therefore, constraints (4.19) through (4.23) are used for both units R4 and R6, while constraints (4.26) through (4.30) are not needed in this case study.

The hot and cold streams for heat exchange can also be identified from Figure 4.4. There are three potential hot streams ($s4$outR1, $s8$outR5 and $s11$outR5) and five

TABLE 4.2

Material and Utility Data for the Case Study

State	Storage Capacity (kg)	Initial Inventory (kg)	Supply Temperature (°C)	Product Storage Temperature (°C)	Heat Capacity (kJ/kg °C)	Price ($/kg)
$s1$	UL	AA	80	—	2.5	0.5
$s2$	UL	AA	40	—	3.4	0.5
$s3$	160	0	0	—	3.3	—
$s4$	240	0	0	—	3.7	—
$s5$	160	0	0	—	4.0	—
$s6$	400	0	0	—	3.6	—
$s7$	UL	AA	60	—	3.5	0.5
$s8$	280	0	0	—	4.0	—
$s9$	UL	0	0	40	3.2	10
$s10$	UL	0	0	40	3.3	5
$s11$	UL	0	0	80	3.7	10

Material transfer time	10 min
Minimum temperature difference for heat transfer	10°C
Cooling water cost	$0.6/MJ
Cooling water inlet/outlet temperature	20/30°C
Heat capacity of cooling water	4.2 kJ/kg °C
Mass flow rate of cooling water	1800 kg/h
Steam cost	$3/MJ
Saturated steam temperature	200°C
Latent heat of steam	1940 kJ/kg
Mass flow rate of steam	40 kg/h

Note: UL, unlimited; AA, available as/when required.

potential cold streams ($s1$inR1, $s3$outR2, $s5$outR3, $s6$outR4 and $s7$inR5). Note that the feed stream to R2 ($s2$inR2) and the product streams from R6 ($s9$outR6 and $s10$outR6) are left out; the former has no need for heating or cooling, while storage temperatures of the latter can only be achieved with cooling water. The objective is to maximize the profit given by the difference between product revenue and operating cost associated with raw materials and external utilities.

Batch operations are normally carried out with fixed batch sizes in the process industries in order to minimize changes. Therefore, batch sizes of all processing tasks are fixed at the maximum capacities given in Table 4.1. Prior to exploring heat integration opportunities, the batch plant produces 140 kg of $s9$, 140 kg of $s10$ and 120 kg of $s11$ within the 16-h time horizon, while consuming 161.76 MJ of steam and 193.56 MJ of cooling water. This gives a profit of $2498.58, obtained from the presented model by setting all the binary variables associated with heat exchange to zero. Figure 4.5 shows the Gantt chart of production without heat integration. Note that the heating and cooling required in R5 increase the material residence time to

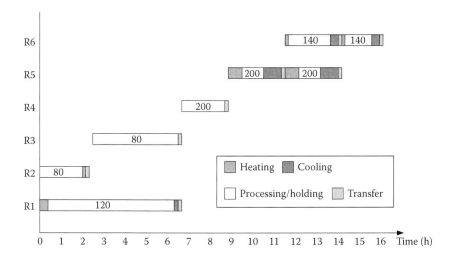

FIGURE 4.5 Gantt chart of production without heat integration.

at least 2.47 h, which is much longer than the task duration of 1 h. Thus, there is not enough time to process one more batch in R5.

When heat exchange between outlet streams (three hot, s4outR1, s8outR5 and s11outR5; three cold, s3outR2, s5outR3 and s6outR4) is considered, the plant can produce 140 kg of s9, 140 kg of s10 and one batch more of s11 (i.e. 180 kg) within the same time horizon. However, a higher utility cost of $661.97 for 188.16 MJ of steam and 162.48 MJ of cooling water is involved with increased production. The profit in this case is $2808.03, corresponding to a 12.38% improvement compared to the case without heat integration. Figure 4.6 shows the optimal schedule with

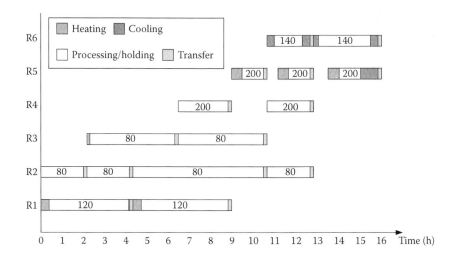

FIGURE 4.6 Optimal schedule with heat exchange between outlet streams.

heat exchange between outlet streams. It can be seen that almost double the number of batches are processed (increased from 8 to 15), but some are only produced for heat exchange, namely the last two batches in R2. These surplus intermediates may be used later over an extended time horizon. The detailed heat exchange results are shown in Tables 4.3 and 4.4. There are six matches of heat exchange to reduce the heating and cooling required, especially in R1 and R5 where the cooling needs can even be eliminated.

When raw material streams (two additional cold, $s1inR1$ and $s7inR5$) are also considered for heat exchange, the product revenue can be increased to $4500, with the production of another batch of $s11$, and the utility cost reduced to $525.24, giving a profit of $3534.76. This corresponds to a further 29.09% improvement compared to the previous cases. Figure 4.7 shows the optimal schedule with heat exchange among raw material and outlet streams, with Tables 4.5 and 4.6

TABLE 4.3
Heat Exchange between Outlet Streams

Stream match	Time Point			
(Hot-cold)	$p = 2$	$p = 3$	$p = 4$	$p = 5$
$s4outR1$-$s3outR2$	10,560			
$s4outR1$-$s6outR4$		22,200		
$s8outR5$-$s5outR3$			8,360	
$s8outR5$-$s6outR4$				16,800
$s11outR5$-$s3outR2$			17,760	17,760

Note: All values given in kJ.

TABLE 4.4
Initial-Final Temperatures of Outlet Streams for Heat Exchange

Stream	$p = 1$	$p = 2$	$p = 3$	$p = 4$	$p = 5$	$p = 6$
$s3outR2$	40–40	40–80		40–107.27	40–107.27	
$s4outR1$		172.61–148.83	180–130			
$s5outR3$		80–80	80–106.13			
$s6outR4$			120–150.83		120–143.33	
$s8outR5$				160–145.07	160–130	80–80
$s11outR5$				160–80	160–80	80–80

Note: All values given in °C.

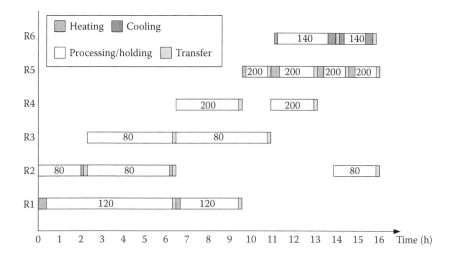

FIGURE 4.7 Optimal schedule with heat exchange among raw material and outlet streams.

TABLE 4.5
Heat Exchange among Raw Material and Outlet Streams

Stream Match (Hot-cold)	Time Point					
	$p = 2$	$p = 3$	$p = 4$	$p = 5$	$p = 6$	$p = 7$
s4outR1-s1inR1	13,840					
s4outR1-s7inR5		36,240				
s8outR5-s5outR3			22,400			
s8outR5-s6outR4				16,800		
s11outR5-s7inR5			17,760	17,760	17,760	
s11outR5-s3outR2						17,760

Note: All values given in kJ.

presenting the detailed heat exchange results. Note that raw material streams take priority over the other cold streams as the former involve larger enthalpy changes. Consequently, streams s3outR2 and s6outR4 are less involved in heat integration, and there are fewer batches produced in R2. The computational results for this case study, including model size and solution time, are summarized in Table 4.7 An overall comparison shows that heat integration of material transfer streams enables the batch plant to achieve higher production with lower utility consumption.

TABLE 4.6
Initial-Final Temperatures of Raw Material and Outlet Streams for Heat Exchange

Stream				Time Point			
	$p = 1$	$p = 2$	$p = 3$	$p = 4$	$p = 5$	$p = 6$	$p = 7$
s1inR1		80–80	80–126.13				
s3outR2	80–80	80–80					40–107.27
s4outR1		180–148.83	180–98.38				
s5outR3		80–80		80–150			
s6outR4			120–120	120–143.33			
s7inR5				60–163.54	60–110.74	60–110.74	60–110.74
s8outR5				160–120	160–130	160–160	160–160
s11outR5				160–80	160–80	160–80	160–80

Note: All values given in °C.

TABLE 4.7

Computational Results for the Case Study

	Without Heat Integration	Heat Integration of	
		Outlet Streams	Raw Material/Outlet Streams
Profit ($)	2498.58	2808.03	3534.76
Improvement	—	12.38%	41.47%
Product revenue ($)	3300	3900	4500
$s9$ production (kg)	140	140	140
$s10$ production (kg)	140	140	140
$s11$ production (kg)	120	180	240
Raw material cost ($)	200	430	440
Utility cost ($)	601.42	661.97	525.24
Steam (MJ)	161.76	188.16	158.56
Cooling water (MJ)	193.56	162.48	82.6
Model statistics			
No. of time points	6	6	7
No. of constraints	1492	1492	1966
No. of variables	1029	1029	1296
No. of binaries	150	204	274
Solution time (CPU s)	62	166	2311

4.7   SUMMARY

A superstructure-based mathematical model for simultaneous process scheduling and heat integration of multipurpose batch plants has been presented in this chapter. Based on a robust scheduling framework adapted from Seid and Majozi (2012), the MINLP formulation ensures proper sequencing of tasks over the time horizon of interest, with the aim of synchronizing the material transfer times in order to maximize heat recovery. An illustrative case study was solved to demonstrate the application of the proposed model. The results indicate that heat integration of process streams allows batch plants to achieve considerably higher production (+36.36%) at a much lower utility cost (−12.67%). It should be noted that the presented formulation assumes intermediate streams to exchange heat only before storage, and external utilities to be used only to the unit jacket. These simplifying assumptions can be relaxed for even better results. On the other hand, direct and indirect heat integration between processing units may be worth considering in future investigation.

4A   APPENDIX

The necessary assignment, sequence and time horizon constraints for scheduling are adapted from Seid and Majozi (2012) and presented here.

4A.1 ASSIGNMENT CONSTRAINT

Constraint (4A.1) states that at any time point, only one task can be performed in a unit. This constraint is only needed if more than one task can be carried out in a given unit.

$$\sum_{s_{in,j} \in S_{in,j}^{eff}} y\left(s_{in,j}, p\right) \leq 1 \quad \forall j \in J, \quad p \in P \tag{4A.1}$$

4A.2 SEQUENCE CONSTRAINT FOR THE SAME TASK IN THE SAME UNIT

Constraint (4A.2) states that a new task can only start in a processing unit after the previous task is complete and the product material removed from the unit.

$$t_u\left(s_{in,j}, p\right) \geq t_p\left(s_{in,j}, p-1\right) + tt\left(y\left(s_{in,j}, p-1\right)\right) \quad \forall j \in J, \quad p \in P, \quad p > 1, \quad s_{in,j} \in S_{in,j}^{eff} \tag{4A.2}$$

4A.3 SEQUENCE CONSTRAINTS FOR STORAGE

Constraints (4A.3) through (4A.6) ensure that the times at which a state is transferred to and from the storage at a given time point are later than those at the previous time point.

$$tm_{in}\left(s, p\right) \geq tm_{in}\left(s, p-1\right) \quad \forall p \in P, \quad p > 1, \quad s \notin S^r \tag{4A.3}$$

$$tm_{in}\left(s, p\right) \geq tm_{out}\left(s, p-1\right) \quad \forall p \in P, \quad p > 1, \quad s \in S^i \tag{4A.4}$$

$$tm_{out}\left(s, p\right) \geq tm_{in}\left(s, p-1\right) \quad \forall p \in P, \quad p > 1, \quad s \in S^i \tag{4A.5}$$

$$tm_{out}\left(s, p\right) \geq tm_{out}\left(s, p-1\right) \quad \forall p \in P, \quad p > 1, \quad s \notin S^p \tag{4A.6}$$

4A.4 SEQUENCE CONSTRAINTS FOR PROCESSING UNITS AND STORAGE

The upper bound for the amount of material sent to storage is given by Constraint (4A.7). When a state is transferred from its producing unit to the storage, Constraints (4A.8) and (4A.9) ensure that the material transfer takes place upon completion of the producing task.

$$ms_{in}\left(s, j, p\right) \leq V_j^U w_{in}\left(s, j, p\right) \quad \forall j \in J_s^p, \quad p \in P, \quad s \notin S^r \tag{4A.7}$$

$$t_p\left(s_{in,j}, p\right) \geq tm_{in}\left(s, p\right) - tt\left(w_{in}\left(s, j, p\right)\right) - H\left(1 - w_{in}\left(s, j, p\right)\right)$$
$$\forall j \in J_s^p, \quad p \in P, \quad s \notin S^r, \quad s_{in,j} \in S_{in,j}^{eff} \tag{4A.8}$$

$$t_p\left(s_{in,j}, p\right) \leq tm_{in}\left(s, p\right) - tt\left(w_{in}\left(s, j, p\right)\right) + H\left(1 - w_{in}\left(s, j, p\right)\right)$$
$$\forall j \in J_s^p, \quad p \in P, \quad s \notin S^r, \quad s_{in,j} \in S_{in,j}^{eff} \tag{4A.9}$$

The upper bound for the amount of material sent from storage is given by Constraint (4A.10). When a state is transferred from the storage to the consuming unit, Constraints (4A.11) and (4A.12) ensure that the consuming task starts right after the material transfer.

$$ms_{out}\left(s,j,p\right)\le V_j^U w_{out}\left(s,j,p\right) \quad \forall j \in J_s^c, \quad p \in P, \quad s \notin S^p \tag{4A.10}$$

$$t_u\left(s_{in,j},p\right)\ge tm_{out}\left(s,p\right)+tt\left(w_{out}\left(s,j,p\right)\right)-H\left(1-w_{out}\left(s,j,p\right)\right)$$
$$\forall j \in J_s^c, \quad p \in P, \quad s \notin S^p, \quad s_{in,j} \in S_{in,j}^{eff} \tag{4A.11}$$

$$t_u\left(s_{in,j},p\right)\le tm_{out}\left(s,p\right)+tt\left(w_{out}\left(s,j,p\right)\right)+H\left(1-w_{out}\left(s,j,p\right)\right)$$
$$\forall j \in J_s^c, \quad p \in P, \quad s \notin S^p, \quad s_{in,j} \in S_{in,j}^{eff} \tag{4A.12}$$

4A.5   SEQUENCE CONSTRAINTS FOR PRODUCING AND CONSUMING UNITS

The upper bound for the amount of material sent between processing units is given by Constraint (4A.13). When a state is transferred from its producing unit to its consuming unit, Constraints (4A.14) and (4A.15) ensure that the material transfer is carried out between the end of the producing task and the start of the consuming task.

$$mt\left(s,j,j',p\right)\le V_j^U z\left(j,j',p\right) \quad \forall j \in J_s^p, \quad j' \in J_s^c, \quad p \in P, \quad s \in S^i \tag{4A.13}$$

$$t_u\left(s_{in,j'},p\right)\ge t_p\left(s_{in,j},p-1\right)+tt\left(z\left(j,j',p-1\right)\right)-H\left(1-z\left(j,j',p-1\right)\right)$$
$$\forall j \in J_s^p, \quad j' \in J_s^c, \quad p \in P, \quad p > 1, \quad s_{in,j}, s_{in,j'} \in S_{in,j}^{eff} \tag{4A.14}$$

$$t_u\left(s_{in,j'},p\right)\le t_p\left(s_{in,j},p-1\right)+tt\left(z\left(j,j',p-1\right)\right)+H\left(1-z\left(j,j',p-1\right)\right)$$
$$\forall j \in J_s^p, \quad j' \in J_s^c, \quad p \in P, \quad p > 1, \quad s_{in,j}, s_{in,j'} \in S_{in,j}^{eff} \tag{4A.15}$$

4A.6   TIME HORIZON CONSTRAINTS

The use, production and transfer of states should all be finished within the time horizon of interest, as given in Constraints (4A.16) through (4A.19). Note that the material transfer time is taken into account.

$$t_u\left(s_{in,j},p\right)\le H-tt \quad \forall j \in J, \quad p \in P, \quad s_{in,j} \in S_{in,j} \tag{4A.16}$$

$$t_p\left(s_{in,j},p\right)\le H-tt \quad \forall j \in J, \quad p \in P, \quad s_{in,j} \in S_{in,j} \tag{4A.17}$$

$$tm_{in}\left(s,p\right)\le H \quad \forall p \in P, \quad s \notin S^r \tag{4A.18}$$

$$tm_{out}\left(s,p\right)\le H-tt \quad \forall p \in P, \quad s \notin S^p \tag{4A.19}$$

NOMENCLATURE

INDICES AND SETS

$j \in J$	Processing units
$j \in J_{in}^c$	Processing units to have cooling for input material
$j \in J_{out}^c$	Processing units to have cooling for output material
$j \in J_s^c$	Processing units consuming state s
$j \in J_{in}^h$	Processing units to have heating for input material
$j \in J_{out}^h$	Processing units to have heating for output material
$j \in J_s^p$	Processing units producing state s
$j \in J_{in}^*$	Processing units to have heating or cooling for input material
$j \in J_{out}^*$	Processing units to have heating or cooling for output material
$p \in P$	Time points
$s \in S$	States
$s \in S_j^c$	States consumed in unit j
$s \in S^i$	Intermediate states
$s \in S^p$	Product states
$s \in S^r$	Raw material states
$s_{in,j} \in S_{in,j}$	Inlet streams
$s_{in,j} \in S_{in,j}^{eff}$	Effective states representing tasks
$s_{out,j} \in S_{out,j}$	Outlet streams
$s_{in,j}^r \in S_{in,j}^r$	Raw material streams
$s_*^c \in S_*^c \subset S_{out,j} \cup S_{in,j}^r$	Cold streams
$s_*^h \in S_*^h \subset S_{out,j}$	Hot streams

PARAMETERS

CCW	Cooling water cost
$Cp(s)$	Specific heat capacity of state s
$Cp_{in}(s_{in,j})$	Specific heat capacity of the input material to the task
$Cp_{out}(s_{in,j})$	Specific heat capacity of the output material from the task
Cp^{cw}	Specific heat capacity of cooling water
CST	Steam cost
H	Time horizon of interest
$\dot{M}^{cw}$	Constant cooling water flow rate through the jacket
$\dot{M}^{st}$	Constant steam flow rate through the jacket
$Q_0(s)$	Initial amount of state s in the storage
Q^U	Upper bound for heat loads

Q_s^U	Maximum storage capacity for state s
$T_{op}(s_{in,j})$	Operating temperature for the task
$T_{stor}(s)$	Storage temperature of product state s
T_{in}^{cw}	Inlet temperature of cooling water
T_{out}^{cw}	Outlet temperature of cooling water
T_{sat}^{st}	Saturated steam temperature
$Ts_0(s)$	Initial temperature of state s in the storage
tt	Material transfer time
V_j^L	Minimum capacity of unit j
V_j^U	Maximum capacity of unit j
ΔT_{min}	Minimum temperature difference
Γ	A large enough positive value
λ^{st}	Steam latent heat
$\rho_s^c(s_{in,j})$	Fraction of state s in the input consumed by the task
$\rho_s^p(s_{in,j})$	Fraction of state s in the output produced by the task
$\tau(s_{in,j})$	Constant duration of the task

VARIABLES

$ms_{in}(s,j,p)$	Amount of state s sent to storage from unit j at time point p
$ms_{out}(s,j,p)$	Amount of state s sent from storage to unit j at time point p
$mt(s,j,j',p)$	Amount of state s sent from unit j to unit j' at time point p
$m_u(s_{in,j},p)$	Amount of material used for the task at time point p
$q_{ex}(s_{**}^h, s_*^c, p)$	Heat exchanged between the hot and cold streams at time point p
$q_{in}^c(s_{in,j},p)$	Amount of cooling for the input material at time point p
$q_{out}^c(s_{in,j},p)$	Amount of cooling for the output material at time point p
$q_{in}^h(s_{in,j},p)$	Amount of heating for the input material at time point p
$q_{out}^h(s_{in,j},p)$	Amount of heating for the output material at time point p
$qs(s,p)$	Amount of state s stored at time point p
$T_f(s_{out,j},p)$	Final temperature of the outlet stream at time point p
$T_f(s_{in,j}^r,p)$	Final temperature of the raw material stream at time point p
$T_i(s_{out,j},p)$	Initial temperature of the outlet stream at time point p
$T_{in}(s_{in,j},p)$	Temperature of the input material at time point p
$T_{out}(s_{in,j},p)$	Temperature of the output material at time point p
$t_p(s_{in,j},p)$	End time of the task at time point p
$t_u(s_{in,j},p)$	Start time of the task at time point p

$t_{in}^r\left(s_{in,j}, p\right)$	Time required for heating or cooling for the input material at time point p
$t_{out}^r\left(s_{in,j}, p\right)$	Time required for heating or cooling for the output material at time point p
$tm_{in}\left(s, p\right)$	Time at which state s is sent to storage at time point p
$tm_{out}\left(s, p\right)$	Time at which state s is sent from storage at time point p
$Ts_{out}\left(s, p\right)$	Outlet temperature of state s from storage at time point p
$w_{in}\left(s, j, p\right)$	Binary variable indicating if state s is sent to storage from unit j at time point p
$w_{out}\left(s, j, p\right)$	Binary variable indicating if state s is sent from storage to unit j at time point p
$x\left(s_*^h, s_*^c, p\right)$	Binary variable indicating if there is heat exchange between the hot and cold streams at time point p
$y\left(s_{in,j}, p\right)$	Binary variable indicating if the task is active at time point p
$y_{in}^c\left(s_{in,j}, p\right)$	Binary variable indicating if there is cooling for the input material at time point p
$y_{out}^c\left(s_{in,j}, p\right)$	Binary variable indicating if there is cooling for the output material at time point p
$y_{in}^c\left(s_{in,j}, p\right)$	Binary variable indicating if there is heating for the input material at time point p
$y_{out}^h\left(s_{in,j}, p\right)$	Binary variable indicating if there is heating for the output material at time point p
$z\left(j, j', p\right)$	Binary variable indicating if there is material sent from unit j to unit j' at time point p

REFERENCES

Fernández, I., Renedo, C.J., Pérez, S.F., Ortiz, A., Mañana, M., 2012. A review: Energy recovery in batch processes. *Renewable and Sustainable Energy Reviews*. 16, 2260–2277.

Harjunkoski, I., Maravelias, C.T., Bongers, P., Castro, P.M., Engell, S., Grossmann, I.E., Hooker, J., Méndez, C., Sand, G., Wassick, J., 2014. Scope for industrial applications of production scheduling models and solution methods. *Computers and Chemical Engineering*. 62, 161–193.

Majozi, T., 2010. *Batch Chemical Process Integration: Analysis, Synthesis and Optimization*. Dordrecht, the Netherlands: Springer.

Pinto, T., Barbósa-Póvoa, A.P.F.D., Novais, A.Q., 2008. Design of multipurpose batch plants: a comparative analysis between the STN, m-STN, and RTN representations and formulations. *Industrial and Engineering Chemical Research*. 47, 6025–6044.

Rosenthal, R.E., 2015. *GAMS—A User's Guide*. Washington, DC: GAMS Development Corporation.

Seid, R., Majozi, T., 2012. A robust mathematical formulation for multipurpose batch plants. *Chemical Engineering Science*. 68, 36–53.

Tawarmalani, M., Sahinidis, N.V., 2005. A polyhedral branch-and-cut approach to global optimization. *Mathematical Programmes*. 103, 225–249.

5 Heat Integration in Multipurpose Batch Plants

5.1 INTRODUCTION

Energy saving is becoming increasingly important in batch processing facilities. Multipurpose batch plants have become more popular than ever in the processing environment due to their inherent flexibility and adaptability to market conditions, even though the same flexibility may lead to complexities, such as the need to schedule process tasks. These are important features to producing high value–added products such as agrochemicals, pharmaceuticals, polymers, food and specialty chemicals where the demand has grown in recent decades. Many current heat integration methods for multipurpose batch plants use a sequential methodology where the schedule is solved first, followed by heat integration. This can lead to suboptimal results. In this chapter, the heat integration model is built upon a robust scheduling framework. This scheduling formulation has proven to lead to better results in terms of better objective values, fewer required time points and reduced computational time. This is important as inclusion of heat integration into a scheduling model invariably complicates the solution process. The improved scheduling model allows the consideration of industrial-sized problems to simultaneously optimize both the process schedule and energy usage. Both direct and indirect heat integration are considered, as well as fixed and variable batch sizes.

5.2 NECESSARY BACKGROUND

Batch processing is commonly used, when products are required in small quantities or when the processes are complex or specialized, to manufacture high value–added products. Examples include food, pharmaceuticals, fine chemicals, biochemicals and agrochemicals. Batch operations, even though they are becoming increasingly popular, are generally run on a smaller scale compared to continuous operations, and utility requirements are therefore considered less significant. However, utility requirements for the food industry, breweries, dairies, meat processing facilities, biochemical plants and agrochemical facilities contribute largely to their overall cost (Stamp and Majozi, 2011). Energy savings have often been neglected in batch processes in the past and hence significant savings are possible. The literature review is organized into two major sections and includes methods developed in the twentieth and twenty-first centuries.

5.2.1 Developments in the Twentieth Century

Early work that aimed at reducing energy consumption in batch plants was by adopting methods that had been developed for continuous processing plants. One of the very first contributions in this regard for energy integration in batch plants was through applying the principle of time average model (TAM) (Clayton, 1986). However, this approach does not consider the process schedule and assumes that the hot and cold streams exist simultaneously as in continuous processes. The identified energy targets could not be achieved when only direct heat integration was applied since the time schedule was not considered. The TAM formulation was further extended to incorporate heat storage to achieve the maximum energy saving targets (Stolze et al., 1995). Vaselenak et al. (1986) explored the possibility of heat recovery between a number of tanks which required heating and cooling through the consideration of co-current, countercurrent and a combination of the two. The authors did not account for the time schedule; they assumed that all tanks are available at the same time. In order to address the limitations of the TAM, various works considered the time schedule in the heat integration analysis. The work on the design of batch processes based on pinch technology showed that the cost structure should consider the interaction between capital, energy, scheduling and yields in order to address the wider scope (Obeng and Ashton, 1988). The cascade analysis for maximum heat recovery based on time–temperature cascade tables that consider both direct and indirect heat exchange between streams was developed (Kemp and Deakin, 1989). A strategy for creating 'schedules of maximum power of heat integration' which ensures, on the one hand, production targets and, on the other hand, optimal condition for heat integration was demonstrated (Ivanov and Bancheva, 1994), and a methodology was presented to address the important aspect of rescheduling for maximum energy recovery (Corominas, 1993). Different theoretical aspects of optimal energy integration in batch chemical plants have been considered in the previous literature, with little experience on real case studies. This gap was addressed by applying optimal energy integration formulation for an existing antibiotics plant where implementation of the formulation gave an overall energy cost reduction of 39% (Boyadjiev et al., 1996). Most of the research work mentioned here is based on fixed production schedules that already achieve all the plant objectives and minimize the energy requirement afterward. However, in general, even optimal production schedules tend to be quite degenerate, in the sense that there often exists a large number of different schedules, all of which can achieve a given set of production requirements. Nevertheless, the potential for heat integration could vary significantly from one such schedule to another. Consequently, heat integration should be considered as an integral part of the problem of scheduling the production in a given plant. The cost of utilities should be incorporated within the overall economic objective of maximizing the net value of the production over a given time horizon and solved simultaneously (Papageorgiou et al., 1994). There were other studies that resolved heat integration in batch processes by establishing an objective that was to minimize the combined operating and annualized capital costs of the heat-exchanger networks (HENs) for a class of multipurpose batch plants. A mathematical formulation that selects the production campaigns and designs HENs simultaneously has been presented (Bancheva et al., 1996a). However, the formulation limits a heat exchanger to be used for a specific pair of processing units. Design of HENs for a system of batch vessels that exploits heat integration potential with minimum cost has also been developed (Pozna et al., 1998).

5.2.2 Developments in the Twenty-First Century

The research conducted in this century has been concentrated on improving the models and methodologies that had been proposed previously as well as the optimization of heat integration in batch processes through the development of new tools such as genetic algorithms and network evolution techniques. Moreover, there has been a significant increase in the utilization of thermal storage to improve heat integration in a discontinuous process.

5.2.2.1 Models Developed for Direct Heat Integration

The developed models in this category require the hot and cold task should be operated in the same time interval for the heat exchange to occur. Uhlenbruck et al. (2000) improved OMNIUM, which is a tool developed for heat exchanger network synthesis by Hellwig and Thone (1994). The improved OMNIUM tool increased energy recovery by 20%. A single step, interactive computer program (BatcHEN) used for the determination of campaigns (i.e. the set of products which can be produced simultaneously), the heat exchange areas of all possible heat exchangers in the campaigns and the heat exchanger network were presented (Bozan et al., 2001). This work addressed the limitation of the graph theory method (Bancheva et al., 1996b) for the determination of the campaign where it is very complex for handling large numbers of products and process units. A heat integration model based on S-graph scheduling framework approaches was developed (Adonyi et al., 2003). The results of this work showed how utility usage can be reduced considerably with just a slight increase in production makespan. Morrison et al. (2007) developed a user-friendly software package known as Optimal Batch Integration (OBI). Chen and Chang (2009) integrated the task scheduling and heat recovery problems into a unified framework for multipurpose batch processes. The batch scheduling formulation is extended from the continuous Resource Task Network (RTN) formulation, which was originally proposed by Castro et al. (2004). Halim and Srinivasan (2009) discussed a sequential method using direct heat integration. A number of optimal schedules with minimum makespan were found and heat integration analysis was performed on each. The schedule with minimum utility requirement was chosen as the best. Later, Halim and Srinivasan (2011) extended their technique to carry out water reuse network synthesis simultaneously. One key feature of this method is its ability to find the heat integration and water reuse solution without much sacrificing the quality of the scheduling solution as compared to other sequential techniques for heat integration for multipurpose batch plants.

There are other works that considered energy saving and the capital cost associated to heat exchanger to achieve the minimum utility target. The technique for design and synthesis of batch plant (Barbosa-Póvoa et al., 2001) was later extended to incorporate economic savings in utility requirements, while considering the cost of both auxiliary structures (i.e. heat exchangers through their transfer area) and the design of the utility circuits and associated piping costs (Pinto et al., 2003). Liu et al. (2011) formulated a batch heat exchanger network that results in nonlinear programming (NLP). The application of the formulation has demonstrated that it can effectively reduce the annual capital cost and annual total cost of HEN with the employment of common heat exchangers. Maiti et al. (2011) developed a novel heat integrated batch distillation column in order to improve thermal efficiency and reduce the total annual cost. The potential energy integration leads to achieving approximately 56.10% energy savings and 40.53% savings in total annual cost.

5.2.2.2 Models Developed for Indirect Heat Integration

Models developed in this category used heat storage for a more heat recovery and flexible schedule compared to direct heat integration methods. The use of heat storage allows the exchange of heat from hot task to cold task to takes place in different time interval. A systematic procedure based on pinch analysis, backed with a graphical representation, allows the determination of the minimum number of heat storage units and their range of feasible operation as a function of the amount of heat recovery was presented (Krummenacher and Favrat, 2001). Chen and Ciou (2008) formulated a method to design and optimize indirect energy storage systems for batch processes. Their work aimed at simultaneously solving the problem of indirect heat exchange network synthesis and its associated thermal storage policy for recirculated hot/cold heat storage medium (HEN). Most of the previous work solved this sequentially. The BatchHeat software that aimed to highlight the energy inefficiencies in the process thereby enabling the scope for possible heat recovery to be established through direct heat exchange or storage through implementation of cascade analysis was developed (Pires et al., 2005).

5.2.2.3 Models Applied for Real Industrial Case Studies

This section presents methodologies developed for energy recovery taking existing batch processing plans. The application of cascade analysis proposed by Kemp and Macdonald (1987) to reduce utility requirement was applied to a case study about an industry that produces oleic acid from palm olein using immobilized lipase (Chew et al., 2005). The result obtained showed savings of 71.4% and 62.5% for hot and cold utilities, respectively. Application of process integration to investigate the potential to decrease energy usage in the slaughtering and meat processing industry was studied. Above ambient temperatures, heating of water with different target temperatures is a large heat demand in a plant, while at sub-ambient temperatures, the refrigeration plant needs almost all of the shaftwork used at the site. Interaction between, on one hand, energy demands above ambient temperature and, on the other hand, cooling needs below ambient temperature can take place with freezing compressors or heat pumps. The result obtained illustrates that 30% of the external heat demand and more than 10% of the shaftwork used can be saved (Fritzson and Berntsson, 2006).

Majozi (2006) presented a heat integration model for multipurpose batch plants based on the continuous-time scheduling frame work (Majozi and Zhu, 2001). The formulation results in smaller problems compared to the discrete-time formulation, which renders it applicable to large-scale problems. Application of the formulation to an agro-chemical industrial case study showed an 18.5% improvement in profit. The direct heat integration model (Majozi, 2006) was extended to incorporate heat storage for more flexible schedules and utility savings in a later work by Majozi (2009). However, storage size is a parameter in his formulation which is addressed later by Stamp and Majozi (2011), where the storage size is determined by an optimization exercise. Foo et al. (2008) extended the minimum units targeting and network evolution techniques that were developed for batch mass exchange network (MEN) into batch HEN. They applied the technique for energy integration of oleic acid production from palm olein using immobilized lipase. Atkins et al. (2010) applied indirect heat integration using heat storage for a milk powder plant in New Zealand. The traditional composite curves have been used to estimate the maximum heat recovery and to determine the optimal temperatures of the stratified tank. Tokos et al. (2010) applied a batch heat integration technique to a large beverage plant. The opportunities of heat integration between batch operations were analysed by a mixed integer linear programming (MILP) model, which was slightly modified by considering specific industrial circumstances. Muster-Slawitsch et al. (2011) came up with the Green Brewery concept to demonstrate the potential for reducing thermal energy consumption in breweries. Three detailed case studies have been performed. The 'Green Brewery' concept has shown a saving potential of more than 5000 t/y fossil CO_2 emissions from thermal energy supply for the three breweries that were closely considered. Becker et al. (2012) applied time average energy integration approach to a real case study of a cheese factory with non-simultaneous process operations. Their work addressed appropriate heat pump integration. A cost saving of more than 40% was reported. Integration of solar thermal energy in a batch fish tinning process was investigated. The work demonstrated the most favourable heat integration option for the thermosolar and heat pump (José et al., 2013). Recently, a design model for heat recovery using heat storage and integration of industrial solar for a case study of dairy processing was developed. Application of the model gave 37% heat recovery (Timothy et al., 2014). The reader can get a more comprehensive and detailed review on energy recovery for batch processes in the chapter by Fernández et al. (2012).

Many heat integration techniques are applied to predefined schedules, which may lead to suboptimal results. For a more optimal solution, scheduling and heat integration should be combined into an overall problem and solved simultaneously. This work aims to improve the efficiency of energy integration techniques by developing a single framework that contains scheduling and heat integration models for multipurpose batch plants to be solved simultaneously for an optimal solution. A recent robust scheduling formulation by Seid and Majozi (2012) is used as a platform since it has proven to require fewer time points and reduced computational time compared to other models and may also lead to an improved objective value. Compared to other models based on simultaneous approach for heat integration for multipurpose batch plants, the developed model allows a task to be heat integrated with other tasks in more than one time interval during its starting and finishing times for better heat recovery. Additionally, this work generalizes the heat integration problem with the considerations of temperature change during processing of tasks and heat integration to occur in any interval between the starting and finishing time of a task, which is a limitation of the recent work by Stamp and Majozi (2011). It also caters for the availability of heat storage for indirect heat integration which is a limitation of the more advanced sequential methodology for heat integration by Halim and Srinivasan (2009). Most literature has addressed these problems independently.

The subsequent sections are organized as follows. The problem statement and objectives are given in Section 5.3. The developed mathematical model is then discussed in Section 5.4. The model is then applied to three literature examples and the results are compared to recent literature models in Section 5.5. Conclusions are then drawn to highlight the value of the contribution in Section 5.6.

5.3   PROBLEM STATEMENT AND OBJECTIVES

The problem addressed in this work can be stated as follows:

Given:

1. Production scheduling data, including equipment capacities, durations of tasks, time horizon of interest, product recipes, cost of starting materials and selling price of final products
2. Hot duties for tasks requiring heating and cold duties for tasks that require cooling
3. Costs of hot and cold utilities
4. Operating temperatures of tasks requiring heating and cooling
5. Minimum allowable temperature differences
6. Design capacity limits on heat storage

Determine:

1. An optimal production schedule where the objective is to maximize profit, defined as the difference between revenue and the cost of hot and cold utilities
2. The size of heat storage as well as the initial temperature of heat storage

5.4   MATHEMATICAL FORMULATION

5.4.1   Model Constraints

The scheduling model by Seid and Majozi (2012) is adopted since it has proven to result in fewer binary variables, reduced CPU time and a better optimal objective value compared to other scheduling models.

The mathematical model is based on the superstructure in Figure 5.1. Each task may operate using either direct or indirect heat integration. Tasks may also operate in standalone mode using only external utilities. This may be required for control reasons or when thermal driving forces or time do not allow for heat integration. If either direct or indirect heat integration is not sufficient to satisfy the required duty, external utilities may be used to make up the deficit.

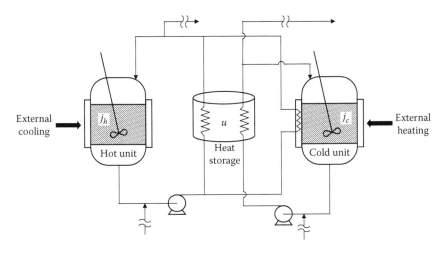

FIGURE 5.1 Superstructure for mathematical model.

Constraints (5.1) and (5.2) are active simultaneously and ensure that one hot unit can only be integrated with one cold unit when direct heat integration takes place in order to avoid operational complexity of the process. However, it is also possible for one unit to integrate with more than one unit at a given time point simultaneously when the summation notation in the equations are not used. Also, if two units are to be heat integrated at a given time point, they must both be active at that time point.

$$\sum_{s_{inj_c}} x\left(s_{inj_c}, s_{inj_h}, p, pp\right) \le y\left(s_{inj_h}, p\right), \quad \forall p, pp \in P, \quad s_{inj_h} \in S_{inJ_h}, \quad s_{inj_c} \in S_{inJ_c} \quad (5.1)$$

$$\sum_{s_{inj_h}} x\left(s_{inj_c}, s_{inj_h}, p, pp\right) \le y\left(s_{inj_c}, p\right), \quad \forall p, pp \in P, \quad s_{inj_h} \in S_{inJ_h}, \quad s_{inj_c} \in S_{inJ_c} \quad (5.2)$$

For better understanding, the difference between time point p and extended time point pp. is explained using Figure 5.2. If a unit j that is active at time point p is integrated with more than one unit in different temperature and time intervals, an extended time point pp. must be defined. Unit $j1$ active at time point p can be integrated with units $j2$ and $j3$ in different time and temperature intervals. At the beginning, unit $j1$ is integrated with unit $j2$ at time point p and the extended time point pp. is the same as time point p. Later, $j1$ is integrated with unit $j3$ in another time interval where extended time point pp. equals to $p + 1$. pp. is equal to or greater than time point p and less than or equal to $n + p$, where n is a parameter which is greater than or equal to zero. If n equals 2, then a unit that is active at time point p can be integrated in three different time intervals. The model should be solved starting from n equals zero and adding one at a time until no better objective value is achieved. This concept is the novelty of this work which addresses the limitation of models based on simultaneous approach for heat integration for multipurpose batch plants where a

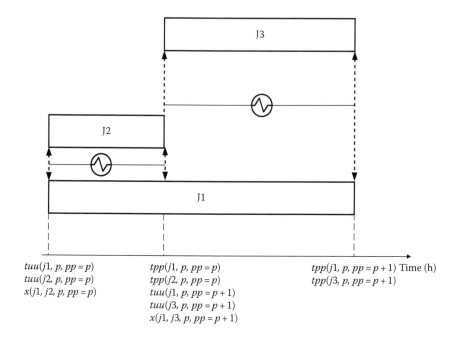

$tuu(j1, p, pp = p)$ $tpp(j1, p, pp = p)$ $tpp(j1, p, pp = p + 1)$ Time (h)
$tuu(j2, p, pp = p)$ $tpp(j2, p, pp = p)$ $tpp(j3, p, pp = p + 1)$
$x(j1, j2, p, pp = p)$ $tuu(j1, p, pp = p + 1)$
                             $tuu(j3, p, pp = p + 1)$
                             $x(j1, j3, p, pp = p + 1)$

FIGURE 5.2   Differentiating time point p and extended time point pp.

task is allowed to be heat integrated with other tasks only for one time interval during the starting and finishing time of the task.

Constraints (5.3) and (5.4) ensure that a unit cannot undergo direct and indirect heat integration simultaneously. This condition simplifies the operation of the
process.

$$\sum_{s_{injh}} x\left(s_{inj_c}, s_{inj_h}, p, pp\right) + z\left(s_{inj_c}, u, p, pp\right) \leq 1,$$

$$\forall p, pp \in P, \quad s_{injh} \in S_{inJ_h}, \quad s_{inj_c} \in S_{inJ_c} \quad u \in U \qquad (5.3)$$

$$\sum_{s_{injc}} x\left(s_{inj_c}, s_{inj_h}, p, pp\right) + z\left(s_{injh}, u, p, pp\right) \leq 1,$$

$$\forall p, pp \in P, \quad s_{injh} \in S_{inJ_h}, \quad s_{inj_c} \in S_{inJ_c} \quad u \in U \qquad (5.4)$$

Constraint (5.5) describes the amount of cooling load required by the hot unit to
reach from its initial temperature to its target temperature. On an occasion where
the temperature in the reactor unit is to be fixed during exothermic reaction, the
heat load becomes the product of the amount of mass that undergoes reaction and
the heat of reaction.

$$cl\left(s_{injh}, p\right) = mu\left(s_{injh}, p\right) cp\left(s_{injh}\right)\left(T_{s_{injh}}^{in} - T_{s_{injh}}^{out}\right), \quad \forall p \in P, \quad s_{injh} \in S_{inJ_h} \qquad (5.5)$$

Constraint (5.6) describes the heating load required by the cold unit to reach from its initial temperature to its target temperature. On an occasion where the temperature in the reactor unit is to be fixed during endothermic reaction, the heat load becomes the product of the amount of mass that undergoes reaction and the heat of reaction.

$$hl\left(s_{inj_c}, p\right) = mu\left(s_{inj_c}, p\right) cp\left(s_{inj_c}\right)\left(T_{s_{inj_c}}^{out} - T_{s_{inj_c}}^{in}\right), \quad \forall p \in P, \quad s_{inj_c} \in S_{inJ_c} \quad (5.6)$$

Constraints (5.7) and (5.8) describe the average heat flow for the hot and cold units, respectively, during the processing time, which is the same as TAM, to address the energy balance during heat integration properly.

$$cl\left(s_{inj_h}, p\right) = avcl\left(s_{inj_h}, p\right)\left(tp\left(s_{inj_h}, p\right) - tu\left(s_{inj_h}, p\right)\right), \quad \forall p \in P, \quad s_{inj_h} \in S_{inJ_h} \quad (5.7)$$

$$hl\left(s_{inj_c}, p\right) = avhl\left(s_{inj_c}, p\right)\left(tp\left(s_{inj_c}, p\right) - tu\left(s_{inj_c}, p\right)\right), \quad \forall p \in P, \quad s_{inj_c} \in S_{inJ_c} \quad (5.8)$$

Constraints (5.9) and (5.10) define the heat load at time point p and extended time point pp. for the cold and hot unit.

$$hlp\left(s_{inj_c}, p, pp\right) = avhl\left(s_{inj_c}, p\right)\left(tpp\left(s_{inj_c}, p, pp\right) - tuu\left(s_{inj_c}, p, pp\right)\right),$$
$$\forall p, pp \in P, \quad s_{inj_c} \in S_{inJ_c} \quad (5.9)$$

$$clp\left(s_{inj_h}, p, pp\right) = avcl\left(s_{inj_h}, p\right)\left(tpp\left(s_{inj_h}, p, pp\right) - tuu\left(s_{inj_h}, p, pp\right)\right),$$
$$\forall p, pp \in P, \quad s_{inj_h} \in S_{inJ_h} \quad (5.10)$$

Constraints (5.11) and (5.12) quantify the amount of heat received from and transferred to the heat storage unit, respectively. There will be no heat received or transferred if the binary variable signifying use of the heat storage vessel, $z\left(s_{inj}, u, p, pp\right)$, is zero.

$$Q\left(s_{inj_c}, u, p, pp\right) = W\left(u\right) cp\left(u\right)\left(T_0\left(u, p, pp\right) - T_f\left(u, p, pp\right)\right) z\left(s_{inj_c}, u, p, pp\right),$$
$$\forall p \in P, \quad s_{inj_c} \in S_{inJc}, \quad u \in U \quad (5.11)$$

$$Q\left(s_{inj_h}, u, p, pp\right) = W\left(u\right) cp\left(u\right)\left(T_f\left(u, p, pp\right) - T_0\left(u, p, pp\right)\right) z\left(s_{inj_h}, u, p, pp\right),$$
$$\forall p \in P, \quad s_{inj_h} \in S_{inJ_h}, \quad u \in U \quad (5.12)$$

Constraints (5.13) and (5.14) are used to calculate the temperature of the hot and cold units at the intervals

$$clp\left(s_{inj_h}, p, pp\right) = mu\left(s_{inj_h}, p\right) cp\left(s_{inj_h}\right)\left(T^{in}\left(s_{inj_h}, p, pp\right) - T^{out}\left(s_{inj_h}, p, pp\right)\right),$$
$$\forall p, pp \in P, \quad s_{inj_h} \in S_{inJ_h} \quad (5.13)$$

$$hlp\left(s_{inj_c}, p, pp\right) = mu\left(s_{inj_c}, p\right)cp\left(s_{inj_c}\right)\left(T^{out}\left(s_{inj_c}, p, pp\right) - T^{in}\left(s_{inj_c}, p, pp\right)\right),$$

$$\forall p, pp \in P, \quad s_{inj_c} \in S_{inJ_c} \tag{5.14}$$

Constraint (5.15) states that the amount of heat exchanged between the cold unit and the heat storage should be less than the heat load required by the cold unit during the interval.

$$Q\left(s_{inj_c}, u, p, pp\right) \le hlp\left(s_{inj_c}, p, pp\right), \quad \forall p, pp \in P, \quad s_{inj_c} \in S_{inJ_c}, \quad u \in U \tag{5.15}$$

Constraint (5.16) states that the amount of heat exchanged between the hot unit and the heat storage should be less than the cooling load required by the hot unit during the interval.

$$Q\left(s_{inj_h}, u, p, pp\right) \le clp\left(s_{inj_h}, p, pp\right), \quad \forall p, pp \in P, \quad s_{inj_h} \in S_{inJ_h}, \quad u \in U \tag{5.16}$$

Constraint (5.17) states that the amount of heat exchanged between the hot and the cold unit should be less than the cooling load required by the hot unit during the interval.

$$\sum_{s_{inj_c}} Qe\left(s_{inj_h}, s_{inj_c}, p, pp\right) \le clp\left(s_{inj_h}, p, pp\right), \quad \forall p, pp \in P, \quad s_{inj_h} \in S_{inJ_h}, \quad s_{inj_c} \in S_{inJ_c} \tag{5.17}$$

Constraint (5.18) states that the amount of heat exchanged between the cold unit and the hot unit should be less than the heat load required by the cold unit during the interval.

$$\sum_{s_{inj_h}} Qe\left(s_{inj_h}, s_{inj_c}, p, pp\right) \le hlp\left(s_{inj_c}, p, pp\right), \quad \forall p, pp \in P, \quad s_{inj_h} \in S_{inJ_h}, \quad s_{inj_c} \in S_{inJ_c} \tag{5.18}$$

Constraint (5.19) ensures that if heat integration occurs, the heat load should have a value that is less than the maximum amount of heat exchangeable. When the binary variable associated with heat integration takes a value of zero, no heat integration occurs and the associated heat load is zero.

$$Qe\left(s_{inj_h}, s_{inj_c}, p, pp\right) \le Q^U x\left(s_{inj_c}, s_{inj_h}, p, pp\right), \quad \forall p, pp \in P, \quad s_{inj_h} \in S_{inJ_h}, \quad s_{inj_c} \in S_{inJ_c} \tag{5.19}$$

Constraints (5.20) and (5.21) ensure that if heat integration between the cold and hot units takes place with the heat storage, then the heat load takes a positive value. This happens only when the binary variable associated with the integration of cold and hot units with the heat storage unit takes the value of one.

$$Q\left(s_{inj_h}, u, p, pp\right) \le Q^U z\left(s_{inj_h}, u, p, pp\right), \quad \forall p, pp \in P, \quad s_{inj_h} \in S_{inJ_h}, \quad u \in U \tag{5.20}$$

$$Q\left(s_{inj_c}, u, p, pp\right) \le Q^U z\left(s_{inj_c}, u, p, pp\right), \quad \forall p, pp \in P, \quad s_{inj_c} \in S_{inJ_c}, \quad u \in U \quad (5.21)$$

Constraints (5.22) and (5.23) state that the temperature of the task at the current time interval should be equal to the temperature at the end of the previous time interval.

$$T^{in}\left(s_{inj_h}, p, pp\right) = T^{out}\left(s_{inj_h}, p, pp-1\right), \quad \forall p, pp \in P, \quad s_{inj_h} \in S_{inJ_h} \quad (5.22)$$

$$T^{in}\left(s_{inj_c}, p, pp\right) = T^{out}\left(s_{inj_c}, p, pp-1\right), \quad \forall p, pp \in P, \quad s_{inj_c} \in S_{inJ_c} \quad (5.23)$$

Constraint (5.24) states that the temperature of the heat storage unit at the current time interval should be equal to the temperature at the end of the previous time interval.

$$T_o\left(u, p, pp\right) = T_f\left(u, p, pp-1\right), \quad \forall p, pp \in P, \quad u \in U \quad (5.24)$$

Constraint (5.25) states that the initial temperature of the heat storage unit at the current time point p should be equal to the final temperature at the previous time point $p-1$.

$$T_o\left(u, p, pp = p\right) = T_f\left(u, p-1, pp = p-1+n\right), \quad \forall p, pp \in P \quad (5.25)$$

Constraints (5.26) and (5.27) state that the temperature at the start of the first time interval, which is time point p, and also pp, should be equal to the initial temperature of the task.

$$T^{in}\left(s_{inj_h}, p, pp\right) = T^{in}_{s_{inj_h}}, \quad \forall p, pp \in P, \quad s_{inj_h} \in S_{inJ_h} \quad (5.26)$$

$$T^{in}\left(s_{inj_c}, p, pp\right) = T^{in}_{s_{inj_c}}, \quad \forall p, pp \in P, \quad s_{inj_c} \in S_{inJ_c} \quad (5.27)$$

Constraints (5.28) and (5.29) ensure that the minimum thermal driving forces are obeyed when there is direct heat integration between a hot and a cold unit.

$$T^{in}\left(s_{inj_h}, p, pp\right) - T^{out}\left(s_{inj_c}, p, pp\right) \ge \Delta T - \Delta T^U \left(1 - x\left(s_{inj_c}, s_{inj_h}, p, pp\right)\right),$$
$$\forall p, pp \in P, \quad s_{inj_h} \in S_{inJ_h}, \quad s_{inj_c} \in S_{inJ_c} \quad (5.28)$$

$$T^{out}\left(s_{inj_h}, p, pp\right) - T^{in}\left(s_{inj_c}, p, pp\right) \ge \Delta T - \Delta T^U \left(1 - x\left(s_{inj_c}, s_{inj_h}, p, pp\right)\right),$$
$$\forall p, pp \in P, \quad s_{inj_h} \in S_{inJ_h}, \quad s_{inj_c} \in S_{inJ_c} \quad (5.29)$$

Constraints (5.30) and (5.31) ensure that the minimum thermal driving forces are obeyed when there is direct heat integration between a hot task and a heat storage unit.

$$T^{in}\left(s_{inj_h}, p, pp\right) - T_f\left(u, p, pp\right) \ge \Delta T - \Delta T^U \left(1 - z\left(s_{inj_h}, u, p, pp\right)\right),$$
$$\forall p, pp \in P, \quad s_{inj_h} \in S_{inJ_h}, \quad u \in U \quad (5.30)$$

$$T^{out}\left(s_{inj_h}, p, pp\right) - T_o\left(u, p, pp\right) \geq \Delta T - \Delta T^U\left(1 - z\left(s_{inj_h}, u, p, pp\right)\right),$$
$$\forall p, pp \in P, \quad s_{inj_h} \in S_{inJ_h} \tag{5.31}$$

Constraints (5.32) and (5.33) ensure that the minimum thermal driving forces are obeyed when there is direct heat integration between a cold task and a heat storage unit.

$$T_o\left(u, p, pp\right) - T^{out}\left(s_{inj_c}, p, pp\right) \geq \Delta T - \Delta T^U\left(1 - z\left(s_{inj_c}, u, p, pp\right)\right),$$
$$\forall p, pp \in P, \quad s_{inj_c} \in S_{inJ_c}, \quad u \in U \tag{5.32}$$

$$T_f\left(u, p, pp\right) - T^{in}\left(s_{inj_c}, p, pp\right) \geq \Delta T - \Delta T^U\left(1 - z\left(s_{inj_c}, u, p, pp\right)\right),$$
$$\forall p, pp \in P, \quad s_{inj_c} \in S_{inJ_c} \tag{5.33}$$

Constraints (5.34) and (5.35) state that temperatures change in the heating and cooling unit when the binary variables associated with heating and cooling are active.

$$T^{in}\left(s_{inj_h}, p, pp\right) - T^{out}\left(s_{inj_h}, p, pp\right) \leq \Delta T^U v\left(s_{inj_h}, p, pp\right), \quad \forall p, pp \in P, \quad s_{inj_h} \in S_{inJ_h} \tag{5.34}$$

$$T^{out}\left(s_{inj_c}, p, pp\right) - T^{in}\left(s_{inj_c}, p, pp\right) \leq \Delta T^U v\left(s_{inj_c}, p, pp\right), \quad \forall p, pp \in P, \quad s_{inj_c} \in S_{inJ_c} \tag{5.35}$$

Constraints (5.36) and (5.37) state that temperatures change in the heat storage unit when the binary variables associated with heating and cooling with the heat storage unit are active.

$$T^{out}\left(u, p, pp\right) - T^{in}\left(u, p, pp\right) \leq \Delta T^U\left(\sum_{s_{inj_c}} z\left(s_{inj_c}, u, p, pp\right) + \sum_{s_{inj_h}} z\left(s_{inj_h}, u, p, pp\right)\right),$$
$$\forall p, pp \in P, \quad s_{inj_h} \in S_{inJ_h}, \quad s_{inj_c} \in S_{inJ_c}, \quad u \in U \tag{5.36}$$

$$T^{out}\left(u, p, pp\right) - T^{in}\left(u, p, pp\right) \geq -\Delta T^U\left(\sum_{s_{inj_c}} z\left(s_{inj_c}, u, p, pp\right) + \sum_{s_{inj_h}} z\left(s_{inj_h}, u, p, pp\right)\right),$$
$$\forall p, pp \in P, \quad s_{inj_h} \in S_{inJ_h}, \quad s_{inj_c} \in S_{inJ_c}, \quad u \in U \tag{5.37}$$

Constraints (5.38) through (5.41) ensure that the times at which units are active are synchronized when direct heat integration takes place.

$$tuu\left(s_{inj_h}, p, pp\right) \geq tuu\left(s_{inj_c}, p, pp\right) - M\left(1 - x\left(s_{inj_c}, s_{inj_h}, p, pp\right)\right),$$
$$\forall p, pp \in P, \quad s_{inj_h} \in S_{inJ_h}, \quad s_{inj_c} \in S_{inJ_c} \tag{5.38}$$

$$tuu\left(s_{inj_h}, p, pp\right) \le tuu\left(s_{inj_c}, p, pp\right) + M\left(1 - x\left(s_{inj_c}, s_{inj_h}, p, pp\right)\right),$$
$$\forall p, pp \in P, \quad s_{inj_h} \in S_{inJ_h}, \quad s_{inj_c} \in S_{inJ_c} \tag{5.39}$$

$$tpp\left(s_{inj_h}, p, pp\right) \ge tpp\left(s_{inj_c}, p, pp\right) - M\left(1 - x\left(s_{inj_c}, s_{inj_h}, p, pp\right)\right),$$
$$\forall p, pp \in P, \quad s_{inj_h} \in S_{inJ_h}, \quad s_{inj_c} \in S_{inJ_c} \tag{5.40}$$

$$tpp\left(s_{inj_h}, p, pp\right) \le tpp\left(s_{inj_c}, p, pp\right) + M\left(1 - x\left(s_{inj_c}, s_{inj_h}, p, pp\right)\right),$$
$$\forall p, pp \in P, \quad s_{inj_h} \in S_{inJ_h}, \quad s_{inj_c} \in S_{inJ_c} \tag{5.41}$$

Constraints (5.42) through (5.49) ensure that the times at which the cold and hot units are synchronized with the time of the heat storage unit are when heat integration takes place.

$$tuu\left(s_{inj_h}, p, pp\right) \ge tuu\left(u, p, pp\right) - M\left(2 - v\left(s_{inj_h}, p, pp\right) - z\left(s_{inj_h}, u, p, pp\right)\right),$$
$$\forall p, pp \in P, \quad s_{inj_h} \in S_{inJ_h}, \quad u \in U \tag{5.42}$$

$$tuu\left(s_{inj_h}, p, pp\right) \le tuu\left(u, p, pp\right) + M\left(2 - v\left(s_{inj_h}, p, pp\right) - z\left(s_{inj_h}, u, p, pp\right)\right),$$
$$\forall p, pp \in P, \quad s_{inj_h} \in S_{inJ_h}, \quad u \in U \tag{5.43}$$

$$tpp\left(s_{inj_h}, p, pp\right) \ge tpp\left(u, p, pp\right) - M\left(2 - v\left(s_{inj_h}, p, pp\right) - z\left(s_{inj_h}, u, p, pp\right)\right),$$
$$\forall p, pp \in P, \quad s_{inj_h} \in S_{inJ_h}, \quad u \in U \tag{5.44}$$

$$tpp\left(s_{inj_h}, p, pp\right) \le tpp\left(u, p, pp\right) + M\left(2 - v\left(s_{inj_h}, p, pp\right) - z\left(s_{inj_h}, u, p, pp\right)\right),$$
$$\forall p, pp \in P, \quad s_{inj_h} \in S_{inJ_h}, \quad u \in U \tag{5.45}$$

$$tuu\left(s_{inj_c}, p, pp\right) \ge tuu\left(u, p, pp\right) - M\left(2 - v\left(s_{inj_c}, p, pp\right) - z\left(s_{inj_c}, u, p, pp\right)\right),$$
$$\forall p, pp \in P, \quad s_{inj_c} \in S_{inJ_c}, \quad u \in U \tag{5.46}$$

$$tuu\left(s_{inj_c}, p, pp\right) \le tuu\left(u, p, pp\right) + M\left(2 - v\left(s_{inj_c}, p, pp\right) - z\left(s_{inj_c}, u, p, pp\right)\right),$$
$$\forall p, pp \in P, \quad s_{inj_c} \in S_{inJ_c}, \quad u \in U \tag{5.47}$$

$$tpp\left(s_{inj_c}, p, pp\right) \ge tpp\left(u, p, pp\right) - M\left(2 - v\left(s_{inj_c}, p, pp\right) - z\left(s_{inj_c}, u, p, pp\right)\right),$$
$$\forall p, pp \in P, \quad s_{inj_c} \in S_{inJ_c}, \quad u \in U \tag{5.48}$$

$$tpp\left(s_{inj_c}, p, pp\right) \leq tpp\left(u, p, pp\right) + M\left(2 - v\left(s_{inj_c}, p, pp\right) - z\left(s_{inj_c}, u, p, pp\right)\right),$$

$$\forall p, pp \in P, \quad s_{inj_c} \in S_{inJ_c}, \quad u \in U \tag{5.49}$$

Constraints (5.50) and (5.51) state that the starting time of the heating load required for the cold unit and the cooling load required for the hot unit at the first time interval should be equal to the starting time of the hot and cold unit.

$$tuu\left(s_{inj_h}, p, pp = p\right) = tu\left(s_{inj_h}, p\right), \quad \forall p, pp \in P, \quad s_{inj_h} \in S_{inJ_h} \tag{5.50}$$

$$tuu\left(s_{inj_c}, p, pp = p\right) = tu\left(s_{inj_c}, p\right), \quad \forall p, pp \in P, \quad s_{inj_c} \in S_{inJ_c} \tag{5.51}$$

Constraints (5.52) and (5.53) state that the starting time of heating and cooling in the current time interval should be equal to the finishing time at the previous time interval.

$$tuu\left(s_{inj_h}, p, pp\right) = tpp\left(s_{inj_h}, p, pp - 1\right), \quad \forall p, pp \in P, \quad s_{inj_h} \in S_{inJ_h} \tag{5.52}$$

$$tuu\left(s_{inj_c}, p, pp\right) = tpp\left(s_{inj_c}, p, pp - 1\right), \quad \forall p, pp \in P, \quad s_{inj_c} \in S_{inJ_c} \tag{5.53}$$

Constraint (5.54) states that the starting time of the heat storage unit at the current time interval should be equal to the finishing time at the previous time interval.

$$tuu\left(u, p, pp\right) = tpp\left(u, p, pp - 1\right), \quad \forall p, pp \in P, \quad u \in U \tag{5.54}$$

Constraint (5.55) states that the finishing time of a heat storage unit in a time interval should be equal to or greater than the starting time of the same time interval.

$$tpp\left(u, p, pp\right) \geq tuu\left(u, p, pp\right), \quad \forall p, pp \in P, \quad u \in U \tag{5.55}$$

Constraint (5.56) states that the starting time of the heat storage unit at time point p should be greater than or equal to the finishing time at the previous time point $p - 1$.

$$tuu\left(u, p, pp = p\right) = tpp\left(u, p - 1, pp = p - 1 + n\right), \quad \forall p, pp \in P, \quad u \in U \tag{5.56}$$

Constraints (5.57) and (5.58) state that if the binary variable associated with heat integration is active, then the binary variable associated with heating and cooling must be active.

$$x\left(s_{inj_c}, s_{inj_h}, p, pp\right) \leq v\left(s_{inj_h}, p, pp\right), \quad \forall p, pp \in P, \quad s_{inj_h} \in S_{inJ_h}, \quad s_{inj_c} \in S_{inJ_c} \tag{5.57}$$

$$x\left(s_{inj_c}, s_{inj_h}, p, pp\right) \leq v\left(s_{inj_c}, p, pp\right), \quad \forall p, pp \in P, \quad s_{inj_h} \in S_{inJ_h}, \quad s_{inj_c} \in S_{inJ_c} \tag{5.58}$$

Constraints (5.59) and (5.60) state that the heating and cooling loads take on a value for a certain duration when the binary variables associated with heating and cooling are active.

$$tpp\left(s_{inj_h}, p, pp\right) - tuu\left(s_{inj_h}, p, pp\right) \leq H * v\left(s_{inj_h}, p, pp\right), \quad \forall p, pp \in P, \quad s_{inj_h} \in S_{inJ_h}$$
(5.59)

$$tpp\left(s_{inj_c}, p, pp\right) - tuu\left(s_{inj_c}, p, pp\right) \leq H * v\left(s_{inj_c}, p, pp\right), \quad \forall p, pp \in P, \quad s_{inj_c} \in S_{inJ_c}$$
(5.60)

Constraint (5.61) states that the cooling of a hot unit will be satisfied by direct heat integration, indirect heat integration or external cooling utility, if required.

$$cl\left(s_{inj_h}, p\right) = cw\left(s_{inj_h}, p\right) + \sum_{S_{inj_c}} \sum_{pp=p}^{pp=p+n} Qe\left(s_{inj_h}, s_{inj_c}, p, pp\right) + \sum_{pp=p}^{pp=p+n} Q\left(s_{inj_h}, u, p, pp\right),$$

$$\forall p, pp \in P, \quad s_{inj_h} \in S_{inJ_h}, \quad s_{inj_c} \in S_{inJ_c}$$
(5.61)

Constraint (5.62) states that the heating of a cold unit will be satisfied by direct heat integration, indirect heat integration or external heating utility, if required.

$$hl\left(s_{inj_c}, p\right) = st\left(s_{inj_c}, p\right) + \sum_{S_{inj_h}} \sum_{pp=p}^{pp=p+n} Qe\left(s_{inj_h}, s_{inj_c}, p, pp\right) + \sum_{pp=p}^{pp=p+n} Q\left(s_{inj_c}, u, p, pp\right),$$

$$\forall p, pp \in P, \quad s_{inj_h} \in S_{inJ_h}, \quad s_{inj_c} \in S_{inJ_c}$$
(5.62)

5.4.2 MODEL OBJECTIVE FUNCTIONS

Equation 5.63 is the objective function in terms of profit maximization, with profit defined as the difference between revenue from product and cost of utility.

$$\max\left(\sum_{s^p} price\left(s^p\right) qs\left(s^p\right) - \sum_{p} \sum_{S_{inj_h}} costcw * cw\left(s_{inj_h}, p\right) - \sum_{p} \sum_{S_{inj_c}} costst * st\left(s_{inj_c}, p\right)\right),$$

$$\forall p, pp \in P, \quad s_{inj_h} \in S_{inJ_h}, \quad s_{inj_c} \in S_{inJ_c}, \quad s_{inj} \in S_{inJ}$$
(5.63)

Equation 5.64 defines the minimization of utility if the product demand is known.

$$\min\left(\sum_{p} \sum_{S_{inj_h}} costcw * cw\left(s_{inj_h}, p\right) + \sum_{p} \sum_{S_{inj_c}} costst * st\left(s_{inj_c}, p\right)\right),$$

$$\forall p, pp \in P, \quad s_{inj_h} \in S_{inJ_h}, \quad s_{inj_c} \in S_{inJ_c}, \quad s_{inj} \in S_{inJ}$$
(5.64)

5.5 CASE STUDIES

In order to demonstrate the application of the proposed model, three literature case studies are presented. The results in all the case studies for the proposed model were obtained using a Pentium 4 with 3.2 GHz processor and 512 MB RAM. CPLEX and CONOPT 2 in GAMS 22.0 were used to solve the MILP and NLP problems, respectively. DICOPT 2 was used as the interface for solving the MINLP problem. The computational results of this work are compared with results from the literature.

5.5.1 CASE STUDY I

This case study, obtained from Kondili et al. (1993), has become one of the most commonly used examples in literature. However, this case study has been adapted by Halim and Srinivasan (2009) to include energy integration. The batch plant produces two different products sharing the same processing units, where Figure 5.3 shows the plant flowsheet. The unit operations consist of preheating, three different reactions and separation. The plant accommodates many common features of multipurpose batch plants, such as units performing multiple tasks, multiple units

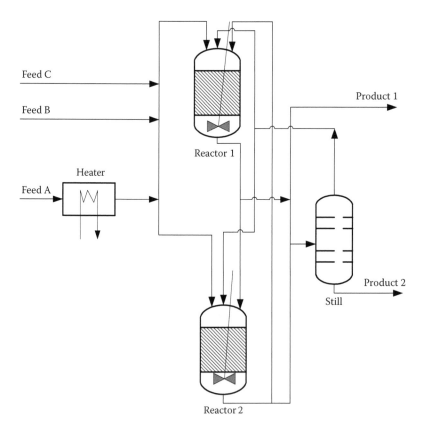

FIGURE 5.3   Flowsheet for Case Study I.

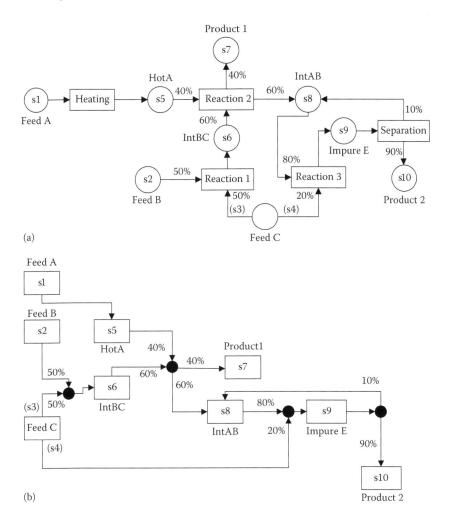

FIGURE 5.4   STN (a) and (b) SSN representations for Case Study I.

suitable for a task and dedicated units for specific tasks. The STN and state sequence network (SSN) representations of the flowsheet are shown in Figure 5.4. The production recipe is as follows:

1. Raw material Feed A is heated from 50°C to 70°C to form HotA used in reaction (5.2).
2. Reactant materials 50% Feed B and 50% Feed C are used in reaction (5.1) to produce IntBC. During the reaction, the material has to be cooled from 100°C to 70°C.
3. 60% of the intermediate material IntBC and 40% of HotA are used in reaction (5.2) to produce product 1 and IntAB. The process needs to be heated from 70°C to 100°C during its operation.

4. 20% Feed C and 80% of the intermediate IntAB from reaction (5.2) are used in reaction (5.3) to produce Impure E. The reaction needs its temperature to be raised from 100°C to 130°C during its operation.
5. The separation process produces 90% product 2 and 10% IntAB from Impure E. Cooling water is used to lower its temperature from 130°C to 100°C.

The processing time of a task i in unit j is assumed to be linearly dependent, $\alpha_i + \beta_i B$, on its batch size B, where α_i is a constant term of the processing time of task i and β_i is a coefficient of variable processing time of task i. The batch-dependent processing time makes this case study more complex. Table 5.1 gives the relevant data on coefficients of processing times, the capacity of the processing units, initial inventory of raw materials, storage capacity and relevant costs. The production demand is given as 200 kg for both Prod1 and Prod2. Table 5.2 gives data pertaining to heat integration. The objective here is optimization with respect to makespan and energy.

5.5.1.1   Results and Discussion

The computational results obtained using the proposed model as well as those of Halim and Srinivasan (2009) are presented in Table 5.3. The model of Stamp and Majozi (2011) is not included in the comparison since the model does not cater for temperature variation during processing of tasks. The minimum temperature driving force for heat exchange is assumed 10°C. For makespan minimization, an objective value of 19.5 h was obtained using the proposed model, which is much better than 19.96 h obtained by Halim and Srinivasan (2009) as a result of using the recent robust scheduling model. Using the makespan obtained, the case study was solved to minimize the energy demand by setting customer requirement for Product 1 and Product 2. The total energy required for the standalone operation was 125.5 MJ.

By using energy integration, the total energy requirement was reduced from 125.5 MJ in standalone operation to 51.4 MJ (38.7 MJ hot utility and 12.8 MJ cold utility) resulting in a 59% energy saving. The performance of the proposed model was also compared to the technique by Halim and Srinivasan (2009). They solved this case study based on a three-step sequential framework. First, the schedule is optimized to meet the economic objective of makespan minimization. Next, alternate schedules are generated through a stochastic search-based integer cut procedure. Finally, the heat integration model is solved for each of the resulting schedules to establish the minimum utility targets. In generating alternate optimal solutions, they used integer cut variables to 6, that is, in each iteration, six tasks were preassigned to different units at different slots. After 1000 MILP iteration of stochastic search, three sets of solutions at three different makespans (20.03, 20.02, 19.96) were obtained. Based on this set of schedules, the heat integration model was solved. The makespan that gave the minimum utility target was 19.96, with the corresponding utility requirement of 40.8 MJ hot utility and 15.6 MJ cold utility. Figure 5.5 shows the optimization results after 1000 iteration of the stochastic search.

The developed model gives a better makespan and a further utility saving of 8.8% compared to the results of Halim and Srinivasan (2009). An additional feature of the proposed model is its efficient solution technique. The optimal solution was found

TABLE 5.1
Scheduling Data for Case Study I

Task (i)	Unit (j)	Max Batch Size (kg)	$\alpha(s_{inj})$	$\beta(s_{inj})$	Material State (s)	Initial Inventory	Max Storage (kg)	Revenue or Cost ($/kg or $/MJ)
Heating (H)	HR	100	0.667	0.007	Feed A	1000	1000	0
Reaction-1 (R1)	RR1	50	1.334	0.027	Feed B	1000	1000	0
	RR2	80	1.334	0.017	Feed C	1000	1000	0
Reaction-2 (R2)	RR1	50	1.334	0.027	HotA	0	100	0
	RR2	80	1.334	0.017	IntAB	0	200	0
Reaction-3 (R3)	RR1	50	0.667	0.013	IntBC	0	150	0
	RR2	80	0.667	0.008	ImpureE	0	200	0
Separation (S)	SR	200	1.334	0.007	Prod1	0	1000	20
					Prod2	0	1000	20
					Cooling water			0.02
					Steam			1

TABLE 5.2

Data Required for Energy Integration for Case Study I

Task (i)	$T_{s_{inj}}^{in}$ (°C)	$T_{s_{inj}}^{out}$ (°C)	Unit (j)	C_p (kJ / kg°C)
Heating (H)	50	70	HR	2.5
Reaction-1	100	70	RR1	3.5
			RR2	3.5
Reaction-2	70	100	RR1	3.2
			RR2	3.2
Reaction-3	100	130	RR1	2.6
			RR2	2.6
Separation	130	100	SR	2.8
Cooling water	20	30		
Steam	170	160		

TABLE 5.3

Computational Results for Case Study I

	Proposed Formulation without Energy Integration	Proposed Formulation with Energy Integration	Halim and Srinivasan (2009) with Energy Integration
Objective ($)	125.5	51.4	56.4
Steam (MJ)	75.2	38.7	40.8
Cooling water (MJ)	50.2	13.5	15.6
Revenue from product ($)	4000	4000	4000
Number of time points/ slots	8	8	N/A
Number of binary variables	92	500	N/A
CPU time (h)	0.39	2	N/A

after solving only one MINLP model, which was solved in a reasonable specified CPU time of 2 h. The authors Halim and Srinivasan (2009) did not report the CPU time required to solve this problem. We believe that it is computationally expensive because of the need to generate alternate optimal schedules.

Figure 5.6 details the amount of heat exchanged between the cold and hot units and the time intervals during which energy integration occurs.

The energy requirements of units RR1, RR2 and HR during the interval 4.976 to 7.660 h are explained to highlight the advantage of the proposed model. The heating load of unit RR2 between 4.976 and 6.660 h is 4.658 MJ. This is partly satisfied through energy integration with unit RR1 in the same time interval resulting in an

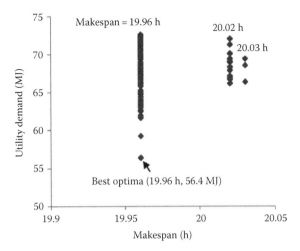

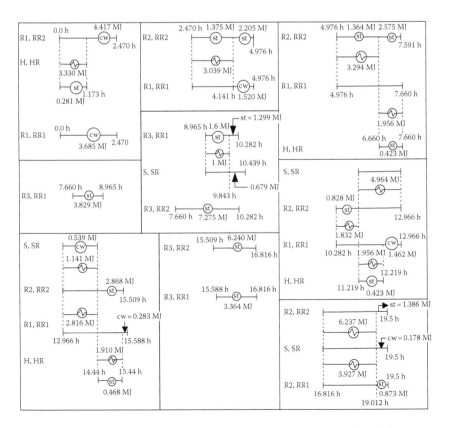

FIGURE 5.5 Optimization results using stochastic search-based integer cut procedure by Halim and Srinivasan (2009).

FIGURE 5.6 Heat exchange network for Case Study I using the proposed model.

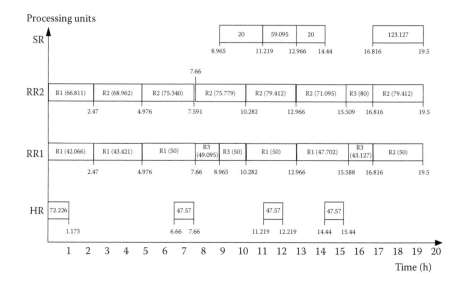

FIGURE 5.7   Gantt chart for Case Study I using the proposed model.

external heating requirement of 1.364 MJ rather than 4.658 MJ if it operated in stand-alone mode. The heating requirement for RR2 between 6.660 and 7.591 h is fully satisfied with external steam (2.575 MJ) since it is not in the heat integrated mode. The cooling requirement of RR1 between 4.976 and 6.660 h is 3.294 MJ. This heating requirement is satisfied by heat integration with unit RR2. For the time interval between 6.660 and 7.660 h, the cooling requirement for RR1 is 1.956 MJ that can be fully satisfied with matching to unit HR. This demonstrates the advantage of the developed model where it allows unit RR1 to be heat integrated with multiple units at different time intervals for better heat recovery. Literature models based on simultaneous approach allow a unit to be heat integrated with other units only for one time interval during its operation. Figure 5.7 shows the Gantt chart related to the optimal usage of resources. It also indicates the types of tasks performed in each piece of equipment and the starting and finishing times of the processing tasks.

5.5.2 CASE STUDY II

This example, which was first examined by Sundaramoorthy and Karimi (2005), is studied extensively in literature. It is a relatively complex problem and is often used in literature to check the efficiency of models in terms of optimal objective value and CPU time required. The plant has many common features of a multipurpose batch plant, with the following features: units performing multiple tasks, multiple units suitable for a task, states shared by multiple tasks and different products produced following different production paths. The STN and SSN representations for this case study are shown in Figure 5.8. The scheduling data are modified to incorporate heat integration opportunities and are presented in Tables 5.4 and 5.5. Data necessary for heat integration are presented in Table 5.6.

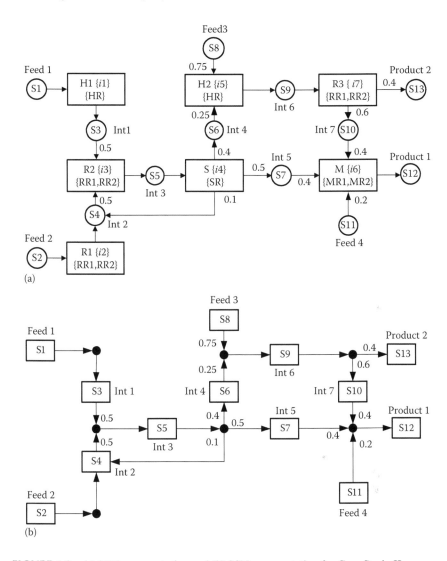

FIGURE 5.8 (a) STN representation and (b) SSN representation for Case Study II.

5.5.2.1 Results and Discussion

The computational results obtained from the different models for this case study are presented in Table 5.7. This case study evaluates the efficiency of the proposed model for indirect heat integration. The recent, more advanced sequential methodology for energy integration by Halim and Srinivasan (2009) is not included in the comparison since their model does not consider the use of heat storage for indirect heat integration.

For indirect heat integration, using a time horizon of 10 h, the proposed model and the model by Stamp and Majozi (2011) give the same optimal objective value of 224,000, while the proposed model gives a better CPU time. It is interesting to

TABLE 5.4
Scheduling Data for Case Study II

Unit	Capacity	Suitability	Mean Processing Time (h)
Heater	100	H1, H2	1, 1.5
Reactor 1	100	RX1, RX2, RX3	2, 1, 2
Reactor 2	150	RX1, RX2, RX3	2, 1, 2
Separator	300	Separation	3
Mixer 1	200	Mixing	2
Mixer 2	200	Mixing	2

TABLE 5.5
Scheduling Data for Case Study II

State	Description	Storage Capacity (Ton)	Initial Amount (Ton)	Revenue (c.u/Ton)
s1	Feed 1	Unlimited	Unlimited	0
s2	Feed 2	Unlimited	Unlimited	0
s3	Int 1	100	0	0
s4	Int 2	100	0	0
s5	Int 3	300	0	0
s6	Int 4	150	50	0
s7	Int 5	150	50	0
s8	Feed 3	Unlimited	Unlimited	0
s9	Int 6	150	0	0
s10	Int 7	150	0	0
s11	Feed 4	Unlimited	Unlimited	0
s12	Product 1	Unlimited	0	1000
s13	Product 2	Unlimited	0	1000

Parameters	Values
Specific heat capacity for heat storage c_p(kJ/kg °C)	4.2
Steam cost (c.u/kWh)	10
Cooling water cost (c.u/kWh)	2
ΔT^{min} (°C)	10
T^L (°C)	20
T^U (°C)	180
W^L (ton)	1
W^U (ton)	3

TABLE 5.6
Heating/Cooling Requirements for Case Study II

Reaction	Type	Heating/Cooling Requirement (kWh)	Operating Temperature (°C)
		Fixed batch size	
RX1	Exothermic	60 (cooling)	100
RX2	Endothermic	80 (heating)	60
RX3	Exothermic	70 (cooling)	140

TABLE 5.7
Computational Results for Case Study II, Fixed Batch Size with Units Operating at 80% Capacity

	Standalone Operation, Stamp and Majozi (2011)	Standalone Operation, Proposed	Indirect Heat Integration, Stamp and Majozi (2011)	Indirect Heat Integration, Proposed
		$H = 10$		
Performance index (cost units)	222,000	222,000	224,000	224,000
External cold duty (kWh)	200	200	0	0
External hot duty (kWh)	160	160	0	0
Heat storage capacity (ton)			1.905	1.905
Initial heat storage temperature (°C)			82.5	82.5
CPU time (s)	5.3	1	68	7.8
Binary variables	66	101	156	209
Time points	7	6	7	6
		$H = 12$		
Performance index (cost units)	285,860	285,860	287,640	287,965
External cold duty (kWh)	270	270	130	17.5
External hot duty (kWh)	160	160	10	0
Heat storage capacity (ton)			5	1.905
Initial heat storage temperature (°C)			87.143	82.5
CPU time (s)	7.7	1.9	238,896	35.2
Binary variables	99	155	206	246
Time points	9	7	11	7

note that by having heat storage, the utility requirement is reduced to 0.0 signifying that operating a heat integrated batch plant with heat storage provides great chance for heat recovery. For a time horizon of 12 h, the model by Stamp and Majozi (Obeng and Ashton, 1988) required more than two days to solve, while the proposed model significantly reduced the required CPU time (35.2 s for the proposed model vs. 238,640 s for Stamp and Majozi's (Obeng and Ashton, 1988)). The reduction in CPU time results from reduced number of time points required (7 for the proposed model vs. 11 for Stamp and Majozi's (2011)). However, the proposed model required a higher number of binary variables. This is due to catering for proper sequencing of tasks for the FIS policy which was inadvertently violated by the scheduling model used by Stamp and Majozi (2011) and is explained well in the next case study. A better optimal objective value of 287,965 was obtained by the proposed model as compared to 287,640 by Stamp and Majozi (2011). This indicates that the use of efficient scheduling techniques as a platform for a heat integration model improves the computational efficiency, both in terms of CPU time and optimal objective value, for heat integration in multipurpose batch plants.

The amount of material processed, the starting and finishing times, the amount of heat exchanged between units and heat storage and the utility requirements of the units for a time horizon of 12 h for indirect heat integration is given in Figure 5.9. The numbers in the boxes represent the amount of batch processed in the unit.

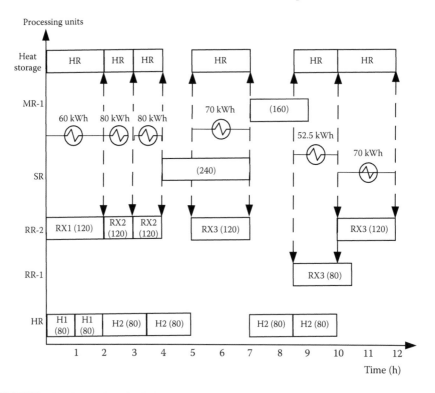

FIGURE 5.9 Gantt chart for indirect heat integration, proposed model.

5.5.3 CASE STUDY III

This case study was taken from the petrochemical plant by Kallrath (2002) and used as a benchmark problem in the scheduling environment for multipurpose batch plants. We adapted this case study to incorporate energy integration. The recipe representation for the plant is presented in Figure 5.10. It is a very complex problem where 10 reactors, 15 storage units and 18 states are considered. There are reactors dedicated to a specific task, while there are also reactors conducting multiple tasks, some capable of executing up to 5 different tasks. The mixed intermediate storage (MIS) policy is assumed in this case study. Scheduling data is given in Table 5.8. The plant is required to satisfy a demand requirement of 50 ton for S14, S17 and S18; 100 for S15 and 400 for S16 within a time horizon of 120 h. The heating and cooling requirements for each task are given in Table 5.9.

5.5.3.1 Results and Discussions

The computational statistics obtained for this case study are presented in Table 5.10. For the standalone operation using the technique of Stamp and Majozi (2011) based on the scheduling techniques of Majozi and Zhu (2001), the model gives an objective value of 6.74×10^6 c.u. The model required 17 time points and 425 binary variables and was solved in a specified CPU time of 5000 s. By implementing the proposed model on the same computer, an objective value of 6.5899×10^6 c.u. was obtained. This model required 550 binary variables and was solved in a specified CPU time of 5000 s.

It would be expected that the proposed model should give the same or a better objective value since the model uses a more robust scheduling technique as a platform. However, the slightly inferior result obtained is due to the need to catering for proper sequencing for the FIS policy. This is better described by the Gantt chart depicted in Figure 5.11. The maximum storage capacity for state S1 is 100 ton, but this limit is violated a number of times during the time horizon using the model of Stamp and Majozi (2011). The numbers inside the boxes in Figure 5.11 represent the amount of batch processed. It is clear from this figure that the state S1 produced from reactor R1 is consumed in reactor R2 in consecutive time points. For example, the state S1 consumed in reactor 1 at time points P1, P2 and P3 is consumed in reactor 2 at time points P2, P3 and P4, respectively. These indicate that the amount of material produced is consumed immediately in the next time point so that the storage constraint is not violated. However, in real time, the storage capacity is violated because of inadequate sequencing constraints that inproperly match the consuming task, the producing task and storage. The same explanation holds for state S5 where there is no storage available to store state S5. However, from Figure 5.11, it can be seen that this constraint is violated in real time and 50 ton storage is actually required. This inadvertent violation of FIS storage policy is resolved using the proposed model, and the right Gantt chart for this case study is presented in Figure 5.12.

Using the proposed model for direct heat integration, an objective value of 7.9107×10^6 c.u. was obtained, which is much better than the standalone operation,

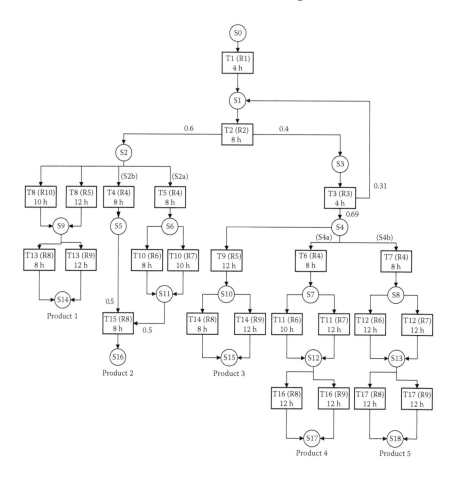

FIGURE 5.10 Recipe representation for Case Study III.

corresponding to a 20% improvement in profit. The utility reduction is also compared for the energy integrated mode 528.7 MJ of utility is required which is much better than the utility requirement of 3053 MJ for standalone operation. This corresponds to an 82.7% utility reduction when the tasks are operated in energy integrated mode. The model required 16 time points, $n = 2$ and 1315 binary variables and was solved in less than 2 h using specified CPU time of 2000 s for the MILP subproblem. The energy integration network for the time horizon of 120 h is depicted in Figure 5.13.

The energy integration in the interval between 12 and 20 h is elaborated for better understanding. The cooling water requirement for Task 2 is 10.625 MW (heat load/duration of the task, 85 MJ/8 h). During the time interval of 12 to 16 h, Task 2 is integrated with the cold task 5 and exchanges a heat load of 33.75 MJ. The cooling requirement for Task 2 in this time interval is 42.5 MJ obtained by

TABLE 5.8

Scheduling Data Required for Case Study III

Unit	Suitability	Capacity (Ton)	Revenue (10^3c.u./Ton)
R1	Task T1	0:100	—
R2	Task T2	0:100	—
R3	Task T3	0:100	—
R4	Task T4, T5, T6, T7	0:100	—
R5	Task T8, T9	0:150	—
R6	Task T10, T11, T12	0:50	—
R7	Task T10, T11, T12	0:50	—
R8	Task T13, T14, T15, T16, T17	0:100	—
R9	Task T13, T14, T16, T17	0:100	—
R10	Task T8	0:100	—
V0	Store S0	0:1000	0
V1	Store S1	0:100	0
V2	Store S2	0:100	0
V3	Store S3	0:100	0
V4	Store S4	0:100	0
V6	Store S6	0:100	0
V7	Store S7	0:100	0
V8	Store S8	0:100	0
V11	Store S11	0:100	0
V13	Store S13	0:100	0
V5, V9, V10 , V12	V5 for S5, V9 for S9, V10 for S10, V12 for S12	0:0	0
V14	Store S14	0:1000	5
V15	Store S15	0:1000	5
V16	Store S16	0:1000	10
V17	Store S17	0:1000	7.5
V18	Store S18	0:1000	7.5

TABLE 5.9

Heating/Cooling Requirement for Case Study III

Task	Heating/Cooling Requirement (MWh)	Operating Temperature (°C)	Cost (c.u/MWh)
T1	$40 + 0.2\beta$ heating	150	—
T2	$70 + 0.15\beta$ cooling	200	—
T3	$50 + 0.3\beta$ cooling	230	—
T4	$50 + 0.3\beta$ heating	120	—
T5	$55 + 0.25\beta$ heating	120	—
T6	$60 + 0.6\beta$ heating	120	—
T7	$45 + 0.45\beta$ heating	120	—
Steam		270	1
Cooling water		30	0.2

TABLE 5.10
Computational Results for Case Study III

	Standalone Operation Using Stamp and Majozi (2011)	Standalone Operation Using Proposed Model	Proposed Model with Direct Energy Integration
Objective value (10^3 c.u.)	6740	6589.9	7910.7
Product 1 (ton)	100	100	90
Product 2 (ton)	600	600	600
Product 3 (ton)	103.6	165.6	120.8
Product 4 (ton)	100	50	55.2
Product 5 (ton)	100	88	100
Cooling water (MJ)	1447	1600	276.75
Steam (MJ)	1488.53	1453	251.95
Binary variable	425	550	1315
CPU time (s)	5000	5000	6008

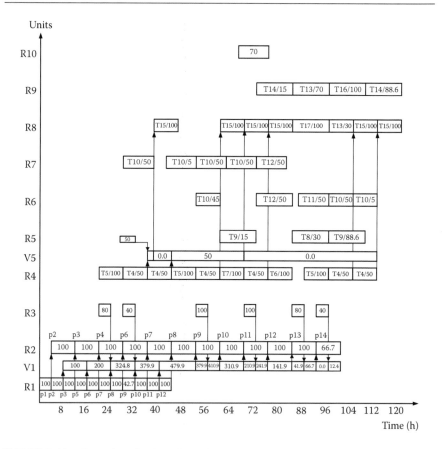

FIGURE 5.11 Gantt chart for Case Study III using Stamp and Majozi (2011) for standalone operation.

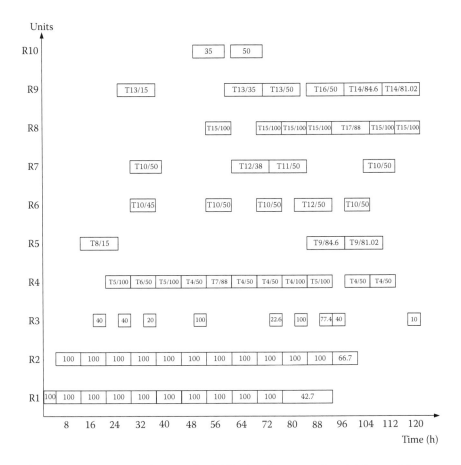

FIGURE 5.12   Gantt chart for Case Study III using the proposed model for standalone operation.

multiplying the energy requirement per hour with the duration of the heat exchange (10.625 MW * 4 = 42.5). The deficit cooling requirement for Task 2, which is (42.5 MJ− 33.75 MJ = 8.75 MJ), is satisfied with external cooling. The heating requirement for Task 5 in this interval is fully satisfied with heat integration of Task 2. The cooling requirement for Task 2 during the interval between 16 and 20 h is fully satisfied with energy integration of Task 1. The heating requirement for Task 1 during this interval is 60 MJ and is satisfied partially with the energy integration of Task 2 and the rest which is 17.5 MJ from external heating. The heating requirement for Task 5, which is 33.5 MJ during the interval between 16 and 20 h, is satisfied fully with external heating since it is not in the energy integrated mode. The same principle is applied to calculate and satisfy the energy requirement for each hot and cold task for the entire time horizon. The Gantt chart that shows the amount of batch processed, the type of task performed in a unit and the starting and finishing times for the energy integrated mode for this case is presented in Figure 5.14.

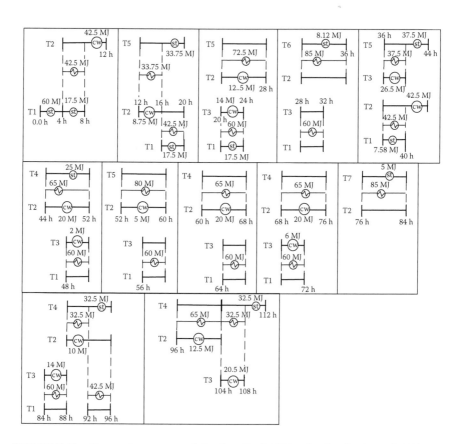

FIGURE 5.13   Heat exchange network for Case Study III for the time horizon of 120 h.

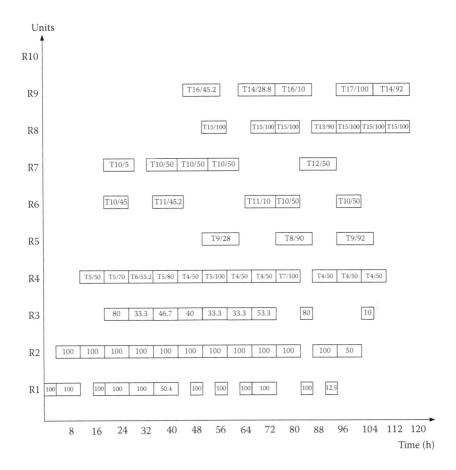

FIGURE 5.14 Gantt chart for Case Study III using direct heat integration.

5.6 CONCLUSIONS

An efficient continuous-time mathematical model for direct and indirect heat integration is presented. Most heat integration models rely on a predefined schedule, which leads to suboptimal results. This chapter incorporates heat integration into the scheduling framework and solved simultaneously. The model is capable of solving for both direct and indirect heat integration. By using a heat storage vessel, a considerable reduction in utility consumption is achieved. When this work is compared to recent existing work, the formulation performs better in terms of both optimal objective value and CPU time. Future comunication will address the developed model to inororate capital cost of heat exchanger.

NOMENCLATURE

SETS

S_{inJ_h} $\{s_{inj_h} | s_{inj_h}$ task which needs cooling$\}$
S_{inJ_c} $\{s_{inj_c} | s_{inj_c}$ task which needs heating$\}$
S_{inJ} $\{s_{inj} | s_{inj}$ any task $\}$
P $\{P | P$ time point$\}$
U $\{u | u$ is a heat storage unit$\}$

PARAMETERS

$cp\left(s_{inj_h}\right)$	Specific heat capacity for the heating task
$cp\left(s_{inj_c}\right)$	Specific heat capacity for the cooling task
$cp\left(u\right)$	Specific heat capacity for the heat storage
$T_{s_{inj_h}}^{in}$	Inlet temperature of the heating task
$T_{s_{inj_h}}^{out}$	Outlet temperature of the heating task
$T_{s_{inj_c}}^{in}$	Inlet temperature of the cooling task
$T_{s_{inj_c}}^{out}$	Outlet temperature of the cooling task
ΔT^U	Maximum thermal driving force
ΔT	Minimum thermal driving force
M	Big-M mostly equivalent to the time horizon
Q^U	Maximum heat requirement from the heating and cooling task
$price\left(s^p\right)$	Price of a product
costst	Cost of steam
costcw	Cost of cooling water
H	time horizon of interest

VARIABLES

$x\left(s_{inj_c}, s_{inj_h}, p, pp\right)$	Binary variable signifying whether heat integration occurs between the hot and cold unit
$z\left(s_{inj}, u, p, pp\right)$	Binary variable signifying whether heat integration occurs between a task and heat storage
$y\left(s_{inj_h}, p\right)$	Binary variable associated to whether the hot state is active at time point p or not
$y\left(s_{inj_c}, p\right)$	Binary variable associated to whether the cold state is active at time point p or not
$v\left(s_{inj}, p, pp\right)$	Binary variable associated to whether the hot and cold states are active at time point p and extended time point pp
$cl\left(s_{inj_h}, p\right)$	Cooling load required by the hot task at time point p
$hl\left(s_{inj_c}, p\right)$	Heating load required by the cold task at time point p
$avcl\left(s_{inj_h}, p\right)$	Average cooling load required by the hot task at time point p using time average model
$avhl\left(s_{inj_c}, p\right)$	Average heating load required by the cold task at time point p using time average model
$mu\left(s_{inj_h}, p\right)$	Amount of material processed by the hot task

$mu\left(s_{inj_c}, p\right)$	Amount of material processed by the cold task
$qs\left(s^p\right)$	Amount of product produced at the end of the time horizon
$tp\left(s_{inj}, p\right)$	Finishing time of a task
$tu\left(s_{inj}, p\right)$	Starting time of a task
$clp\left(s_{inj_h}, p, pp\right)$	Cooling load required by the hot task active at time point p and extended time point pp
$hlp\left(s_{inj_c}, p, pp\right)$	Heating load required by the cold task active at time point p and extended time point pp
$tuu\left(s_{inj}, p, pp\right)$	Starting time of a task at time point p and extended time point pp
$tpp\left(s_{inj}, p, pp\right)$	Finishing time of a task at time point p and extended time point pp
$tuu\left(u, p, pp\right)$	Starting time of a heat storage at time point p and extended time point pp
$tpp\left(u, p, pp\right)$	Finishing time of a heat storage at time point p and extended time point pp
$T^{in}\left(s_{inj}, p, pp\right)$	Inlet temperature of a task active at time point p and extended time point pp
$T^{out}\left(s_{inj}, p, pp\right)$	Outlet temperature of a task active at time point p and extended time point pp
$T_o\left(u, p, pp\right)$	Inlet temperature of a heat storage active at time point p and extended time point pp
$T_f\left(u, p, pp\right)$	Outlet temperature of a heat storage active at time point p and extended time point pp
$Qe\left(s_{inj_h}, s_{inj_c}, p, pp\right)$	Amount of heat load exchanged by the hot and cold unit active at time point p and extended time point pp
$Q\left(s_{inj}, u, p, pp\right)$	Amount of heat load exchanged between a task and heat storage
$cw\left(s_{inj_h}, p\right)$	External cooling water used by the hot task
$st\left(s_{inj_c}, p\right)$	External heating used by the cold task
$W\left(u\right)$	Capacity of heat storage unit

REFERENCES

Adonyi, R., Romero, J., Puigjaner, L., Friedler, F., 2003. Incorporating heat integration in batch process scheduling. *Applied Thermal Engineering*. 23, 1743–1762.

Atkins, M.J., Walmsley, M.R.W., Neale, J.R., 2010. The challenge of integrating non-continuous processes – Milk powder plant case study. *Journal of Cleaner Production*. 18, 927–934.

Bancheva, N., Ivanov, B., Shah, N., Pantelides, C.C., 1996a. Heat exchanger network design for multipurpose batch plants. *Computers and Chemical Engineering*. 20, 989–1001.

Bancheva, N., Ivanov, B., Shah, N., Pantelides, C.C., 1996b. Heat exchanger network design for multipurpose batch plants. *Computers and Chemical Engineering*. 20, 989–1001.

Barbosa-Póvoa, A.P.F.D., Pinto, T., Novais, A.Q., 2001. Optimal design of heat-integrated multipurpose batch facilities: A mixed-integer mathematical formulation. *Computers and Chemical Engineering*. 25, 547–559.

Becker, H., Vuillermoz, A., Maréchal, F., 2012. Heat pump integration in a cheese factory. *Applied Thermal Engineering*. 43, 118–127.

Boyadjiev, C.H.R., Ivanov, B., Vaklieva-Bancheva, N., Pantelides, C.C., Shah, N., 1996. Optimal energy integration in batch antibiotics manufacture. *Computers and Chemical Engineering*. 20, S31–S36.

Bozan, M., Borak, F., Or, I., 2001. A computerized and integrated approach for heat exchanger network design in multipurpose batch plants. *Chemical Engineering Process*. 40, 511–524.

Castro, P.M., Barbosa-Póvóa, A.P., Matos, H.A., Novais, A.Q., 2004. Simple continuous-time formulation for short-term scheduling of batch and continuous processes. *Industrial Engineering and Chemical Research*. 43, 105–118.

Chen, C.L., Chang, C.Y., 2009. A resource-task network approach for optimal short-term/periodic scheduling and heat integration in multipurpose batch plants. *Applied Thermal Engineering*. 29, 1195–1208.

Chen, C.L., Ciou, Y.J., 2008. Design and optimization of indirect energy storage systems for batch process plants. *Industrial and Engineering Chemical Research*. 47, 4817–4829.

Chew, Y.H., Lee, C.T., Foo, C.Y., 2005. Evaluating heat integration scheme for batch production of oleic acid. *Malaysian Science and Technology Congress (MSTC)*. 18–20.

Clayton, R.W., 1986. Cost reductions on an edible oil refinery identified by a process integration study at Van den Berghs and Jurgens Ltd., Report Nr RD14/14. UK: Energy Efficiency Office R&D, Energy Technology Support Unit (ETSU), Harwell Laboratory.

Cerominas, J., Espuña, A., Puigjaner, L., 1993. A new look at energy integration in multi-product batch processes. *Computers and Chemical Engineering*. 17, 15–20.

Fernández, I., Renedo, C.J., Pérez, S.F., Ortiz, A., Mañana, M., 2012. A review: Energy recovery in batch processes. *Renewable and Sustainable Energy Review*. 16, 2260–2277.

Foo, D.C.Y., Chew, Y.H., Lee, C.T., 2008. Minimum units targeting and network evolution for batch heat exchanger network. *Applied Thermal Engineering*. 28, 2089–2099.

Fritzson, A., Berntsson, T., 2006. Efficient energy use in a slaughter and meat processing plant – opportunities for process integration. *Journal of Food Engineering*. 76, 594–604.

Halim, I., Srinivasan, R., 2009. Sequential methodology for scheduling of heat-integrated batch plants. *Industrial Engineering and Chemical Research*. 48(18), 8551–8565.

Halim, I., Srinivasana, R., 2011. Sequential methodology for integrated optimization of energy and water use during batch process scheduling. *Computers and Chemical Engineering*. 35, 1575–1597.

Hellwig, T., Thöne, E., 1994. Omnium: A method for optimization of waste heat utilization in Germany eni verfahern zur optimierung der abwarmenutzunga. *BWK (Brennstoff, Warme, Kraft)*. 46, 393–397 [in German].

Ivanov, B., Bancheva, N., 1994. Optimal reconstruction of batch chemical plants with regard to maximum heat recuperation. *Computers and Chemical Engineering*. 18, 313–317.

José, A.Q., Araceli, G., María, G.A., Jalel, L., 2013. Heat integration options based on pinch and exergy analyses of a thermosolar and heat pump in a fish tinning industrial process. *Energy*. 55, 23–37.

Kallrath, J., 2002. Planning and scheduling in the process industry. *OR Spectrum*. 24(3), 219–250.

Kemp, I.C., Deakin, A.W., 1989. The cascade analysis for energy process integration of batch processes. Part 1. Calculation of energy targets. *Chemical Engineering and Research Design.* 67, 495–509.

Kemp, I.C., Macdonald, E.K., 1987. Energy and process integration in continuous and batch processes. Innovation in process energy utilization. *IChemE Symposium Series.* 105, 185–200.

Kondili, E., Pantelides, C.C., Sargent, R.W.H., 1993. A general algorithm for short-term scheduling of batch operations. I. MILP formulation. *Computer and Chemical Engineering.* 17, 211–227.

Krummenacher, P., Favrat, D., 2001. Indirect and mixed direct–indirect heat integration of batch processes based on Pinch Analysis. *International Journal of Applied Thermodynamics.* 4, 135–143.

Liu, L., Du, J., Xiao, F., Chen, L., Yao, P., 2011. Direct heat exchanger network synthesis for batch process with cost targets. *Applied Thermal Engineering.* 31, 2665–2675.

Maiti, D., Jana, A.K., Samanta, A.N., 2011. A novel heat integrated batch distillation scheme. *Applied Energy.* 88, 5221–5225.

Majozi, T., 2006. Heat integration of multipurpose batch plants using a continuous-time framework. *Applied Thermal Engineering.* 26, 1369–1377.

Majozi, T., 2009. Minimization of energy use in multipurpose batch plants using heat storage: An aspect of cleaner production. *Journal of Cleaner Production.* 17, 945–950.

Majozi, T., Zhu, X.X., 2001. A novel continuous-time MILP formulation for multipurpose batch plants. 1. Short-term scheduling. *Industrial Engineering and Chemical Research.* 40(25), 5935–5949.

Morrison, A.S., Walmsley, M.R.W., Neale, J.R., Burrell, C.P., Kamp, P.J.J., 2007. Non-continuous and variable rate processes: Optimization for energy use. *Asia-Pacific Journal of Chemical Engineering.* 2(5), 380–387.

Muster-Slawitsch, B., Weiss, W., Schnitzer, H., Brunner, C., 2011. The green brewery concept Energy efficiency and the use of renewable energy sources in breweries. *Applied Thermal Engineering.* 31, 2123–2134.

Obeng, E.D.A., Ashton, G.J., 1988. On pinch technology based procedures for the design of batch processes. *Chemical Engineering and Research Design.* 66, 255–259.

Papageorgiou, L.G., Shah, N., Pantelides, C.C., 1994. Optimal scheduling of heat-integrated multipurpose plants. *Industrial and Engineering Chemical Research.* 33(12), 3168–3186.

Pinto, T., Novais, A.Q., Barbosa-Póvoa, A.P.F.D., 2003. Optimal design of heat-integrated multipurpose batch facilities with economic savings in utilities: A mixed integer mathematical formulation. *Annals of the Operation Research.* 120, 201–230.

Pires, A.C., Fernandes, C.M., Nunes, C.P., 2005. An energy integration tool for batch process, sustainable development of energy, water and environment systems. In *Proceedings of the Third Dubrovnik Conference.* pp.175–185.

Pozna, A., Ivanov, B., Bancheva, N., 1998. Design of a heat exchanger network for a system of batch vessels. *Hungarian Journal of Industry and Chemistry.* 26, 203–211.

Seid, R., Majozi, T., 2012. A robust mathematical formulation for multipurpose batch plants. *Chemical Engineering Science.* 68, 36–53.

Stamp, J.D., Majozi, T., 2011. Optimum heat storage design for heat integrated multipurpose batch plants. *Energy.* 36(8), 5119–5131.

Stolze, S., Mikkelsen, J., Lorentzen, B., Petersen, P.M., Qvale, B., 1995. Waste-heat recovery in batch processes using heat storage. *Journal of Energy Resources and Technology.* 117, 142–149.

Sundaramoorthy, A., Karimi, I.A., 2005. A simpler better slot-based continuous-time formulation for short-term scheduling in multipurpose batch plants. *Chemical Engineering Science.* 60, 2679–2702.

Timothy, G, Walmsley, M.R.W., Martin, J., Atkins, M.J., Neale, J.R., 2014. Integration of industrial solar and gaseous waste heat into heat recovery loops using constant and variable temperature storage. *Energy.* 75, 53–67.

Tokos, H., Pintarič, Z.N., Glavič, P., 2010. Energy saving opportunities in heat integrated beverage plant retrofit. *Applied Thermal Engineering.* 30, 36–44.

Uhlenbruck, S., Vogel, R., Lucas, K., 2000. Heat integration of batch processes. *Chemical Engineering and Technology.* 23(3), 226–229.

Vaselenak, J.A., Grossmann, I.E., Westerberg, A.W., 1986. Heat integration in batch processing. *Industrial and Engineering Chemical Process Design Development.* 25(2), 357–366.

6 Design and Synthesis of Heat-Integrated Batch Plants Using an Effective Technique

6.1 INTRODUCTION

The aim of this chapter is to address design, synthesis and scheduling simultaneously with the consideration of economic savings in utility requirements, while considering the cost of both the auxiliary structures. The recent design and synthesis model by Seid and Majozi (2012) is extended to incorporate the design of the associated utility facility, since it is proven to result in better design objectives and computational efficiencies. An additional feature of the proposed model is the determination of optimal pipe connection between processing equipment. The model is implemented in a case study in order to demonstrate its application. From the case study, the profit is increased by 20% and the total utility requirement is reduced by 41.1% for the design and synthesis of energy-integrated batch plant compared to the basic design.

6.2 NECESSARY BACKGROUND

There is a great deal of interest for manufacturing of fine chemicals, pharmaceutical products, polymers and food and beverage using batch operations because of the advantage of producing low-volume, high-quality products, flexibility to adopt complex operations due to fast market change and suitability to manufacture different products using the same facility. In these batch facilities, the different products compete for the available resources like equipment, utilities, manpower and storage, which makes the design and operation of this plant a challenging task. The conceptual design problem must determine the number and capacity of the major processing equipment items, utilities and storage tanks so as to meet these design and production objectives at the lowest possible capital and operating cost. The modelling and solution of multipurpose batch processes has received considerable attention in the last two decades. The literature review which follows covers published work under two major headings: basic design and extended design.

6.2.1 Basic Grass Root Design for Multipurpose Batch Plants

Literatures under this category considered simple choice of equipment and associated scheduling. The formulation of Suhami and Mah (1982) was based on the work of Grossmann and Sargent (1979) for multi-product batch plants. A two-level problem is derived where the upper level is characterized by the definition of production campaigns using a heuristic procedure, and the lower level includes the design problem with additional constraints resulting in a mixed integer nonlinear programming (MINLP) problem deciding allocation of tasks to units during the time of planning. Klossner and Rippin (1984) looked into the unique assignment case and enumerated all possible product configurations by solving a set partitioning problem followed by the solution of a MINLP for each configuration. The set partitioning problem was further studied by Imai and Nishida (1984), who presented an improved heuristic procedure to solve it. However, no comparative results were presented.

Vaselenak et al. (1987) avoided the two-stage nature of these approaches by proposing a superstructure embedding all possible product configurations. However, a major difficulty appears to be associated with the derivation of the limiting set of production time constraints, the so-called horizon constraint. A more complete formulation based on a theory of linear inequalities was presented by Faqir and Karimi (1989) where for every case involving a unique production route the equivalent horizon constraints are identified. This work was later extended (Faqir and Karimi, 1990) to multiple production routes assuming unique task-unit allocations with discrete batch size equipment. Cerda et al. (1989) proposed a more general problem than the ones mentioned earlier. Here, the restrictive assumption that all units within a stage must be allocated to a task if the stage itself is allocated to the same task was relaxed.

Papageorgaki and Reklaitis (1990) proposed a mixed integer linear programming (MILP) formulation which accommodates equipment used in and out of phase, units available with different sizes in a processing stage and tasks that can be processed in different units. Most of the previous published works were restricted to the following key assumptions: (1) There is a pre-specified assignment of equipment items to product tasks. (2) Parallel production is only allowed for products that have no common equipment requirements. (3) All units of a given type are identical and can be used in the out-of-phase mode. (4) All units of a given type are devoted to the production of only one product at a time. Shah and Pantelides (1991) presented a formulation incorporating long-term campaign planning. Previous work was limited to all the processing steps involved in the manufacturing of a product taking place within any campaign producing it. This is relaxed with the use of intermediate storage to decouple the manufacturing of each product into several stages each of which can be run independently in campaign mode.

Voudouris and Grossmann (1996) presented a formulation for a special class of sequential multipurpose batch plants where not all the products use the same processing stages. A cyclic schedule was implemented that has the effect of aggregating the number of batches for each product in order to allow the consideration of problems of practical size. It is shown that the no-wait characteristics of substrains can be exploited with a reduction scheme that has the effect of greatly decreasing the dimensionality of the problem. This reduction scheme can be complemented with a tight formulation of the underlying disjunctions in the MILP problem to reduce the computational expense.

Xia and Macchietto (1997) presented a general formulation considering linear/ nonlinear models for capital cost of processing units/storage vessels and processing task models with fixed, linear and nonlinear processing time in a general form. The formulation resulted in a MINLP solved by a stochastic MINLP optimizer. A short-term scheduling model was implemented based on discrete time representation which allowed batches of a product to be processed in any sequence.

Barbosa-Povoa and Pantelides (1997) developed a model for single campaign structure with fixed product slate is assumed within a non-periodic operation. A resource-task network (RTN) scheduling framework by Pantelides (1994) was used as a platform. The model allows a detailed consideration of the design problem taking into account the trade-offs between capital costs, revenues and operational flexibility. The optimal solution involves the selection of the required processing and storage equipment items. Lin and Floudas (2001) presented a formulation based on the scheduling formulation of Ierapetritou and Floudas (1998) using a unit-specific event-point and continuous-time representation. The formulation addressed a single campaign production approach where multiple batches of a product can be produced in the given time horizon. A comparison with earlier formulations was given. Heo et al. (2003) developed a three-step procedure for the design of multipurpose batch plants using cyclic scheduling. The first MILP model gives the minimum number of equipment units required to produce the products; this configuration is used as an initial plant configuration. The second MILP model determines the minimum cycle time and the third determines the equipment size and scheduling that minimizes the cost. The method gave a better solution compared to the separable programming and the evolutionary design method proposed by Fuchino et al. (1994). Castro et al. (2005) presented a general mathematical formulation for the simultaneous design and scheduling of multipurpose plants. The formulation is based on the RTN process representation and can handle both short-term and periodic problems. The performance of the formulation is illustrated through the solution of two periodic example problems that have been examined in the literature, where the selection and design of the main equipment items and their connecting pipes is considered. They also investigated two example problems by changing demand rates, with the results showing that this set of parameters, which is usually subject to a high degree of uncertainty, affects not only the optimal structure and cost of the plant but also the optimal cycle time. This is a clear indication that both the design and scheduling aspects of the problem should be considered simultaneously. A comparison with an earlier approach of Lin and Floudas (2001) is also presented.

6.2.2 EXTENDED DESIGN

Literature under this category besides the basic design important features, such as plant topology, layout and rational use of utility, has been covered. Barbosa-Povoa and Macchietto (1994) presented for the first time the design of multipurpose batch plants considering plant topology. The formulation is based on the maximal state task network (mSTN) which describes both the recipes and plant possible super-structure. The model optimizes the structural aspect of the plant and the associated production schedule accounting for capital cost of equipment, pipework, operating costs and revenues. Both short-term and cyclic scheduling was studied. Penteado and Ciric (1996) and Barbosa-Povoa et al. (2001a) addressed layout aspects in the design problems. A 2D small case study was analysed.

Georgiadis et al. (1997) used a space discretization technique to consider the allocation of equipment items to floor as well as the block layout of each floor. The main drawback of this formulation is a suboptimal solution may result due to the discretization of the available space. Barbosa-Povoa et al. (2002) proposed a general model where both 2D and 3D space was considered. Irregular shapes for each piece of equipment could also be accounted for in the formulation. Patziatsis et al. (2005) formulated simultaneous layout, design and planning of pipeless batch plants. Barbosa-Povoa (2007) reviewed the design and retrofit of batch plants. The authors included published literature for the last two decades. From the conclusion of the authors, there is still a gap in developing efficient techniques addressing large-scale problems, detail process operation (cleaning in place, change over and utility requirements) and plant layout and topology. Pinto et al. (2008) presented a comparative analysis between the state task network (STN), m-STN and RTN representations for the design of multipurpose batch plants. A number of problems are solved and compared with the formulations based on the different representations.

The increasing importance of utilities rationalization within the design of multi-purpose batch plants was first addressed by Barbosa-Povoa et al. (2001b), who presented a mathematical formulation for the detailed design of multipurpose batch process facilities. This work was later extended by Pinto et al. (2003) where economic savings in utility requirements were obtained while considering both the cost of the auxiliary structures (i.e. heat exchanger through their transfer area) and the design of the utility circuits and associated piping costs.

In this work, an efficient formulation for the design of multipurpose batch plants that considers plant topology and heat integration for sustainable design is developed. The formulation is posed as a MILP in which the binary variables are the structural choice variables. The proposed model is able to accommodate equipment used in and out of phase, units available in two or more sizes within a processing stage, multiple choices of equipment types for each product task, unit dependent processing time and processing time dependent on batch.

6.3   PROBLEM STATEMENT

The optimal plant synthesis and design for multipurpose batch plants can be achieved by developing a model that can solve the following problem:

Given:

1. The product recipes (STNs) describing the production of one or more products over a single campaign structure
2. The plant flowsheet with all possible equipment units to be installed and the involved connectivity

3. The equipment units' suitability to perform the process/storage tasks
4. The connections' suitability to transfer materials
5. The operating and capital cost data involved in the plant
6. The time horizon of planning
7. The production requirements over the time horizon
8. Scheduling data
9. Heat integration data such as operating temperature of a task, heat load of a task, cooling and heating cost

Determine:

1. The optimal plant configuration (i.e. number and type of equipment and the optimal design size of the equipment)
2. Minimum cooling and heating requirement
3. Detail of the heat exchanger network
4. A process schedule that allows the selected resources to achieve the required production (i.e. the starting and finishing times of all tasks, storage policies, batch sizes, amounts transferred and allocation of tasks to the equipment), so as to optimize the economic performance of the plant, measured in terms of the capital expenditure and the operating costs and revenues

6.4 MODEL FORMULATION

In this work, the mathematical modelling is developed and has the following constraints.

6.4.1 EQUIPMENT EXISTENCE CONSTRAINTS

Constraint (6.1) states that in order to execute a task in the unit, the unit must be selected first.

$$\sum_{sin,j \in S^*_{in,j}} y\left(s_{in,j}, p\right) \le e(j), \quad \forall j \in J, \quad p \in P \tag{6.1}$$

6.4.2 UNIT SIZE CONSTRAINTS

Constraint (6.2) implies that if the unit is selected, the design capacity should be between the minimum and maximum design capacity.

$$V^l_j e(j) \le ss(j) \le V^u_j e(j), \quad \forall j \in J \tag{6.2}$$

6.4.3 CAPACITY CONSTRAINTS

Constraint (6.3) implies that the total amount of all the states consumed at time point p is limited by the capacity of the unit which consumes the states and represents the lower and upper bounds in the capacity of a given unit that processes the effective state.

$$\gamma^L_{s_{in,j}} ss(j) - V^L_j \left(1 - y\left(s_{in,j}, p\right)\right) \leq mu\left(s_{in,j}, p\right) \leq \gamma^U_{s_{in,j}} ss(j),$$

$$\forall p \in P, \quad j \in J, \quad s_{in,j} \in S_{in,J} \tag{6.3}$$

Constraints (6.4) and (6.5) ensure the amount of material stored at any time point p is limited by the capacity of the storage.

$$q_s(s, p) \leq s(v), \quad \forall s \in S, \quad p \in P, \quad v \in V \tag{6.4}$$

$$V^L_v eu(v) \leq s(v) \leq V^U_v eu(v), \quad \forall s \in S, \quad p \in P, \quad v \in V \tag{6.5}$$

6.4.4 MATERIAL BALANCE FOR STORAGE

Constraint (6.6) states that the amount of material stored at each time point p is the amount stored at the previous time point adjusted by some amount resulting from the difference between state s produced by tasks at the previous time point $(p - 1)$ and used by tasks at the current time point p. This constraint is used for an intermediate state rather than a product, since the latter is not consumed, but only produced within the process.

$$q_s(s, p) = q_s(s, p-1) - \sum_{j \in J^{sc}_s} muu(s, j, p) + \sum_{j \in J^{sp}_s} muu(s, j, p-1),$$

$$\forall p \in P, \quad p \geq 1, \quad s \in S \tag{6.6}$$

Constraint (6.7) is used for the material balance around storage at the first time point.

$$q_s(s, p) = QO(s) - \sum_{j \in J^{sc}_s} muu(s, j, p), \quad \forall p \in P, \quad p = 1, \quad s \in S \tag{6.7}$$

Constraint (6.8) states that the amount of product stored at time point p is the amount stored at the previous time point and the amount of product produced at time point p.

$$q_s(s^p, p) = q_s(s^p, p-1) + \sum_{s_{in,j} \in s^p_{in,J}} \rho^{sp}_{s_{in,j}} mu\left(s_{in,j}, p\right), \quad \forall p \in P, \quad s^p \in S^p \tag{6.8}$$

6.4.5 MATERIAL BALANCE AROUND THE PROCESSING UNIT

Constraint (6.9) is used to cater for the material processed in the unit and equals the amount of material directly coming from the unit producing it and from the storage.

$$\rho^{sc}_{s_{in,j}} mu\left(s_{in,j}, p\right) = muu(s, j, p) + \sum_{j' \in j^{sp}_s} mux(s, j, j', p),$$

$$\forall p \in P, \quad j, j' \in J, \quad s_{in,j} \in S_{in,J}, \quad s \in S \tag{6.9}$$

Constraint (6.10) is used to define the amount of material produced at time point p which is sent to storage at the same time point p and to units that consume it at time point $p + 1$.

$$\rho_{s_{in,j}}^{sp} mu\left(s_{in,j}, p\right) = muu\left(s, j, p\right) + \sum_{j' \in j_s^{sc}} mux\left(s, j', j, p+1\right),$$

$$\forall p \in P, \quad j, j' \in J, \quad s_{in,j} \in S_{in,J}, \quad s \in S \qquad (6.10)$$

6.4.6 EXISTENCE CONSTRAINTS FOR PIPING

Constraints (6.11) and (6.12) are applicable to ensure whether pipe connections exist between the processing units and storage as well as to ensure the amount of material transferred through the connection is limited by the design capacity of the connection. The material transfer time is assumed fixed.

$$V_{j,v}^{L} z\left(j, v\right) \leq pip\left(j, v\right) \leq V_{j,v}^{U} z\left(j, v\right), \quad \forall j \in J, \quad v \in V \qquad (6.11)$$

$$muu\left(s, j, p\right) \leq pip\left(j, v\right), \quad \forall j \in J, \quad v \in V, \quad p \in P, \quad s \in S \qquad (6.12)$$

Constraints (6.13) and (6.14) are similar to Constraints (6.11) and (6.12) and are used when the connection is between processing units.

$$V_{j,j'}^{L} w\left(j, j'\right) \leq pipj\left(j, j'\right) \leq V_{j,j'}^{U} w\left(j, j'\right), \quad \forall j \in J \qquad (6.13)$$

$$mux\left(s, j, j', p\right) \leq pipj\left(j, j'\right), \quad \forall j, j' \in J, \quad p \in P, \quad s \in S \qquad (6.14)$$

6.4.7 DURATION CONSTRAINTS (BATCH TIME AS A FUNCTION OF BATCH SIZE)

Constraint (6.15) describes the duration constraint modelled as a function of batch size where the processing time is a linear function of the batch size. For zero-wait (ZW), only the equality sign is used.

$$t_p\left(s_{in,j}, p\right) \geq t_u\left(s_{in,j}, p\right) + \tau\left(s_{in,j}\right) y\left(s_{in,j}, p\right) + \beta\left(s_{in,j}\right) mu\left(s_{in,j}, p\right),$$

$$\forall j \in J, \quad p \in P, \quad s_{in,j} \in S_{in,J} \qquad (6.15)$$

6.4.8 SEQUENCE CONSTRAINTS

The following two subsections address the proper allocation of tasks in a given unit to ensure the starting time of a new task is later than the finishing time of the previous task.

6.4.8.1 Same Task in Same Unit

Constraint (6.16) states that a state can only be used in a unit, at any time point, after all the previous tasks are complete.

$$t_u\left(s_{in,j}, p\right) \geq t_p\left(s_{in,j}, p-1\right), \quad \forall j \in J, \quad p \in P, \quad s_{in,j} \in S_{in,j}^{*} \tag{6.16}$$

6.4.8.2 Different Tasks in Same Unit

Constraint (6.17) states that a task can only start in the unit after the completion of all the previous tasks that can be performed in the unit.

$$t_u\left(s_{in,j}, p\right) \geq t_p\left(s'_{in,j}, p-1\right), \quad \forall j \in J, \quad p \in P, \quad s_{in,j} \neq s'_{in,j}, \quad s_{in,j}, s'_{in,j} \in S_{in,j}^{*} \tag{6.17}$$

If the state is consumed and produced in the same unit, where the produced state is unstable then in addition to Constraints (6.17) through (6.19) are used.

$$t_p\left(s_{in,j}^{usp}, p-1\right) \geq t_u\left(s_{in,j}^{usc}, p\right) - H\left(1 - y\left(s_{in,j}^{usp} p-1\right)\right),$$
$$\forall j \in J, \quad p \in P, \quad s_{in,j}^{usc} \in S_{in,j}^{usc}, \quad s_{in,j}^{usp} \in S_{in,j}^{usp} \tag{6.18}$$

$$t_p\left(s_{in,j}^{usp}, p\right) \geq t_p\left(s_{in,j}^{usp}, p-1\right), \quad \forall j \in J, \quad p \in P, \quad s_{in,j}^{usp} \in S_{in,j}^{usp} \tag{6.19}$$

It should be noted that in this particular situation $t_u\left(s_{in,j}^{usc}, p\right) = t_u\left(s_{in,j}, p\right)$ and $t_p\left(s_{in,j}^{usc}, p\right) = t_p\left(s_{in,j}, p\right)$. Consequently, Constraints (6.17) and (6.18) enforce $t_u\left(s_{in,j}^{usc}, p\right) = t_p\left(s_{sin,j}^{usc}, p-1\right)$.

6.4.9 Sequence Constraints for Different Tasks in Different Units

These constraints state that for different tasks that consume and produce the same state, the starting time of the consuming task at time point p must be later than the finishing time of any task at the previous time point $p - 1$ provided that the state is used.

6.4.9.1 If an Intermediate State s Is Produced from One Unit

Constraints (6.20) and (6.21) work together in the following manner:

$$\rho\left(s_{in,j}^{sp}\right) mu\left(s_{in,j}, p-1\right) \leq q_s\left(s, p\right) + V_j^{U} t\left(j, p\right), \quad \forall j \in J, \quad p \in P, \quad s_{in,j} \in S_{in,J}^{sp} \tag{6.20}$$

$$t_u\left(s_{in,j'}, p\right) \geq t_p\left(s_{in,j}, p-1\right) - H\left(\left(2 - y\left(s_{in,j}, p-1\right) - t\left(j, p\right)\right)\right),$$
$$\forall j \in J, \quad p \in P, \quad s_{in,j} \in S_{in,J}^{sp}, \quad s_{in,j'} \in S_{in,J}^{sc} \tag{6.21}$$

Constraint (6.20) states that if the state s is produced from unit j at time point $p - 1$ but is not consumed at time point p by another unit j', that is, $t(j, p) = 0$, then the amount produced cannot exceed allowed storage, that is, $q_s(s, p)$. On the other hand, if state s produced from unit j at time point $p - 1$ is used by another unit j', then the amount of state s stored at time point p, i.e. $q_s(s,p)$, is less than the amount of state s produced at time point $p - 1$. The outcome is that the binary variable $t(j, p)$ becomes 1 in order for Constraint (6.20) to hold. Constraint (6.21) states that the starting time of a task consuming state s at time point p must be later than the finishing time of a task that produces state s at the previous time point $p - 1$, provided that state s is used. Otherwise, the sequence constraint is relaxed.

6.4.9.2 If an Intermediate State Is Produced from More than One Unit

Constraint (6.22) states that the amount of state s used at time point p can either come from storage or from other units that produce the same state, depending on the binary variable $t(j, p)$. If the binary variable $t(j, p)$ is 0, which means that state s produced from unit j at time point $p-1$ is not used at time point p, then Constraint (6.21) is relaxed. If $t(j, p)$ is 1, state s produced from unit j at time point $p-1$ is used; as a result Constraint (6.21) holds. Although Constraint (6.22) is nonlinear, it can be linearized exactly using the Glover transformation[31].

$$\sum_{s_{in,j} \in S_{in,J}^{sc}} \rho_{s_{in,j}}^{sc} mu\left(s_{in,j}, p\right) \le qs(s, p-1) + \sum_{s_{in,j} \in S_{in,J}^{sp}} \rho_{s_{in,j}}^{sp} mu\left(s_{in,j}, p-1\right) t\left(j, p\right)$$

$$\forall j \in J, \quad p \in P \tag{6.22}$$

6.4.9.3 Sequence Constraints for Completion of Previous Tasks

Constraint (6.23) states that a consuming task can start after the completion of the previous task. Constraint (6.23) takes care of proper sequencing time when a unit uses material which is previously stored, that is, when the producing task is active at time point $p - 2$ and later produces and transfers the material to the storage at time point $p - 1$. This available material in the storage at time point $p - 1$ is then used by the consuming task in the next time points. This necessitates that the starting time of the consuming task must be later than the finishing time of the producing task at time point $p - 2$. Consequently, Constraint (6.23) together with Constraint (6.21) results in a feasible sequencing time when the consuming task uses material, which is previously stored or/and material which is currently produced by the producing units.

$$t_u\left(s_{in,j'}, p\right) \ge t_p\left(s_{in,j}, p-2\right) - H\left(1 - y\left(s_{in,j}, p-2\right)\right),$$

$$\forall j \in J, \quad p \in P, \quad s_{in,j} \in S_{in,J}^{sp}, \quad s_{in,j'} \in S_{in,j'}^{sc} \tag{6.23}$$

6.4.10 Sequence Constraints for Finite Intermediate Storage (FIS) Policy

According to Constraints (6.24) and (6.21), the starting time of a task that consumes state s at time point p must be equal to the finishing time of a task that produces state

s at time point $p - 1$, if both consuming and producing tasks are active at time point p and time point $p - 1$, respectively.

$$t_u\left(s_{in,j'},p\right) \leq t_p\left(s_{in,j},p-1\right) + H\left(2 - y\left(s_{in,j'},p\right) - y\left(s_{in,j},p-1\right)\right)$$

$$\forall j \in J, \quad p \in P, \quad s_{in,j} \in S_{in,J}^{sp}, \quad s_{in,j'} \in S_{in,J}^{sc} \tag{6.24}$$

6.4.11 TIME HORIZON CONSTRAINTS

The usage and the production of states should be within the time horizon of interest. These conditions are expressed in Constraints (6.25) and (6.26).

$$t_u\left(s_{in,j},p\right) \leq H, \quad \forall s_{in,j} \in S_{in,J}, \quad p \in P, \quad j \in J \tag{6.25}$$

$$t_p\left(s_{in,j},p\right) \leq H, \quad \forall s_{in,j} \in S_{in,J}, \quad p \in P, \quad j \in J \tag{6.26}$$

6.4.12 ENERGY INTEGRATION CONSTRAINTS

Constraints (6.27) and (6.28) are active simultaneously and ensure that one hot unit will be integrated with one cold unit when direct heat integration takes place.

$$\sum_{i_c} x\left(i_h, i_c, p\right) \leq y\left(i_h, p\right), \quad \forall p \in P, \quad i_h \in I_h \tag{6.27}$$

$$\sum_{i_h} x\left(i_h, i_c, p\right) \leq y\left(i_c, p\right), \quad \forall p \in P, \quad i_c \in I_c \tag{6.28}$$

Constraint (6.29) ensures that minimum thermal driving forces are obeyed.

$$T\left(i_h\right) - T\left(i_c\right) \geq \Delta T^{min} - M\left(1 - x\left(i_h, i_c, p\right)\right), \quad \forall p \in P, \quad i_c, i_h \in I \tag{6.29}$$

Constraint (6.30) states that the cooling of a heat source will be satisfied by direct heat integration as well as external utility if required.

$$\left(E\left(i_h\right)y\left(i_h,p\right) + \xi\left(i_h\right)mu\left(i_h,p\right)\right)\left(tp\left(i_h,p\right) - tu\left(i_h,p\right)\right)$$

$$= cw\left(i_h,p\right) + \sum_{i_c} xx(i_h, i_c, p), \quad \forall p \in P, \quad i_h \in I_h \tag{6.30}$$

Constraint (6.31) ensures that the heating of a heat sink will be satisfied by direct heat integration as well as external utility if required.

$$\left(E\left(i_c\right)y\left(i_c,p\right) + \xi\left(i_c\right)\right)\left(tp\left(i_c,p\right) - tu(i_c,p)\right)$$

$$= st\left(i_c,p\right) + \sum_{i_h} xx(i_h, i_c, p), \quad \forall p \in P, \quad i_c \in I_c \tag{6.31}$$

Constraint (6.32) states that the amount of heat exchanged between the hot and cold unit is limited by the total duration of the cold unit.

$$w\left(i_h, i_c, p\right) \le \left(E\left(i_h\right) y\left(i_h, p\right) + \xi\left(i_h\right) mu\left(i_h, p\right)\right)\left(tp\left(i_c, p\right) - tu\left(i_c, p\right)\right),$$
$$\forall p \in P, \quad i_c \in I_c, \quad i_h \in I_h \tag{6.32}$$

Constraint (6.33) ensures that the amount of heat transferred from the hot unit to the cold unit is limited by the duration of the hot unit.

$$w\left(i_h, i_c, p\right) \le \left(E\left(i_c\right) y\left(i_c, p\right) + \xi\left(i_c\right) mu\left(i_c, p\right)\right)\left(tp\left(i_h, p\right) - tu\left(i_h, p\right)\right),$$
$$\forall p \in P, \quad i_c \in I_c, \quad i_h \in I_h \tag{6.33}$$

Constraint (6.34) states that the amount of heat exchanged between the hot and cold unit is determined by the value of the binary variable $x\left(i_h, i_c, p\right)$.

$$w\left(i_h, i_c, p\right) \le Mx\left(i_h, i_c, p\right), \quad \forall p \in P, \quad i_c \in I_c, \quad i_h \in I_h \tag{6.34}$$

Constraints (6.35) and (6.36) are used to determine the heat exchanger area required.

$$UA\left(T\left(i_h\right) - T\left(i_c\right)\right)\left(tp\left(i_h, p\right) - tu\left(i_h, p\right)\right) \ge w\left(i_h, i_c, p\right), \quad \forall i_h \in I_h, \quad i_c \in I_c, \quad p \in P \tag{6.35}$$

$$UA\left(T\left(i_h\right) - T\left(i_c\right)\right)\left(tp\left(i_c, p\right) - tu\left(i_c, p\right)\right) \ge w\left(i_h, i_c, p\right), \quad \forall i_h \in I_h, \quad i_c \in I_c, \quad p \in P \tag{6.36}$$

Constraints (6.37) and (6.38) are used to synchronize units when heat integration takes place.

$$tp\left(i_h, p\right) \ge tp\left(i_c, p\right) - M\left(1 - x\left(i_h, i_c, p\right)\right), \quad \forall p \in P, \quad i_c \in I_c, \quad i_h \in I_h \tag{6.37}$$

$$tp\left(i_h, p\right) \le tp\left(i_c, p\right) + M\left(1 - x\left(i_h, i_c, p\right)\right), \quad \forall p \in P, \quad i_c \in I_c, \quad i_h \in I_h \tag{6.38}$$

Constraints (6.39) and (6.40) are used to cater for external utilities.

$$u_c = \sum_p \sum_{i_c} cw\left(i_c, p\right), \quad \forall p \in P, \quad i_c \in I_c \tag{6.39}$$

$$u_h = \sum_p \sum_{i_h} st\left(i_h, p\right), \quad \forall p \in P, \quad i_h \in I_h \tag{6.40}$$

6.4.13 Objective Function

Constraint (6.41) is the objective function expressed as maximization of profit. This is obtained from revenue from the sale of products less operating costs for tasks, raw material costs and capital costs from piping and equipment.

$$
\begin{aligned}
\text{maximize} & \left(
\begin{array}{l}
\displaystyle\sum_{s^p} price\left(s^p\right) q_s\left(s^p, p\right) - \sum_p \sum_{s_{inj} \in S_{rm}} mu\left(s_{inj}, p\right) - \\[2ex]
\displaystyle\sum_p \sum_{s_{inj} \in S_{inJ}} \left(FOC * y\left(s_{inj}, p\right) + VOC * mu\left(s_{inj}, p\right)\right) - C_c u_c - C_h u_h
\end{array}
\right) * \left(AWH/H\right) \\[3ex]
& - \sum_{j \in J} \sum_{v \in V} \left(CNC * z\left(j, v\right) + VCN * pip\left(j, v\right)\right) \\[3ex]
& - \sum_{j' \in J} \sum_{j \in J} \left(FCNC * w\left(j, j'\right) + VCNC * pipj\left(j, j'\right)\right) \\[3ex]
& - \sum_{j \in J} \left(FEC * e\left(j\right) + VEQ * ss\left(j\right)\right) \\[3ex]
& - \sum_{v \in V} \left(FECS * eu\left(v\right) + VEQS * s\left(v\right)\right), \quad \forall p = P, \quad s^p \in S^p
\end{aligned}
\tag{6.41}
$$

Constraint (6.42) is the objective function expressed as minimization of capital and operating cost if the demand for the products is known beforehand within the specified time horizon.

$$
\begin{aligned}
\text{minimize} & \left(
\begin{array}{l}
\displaystyle\sum_p \sum_{s_{inj} \in S_{rm}} mu\left(s_{inj}, p\right) + \\[2ex]
\displaystyle\sum_p \sum_{s_{inj} \in S_{inJ}} \left(FOC * y\left(s_{inj}, p\right) + VOC * mu\left(s_{inj}, p\right)\right) + C_c u_c + C_h u_h
\end{array}
\right) * \left(AWH/H\right) \\[3ex]
& \left(
\begin{array}{l}
+ \displaystyle\sum_{j \in J} \sum_{v \in V} \left(CNC * z\left(j, v\right) + VCN * pip\left(j, v\right)\right) \\[3ex]
+ \displaystyle\sum_{j' \in J} \sum_{j \in J} \left(FCNC * w\left(j, j'\right) + VCNC * pipj\left(j, j'\right)\right) \\[3ex]
+ \displaystyle\sum_{j \in J} \left(FEC * e\left(j\right) + VEQ * ss\left(j\right)\right) \\[3ex]
+ \displaystyle\sum_{v \in V} \left(FECS * eu\left(v\right) + VEQS * s\left(v\right)\right)
\end{array}
\right) * CCF, \\[3ex]
& \forall p = P, \quad s^p \in S^p
\end{aligned}
\tag{6.42}
$$

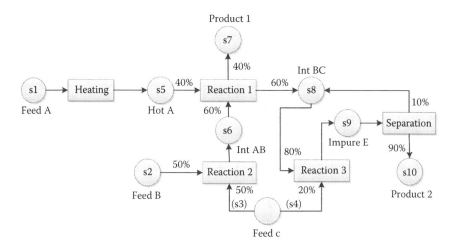

FIGURE 6.1 STN representation for the case study.

6.5 CASE STUDY

This example, which was first examined by Kondili et al. (1993), is studied extensively in literature. It is a relatively complex problem and is often used in literature to check the efficiency of models in terms of optimal objective value and CPU time required. The plant has many common features of a multipurpose batch plant, with the following features: units performing multiple tasks, multiple units suitable for a task, states shared by multiple tasks and different products produced following different production paths. The STN for this case study is depicted in Figure 6.1. Full connectivity between equipment is assumed. The plant superstructure is shown in Figure 6.2. The scheduling and design data are presented in Table 6.1. The raw material costs for state A, B and C are 1, 0.5 and 0.2 c.u. respectively. The selling price for product 1 and 2 are 4 and 7 respectively. Reaction 1 operates at 200°C and it is exothermic which requires 600 kWh/m.u. of cooling load. Reaction 2 operates at 70°C and it is endothermic which requires 800 kWh/m.u. of heating load. Reaction 3 operates at 160°C and it is exothermic which requires 700 kWh/m.u. of cooling load. Steam cost and cooling water cost are 0.7 and 0.15 c.u/MJ respectively.

6.6 RESULTS AND DISCUSSION

The computational results found are depicted in Tables 6.2 and 6.3. An overall profit of 2.6×10^8 was obtained using the model that catered for energy integration which is much better than the objective value of 2.08×10^8 obtained by the model without considering energy integration. Consequently, a 20% increase in profit was achieved by allowing heat integration between the hot and cold tasks. From Table 6.3, a significant saving in hot utility can be attained by integrating the

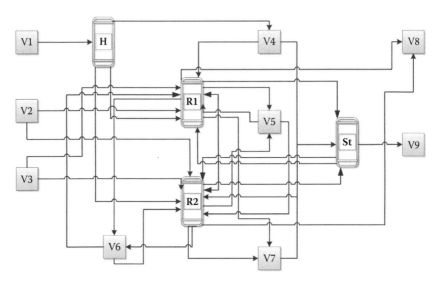

FIGURE 6.2   Superstructure for the case study.

TABLE 6.1
Equipment and Scheduling Data for the Case Study

Unit	Capacity (ton/m²)	Suitability	Processing Time	Cost Model (k c.u.)
Heater	5.0–30.0	Heating	1	200.0 + 10 s
Reactor 1	5.0–50.0	Reaction 1	2	5000.0 + 15 s
		Reaction 2	2	
		Reaction 3	1	
Reactor 2	5.0–50	Reaction 1	2	5000.0 + 10s
		Reaction 2	2	
		Reaction 3	1	
Still	5.0–20.0	Separation	2.5	150.0 + 5 s
Vessel 4	5.0–50.0	(Hot A)		50.0 + 3 s
Vessel 5	5.0–50.0	(Int BC)		70.0 + 3 s
Vessel 6	5.0–50.0	(Int AB)		60.0 + 4 s
Vessel 7	5.0–50.0	(Impure E)		90.0 + 5 s
Heat exchanger 1	0.0–200.0	RX1–RX2		60.0 + 5 s
Heat exchanger 2	0.0–200.0	RX2–Rx3		60.0 + 5 s

TABLE 6.2
Equipment Design Capacity for the Case Study

Unit	Equipment Capacity Range	Design Capacity with Consideration of Energy Integration	Design Capacity without Consideration of Energy Integration
Heater	5–30	19	13.3
Reactor 1	5–50	47.7	20
Reactor 2	5–50	50	20
Still	5–20	20	33.3
Vessel 4	5–50	19	13.3
Vessel 5	5–50	30.6	28.7
Vessel 6	5–50	32.1	14
Vessel 7	5–50	33.3	0
Heat exchanger 1	0–200	25.6	—
Heat exchanger 2	0–200	43.2	—

hot and cold tasks leading to supplanting most of the external utility requirement (160 GJ hot utility required by energy-integrated batch plant vs. 271.9 GJ hot utility required by the basic design without energy integration). Consequently, a saving of 41.1% in hot utility requirement is achieved. This indicates that solving a design and synthesis problem for batch plants by considering scheduling and heat integration results not only in efficient use of equipment resources for maximum production of products but also rational use of energy for sustainable operation. It is worth mentioning that this comprehensive model also determines the optimal heat exchanger area required in order to allow the heat transfer. An optimal area of 25.6 m^2 for heat exchanger 1 and 43.2 m^2 for heat exchanger 2 was obtained. Figure 6.2 details the possible amount of energy integration between the cold and hot units and the time intervals during which energy integration occurred. The energy requirements of reactor 1 and reactor 2 during the interval 4–6 h are highlighted to elaborate on the application of the proposed model. The cooling load of reactor 2 between 5 and 6 h was 25 GJ. This cooling requirement is fully satisfied through energy integration with reactor 1 in the same time interval, resulting in supplanting an external cooling requirement if operated in standalone mode. At the beginning of the operation of reactor 1 from 4 to 5 h, the heating requirement was 25.8 GJ. This value was obtained using the time average model by multiplying the duration (5 – 4 h) and the energy demand per hour (51.6 GJ/2 h (total duration of the task) = 25.8 GJ), where the heating requirement is fully satisfied by external heating. For the rest of its operation between 5 and 6 h, the heating requirement was 25.8 GJ, satisfied partly with energy integration (25 GJ) and the difference by external heating with energy requirement of 0.8 GJ. The amount of material each unit is processing, the type of task each unit is conducting at a specific time point and the starting and finishing times for each task are also presented in Figure 6.3.

TABLE 6.3
Computational Results for the Case Study

Method	Time Point	Objective Value	Binary Variables	Continuous Variables	Constraints	CPU Time (s)	Cold Utility	Hot Utility
Design without energy integration	11	$2.08 * 10^8$	176	793	2079	19	220.4 GJ	271.9 GJ
Design with energy integration	11	$2.6 * 10^8$	222	887	2367	198	34.4 GJ	160 GJ

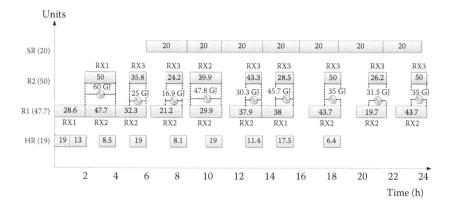

FIGURE 6.3   Gantt chart for the case study.

6.7   CONCLUSIONS

An efficient method for the design and synthesis of batch plants that incorporate energy integration was presented. The developed model simultaneously solved scheduling and energy integration within the same design framework leading to optimal design of equipment and rational use of energy. From the case study, the profit maximization was increased by 20% in the energy-integrated batch plant compared to the basic design where utilities were the only option to satisfy the heating and cooling load requirements demanded by the tasks. The total hot utility requirement was reduced by 41.1% for the energy-integrated batch plant compared to the basic design. From this, it can be concluded that for efficient design of batch plants, energy integration must be incorporated in the design problem.

NOMENCLATURE

SETS

J $\{j|j$ is a piece of equipment including heat exchanger$\}$

J_s^{sc} $\{j_s^{sc}|j_s^{sc}$ is a piece of equipment that consumes state $s\}$

J_s^{sp} $\{j_s^{sp}|j_s^{sp}$ is a piece of equipment that produces state $s\}$

V $\{v|v$ is a storage$\}$

P $\{p|p$ is a time point$\}$

$S_{in,J}^{sc}$ $\{s_{in,j}^{sc}|s_{in,j}^{sc}$ is a task which consumes state $s\}$

$S_{in,j}^{*}$ $\{s_{in,j}^{*}|s_{in,j}^{*}$ is a task performed in unit $j\}$

$S_{in,J}$ $\{s_{in,j}|s_{in,j}$ is an effective state representing a task$\}$

$S_{in,J}^{usc}$ $\{s_{in,j}^{usc}|s_{in,j}^{usc}$ is a task which consumes unstable state $s\}$

$S_{in,J}^{sp}$ $\{s_{in,j}^{sp}|s_{in,j}^{sp}$ is a task which produces state s other than a product$\}$

$S_{in,J}^{usp}$ $\{S_{in,j}^{usp}|S_{in,j}^{usp}$ is a task which produces unstable state $s\}$
S $\{s|s$ is any state$\}$
S^p $\{s|s$ is any state which is a product$\}$
S_{rm} $\{s|s$ is any state which is a raw material$\}$
$S_{in,J}^p$ $\{S_{in,j}^p|S_{in,j}^p$ task which produce state s which is a product$\}$
I_c $\{i_c|i_c$ is a task requires heating$\}$, $I_h = \{i_h|i_h$ is a task requires cooling$\}$

CONTINUOUS VARIABLES

$t_u\left(s_{in,j}, p\right)$	Time at which a task in unit j starts
$t_p\left(s_{in,j}, p\right)$	Time at which a task in unit j finishes
$ss\left(j\right)$	Design capacity of unit j
$s\left(v\right)$	Design capacity of storage unit v
$mu\left(s_{in,j}, p\right)$	Amount of material processed by a task
$muu\left(s, j, p\right)$	Amount of state s transferred between unit j and storage unit at time point p
$mux\left(s, j, j', p\right)$	Amount of state s transferred between unit j and another unit j' at time point p
$q_s\left(s, p\right)$	Amount of state s stored at time point p in storage unit
$pip\left(j, v\right)$	Design capacity of a pipe that connects unit j and storage unit v
$pipj\left(j, j'\right)$	Design capacity of a pipe that connects unit j and another unit j'
$cw\left(i_h, p\right)$	External cooling required by task i_h at time point p
$st\left(i_c, p\right)$	External heating required by task i_c at time point p
$w\left(i_h, i_c, p\right)$	Amount of heat exchanged between the cold and hot task
A	Optimal heat exchanger area required
$mu\left(i_h, p\right)$	Amount of material processed by the hot task
$mu\left(i_c, p\right)$	Amount of material processed by the cold task

BINARY VARIABLES

$y\left(s_{in,j}, p\right)$	1 if state s is used in unit j at time point p; 0 otherwise
$e\left(j\right)$	1 if unit j is selected; 0 otherwise
$eu\left(v\right)$	1 if storage unit v is selected; 0 otherwise
$z\left(j, v\right)$	1 if pipe that connects unit j and storage unit v is selected; 0 otherwise
$w\left(j, j'\right)$	1 if pipe that connects unit j and j' is selected; 0 otherwise
$t\left(j, p\right)$	1 if the state produced by unit j at time point p is consumed; 0 otherwise
$x\left(i_h, i_c, p\right)$	1 if hot task i_h is integrated with the cold task i_c at time point p; 0 otherwise

PARAMETERS

M	Any large number
$t\left(s_{in,j}\right)$	Duration of a task conducted in unit j
$\beta\left(s_{in,j}\right)$	Coefficient of variable term of processing time of a task
V_j^L	Lower bound for unit j
V_j^U	Upper bound for unit j
V_v^L	Lower bound for storage unit v
V_v^U	Upper bound for storage unit v
$\gamma_{s_{in,j}}^L$	Minimum percentage equipment utilization for a task
$\gamma_{s_{in,j}}^U$	Maximum percentage equipment utilization for a task
$QO(s)$	Initial amount of state s stored in unit
$\rho\left(s_{in,j}^{sp}\right)$	Portion of state s produced by a task
$\rho\left(s_{in,j}^{sc}\right)$	Portion of state s consumed by a task
$V_{j,v}^L$	Lower bound for the capacity of the pipe connecting unit j and storage unit v
$V_{j,v}^U$	Upper bound for the capacity of the pipe connecting unit j and storage unit v
$V_{j,j'}^L$	Lower bound for the capacity of the pipe connecting unit j and another unit j'
$V_{j,j'}^U$	Upper bound for the capacity of the pipe connecting unit j and another unit j'
$price\left(s^p\right)$	Selling price for a product
AWH	Annual working hour
H	Time horizon of interest
FOC	Fixed operating cost for a task
VOC	Variable operating cost for a task
CCF	Capital charge factor
FEC	Fixed capital cost of equipment
VEQ	Variable capital cost of equipment
$FECS$	Fixed capital cost of storage
$VEQS$	Variable capital cost of storage
$FCNC$	Fixed capital cost for pipe connection between processing units
$VCNC$	Variable capital cost for pipe connection between processing units
CNC	Fixed capital cost for pipe connection between processing unit and storage
VCN	Variable capital cost for pipe connection between processing unit and storage
C_p	Specific heat capacity of fluid
$E(i)$	Constant coefficient of amount of heat required or removed corresponding to task i

$\xi(i)$	Variable coefficient of amount of heat required or removed corresponding to task i
$T(i)$	Operating temperature for a task
$\Delta T^{\min}$	Minimum allowable thermal driving force
C_c	Cost unit for cooling
C_h	Cost unit for steam
U	Overall heat transfer coefficient

REFERENCES

Barbosa-Povoa, A.P., Macchietto, S., 1994. Detailed design of multipurpose batch plants. *Computers and Chemical Engineering*. 18(11/12), 1013–1042.

Barbosa-Povoa, A.P., Mateus, R., Novais, A.Q., 2001a. Optimal 2D design layout of industrial facilities. *International Journal of Production Research*. 39(12), 2567–2593.

Barbosa-Povoa, A.P., Mateus, R., Novais, A.Q., 2002. Optimal design and layout of industrial facilities: An application to multipurpose batch plants. *Industrial and Engineering Chemistry Research*. 41(15), 3610–3620.

Barbosa-Povoa, A.P., Pantelides, C.C., 1997. Design of multipurpose plants using the resource-task network unified framework. *Computers and Chemical Engineering*. 21, S703–S708.

Barbosa-Povoa, A.P., Pinto, T., Novais, A.Q., 2001b. Optimal design of heat-integrated multipurpose batch facilities: A mixed integer mathematical formulation, *Computers and Chemical Engineering*. 25, 547–559.

Barbosa-Povoa, A.P.F.D., 2007. A critical review on the design and retrofit of batch plants. *Computers and Chemical Engineering*. 31(7), 833–855.

Castro, P.M., Barbosa-Povoa, A.P., Novais, A.Q., 2005. Simultaneous design and scheduling of multipurpose plants using resource task network based continuous-time formulations. *Industrial and Engineering Chemistry Research*. 44(2), 343–357.

Cerda, J., Vicente, M., Gutierrez, J.M., Esplugas, S., Mata, J., 1989. A new methodology for the optimal-design and production schedule of multipurpose batch plants. *Industrial and Engineering Chemistry Research*. 28(7), 988–998.

Faqir, N.M., Karimi, I.A., 1989a. Optimal-design of batch plants with single production routes. *Industrial and Engineering Chemistry Research*. 1989, 28(8), 1191–1202.

Faqir, N.M., Karimi, L.A., 1989b. Design of multipurpose batch plants with multiple production routes. *Conference on Foundations of Computer Aided Process Design*, Snowmass, CO, 451–468.

Fuchino, T., Muraki, M., Hayakawa, T., 1994. Scheduling method in design of multipurpose batch plants with constrained resources. *Journal of Chemical Engineering of Japan*. 27(3), 363–368.

Georgiadis, M.C., Rotstein, G.E., Macchietto, S., 1997. Optimal layout design in multipurpose batch plants. *Industrial and Engineering Chemistry Research*. 36(11), 4852–4863.

Grossman, I.E., Sargent, R.W.H., 1979. Optimum design of multipurpose chemical plants. *Industrial and Engineering Chemistry Process Design and Development*. 18(2), 343–348.

Heo, S.K., Lee, K.H., Lee, H.K., Lee, I.B., Park, J.H., 2003. A new algorithm for cyclic scheduling and design of multipurpose batch plants. *Industrial and Engineering Chemistry Research*. 42(4), 836–846.

Ierapetritou, M.G., Floudas, C.A., 1998. Effective continuous-time formulation for short-term scheduling: 1 Multipurpose batch processes. *Industrial and Engineering Chemistry Research*. 37, 4341–4359.

Imai, M., Nishida, N., 1984. New procedure generating suboptimal configurations to the optimal-design of multipurpose batch plants. *Industrial and Engineering Chemistry Process Design and Development.* 23(4), 845–847.

Klossner, J., Rippin, D.W.T., 1984. Combinatorial problems in the design of multiproduct batch plants—Extension to multiplant and partly parallel operations. In *AIChE Annual Meeting,* San Francisco, CA.

Kondili, E., Pantelides, C.C., Sargent, R.W.H., 1993. A general algorithm for short-term scheduling of batch operations. I. MILP formulation. *Computer and Chemical Engineering.* 17, 211–227.

Lin, X., Floudas, C.A., 2001. Design, synthesis and scheduling of multipurpose batch plants via an effective continuous-time formulation. *Computers and Chemical Engineering.* 25665–25674.

Pantelides, C.C., 1994. Unified frameworks for the optimal process planning and scheduling. *Foundations on Computer Aided Process Operations,* CACHE, Austin, TX, pp. 253–274.

Papageorgaki, S., Reklaitis, G.V., 1990. Optimal design of multipurpose batch plants. 1. Problem formulation. *Industrial and Engineering Chemistry Research.* 29(10), 2054–2062.

Patziatsis, D.I., Xu, G., Papageorgiou, L.G., 2005. Layout aspects of pipeless batch plants. *Industrial and Engineering Chemistry Research.* 44(15), 5672–5679.

Penteado, F.D., Ciric, A.R., 1996. An MINLP approach for safe process plant layout. *Industrial and Engineering Chemistry Research.* 35, 1354–1361.

Pinto, T., Barbosa-Povoa, A.P., Novais, A.Q., 2003. Optimal design of heat-integrated multipurpose batch facilities: A mixed integer mathematical formulation. *Computers and Chemical Engineering.* 25, 547–559.

Pinto, T., Barbosa-Povoa, A.P., Novais, A.Q., 2008. Design of multipurpose batch plants: A comparative analysis between the STN, m-STN, and RTN representations and formulations. *Industrial and Engineering Chemistry Research.* 47, 6025–6044.

Seid, E.R., Majozi, T., 2012. A robust mathematical formulation for multipurpose batch plants. *Chemical Engineering Science.* 68, 36–53.

Shah, N., Pantelides, C.C., 1991. Optimal long-term campaign planning and design of batch-operations. *Industrial and Engineering Chemistry Research.* 30(10), 2308–2321.

Suhami, I., Mah, R.S.H., 1982. Optimal design of multipurpose batch plants. *Industrial and Engineering Chemistry Process Design and Development.* 21(1), 94–100.

Vaselenak, J.A., Grossmann, I.E., Westerberg, A.W., 1987. An embedding formulation for the optimal scheduling and design of multipurpose batch plants. *Industrial and Engineering Chemistry Research.* 26(1), 139–148.

Voudouris, V.T., Grossmann, I.E., 1996. MILP model for scheduling and design of a special class of multipurpose batch plants. *Computers and Chemical Engineering.* 20(11), 1335–1360.

Xia, Q.S., Macchietto, S., 1997. Design and synthesis of batch plants MINLP solution based on a stochastic method. *Computers and Chemical Engineering.* 21, S697–S702.

7 Simultaneous Scheduling and Water Optimization
Reduction of Effluent in Batch Facilities

7.1 INTRODUCTION

Wastewater minimization can be achieved by employing water reuse opportunities. This chapter presents a methodology to address the problem of wastewater minimization by extending the concept of water reuse to include a wastewater regenerator. The regenerator purifies wastewater to such a quality that it can be reused in other operations. This further increases water reuse opportunities in the plant, thereby significantly reducing freshwater demand and effluent generation. The mathematical model determines the optimum batch production schedule that achieves the minimum wastewater generation within the same framework. The model was applied to two case studies involving multiple contaminants, and wastewater reductions of 19.2% and 26% were achieved.

7.2 MOTIVATION FOR THE STUDY

The study carried out in this chapter was motivated by the following reasons. Firstly, earlier work by Majozi and Gouws (2009) did not consider the effect of regeneration while addressing wastewater minimization in the presence of multiple contaminants. This has been identified as a limitation as regeneration will improve water reuse opportunities in a plant.

Secondly, most published works in literature directly or indirectly avoid treating time as a variable. The results obtained from such formulations could very well be suboptimal. An example is the second case study investigated by Liu et al. (2009) which considered wastewater minimization in a single contaminant problem by incorporating regeneration. The case study was based on a given fixed schedule. Figure 7.1a and b highlights the impact of treating time as a variable during the first cycle of operation.

From Figure 7.1, when time is treated as a variable, the freshwater usage is 64.1 tons compared to 68.594 tons obtained by Liu et al. (2009) using a predefined schedule. This amounts to water savings of 6.6%. This improvement was brought about by the flexibility of the schedule. Apart from process A which occurs within the same time interval in both figures, the order of the other processes differs.

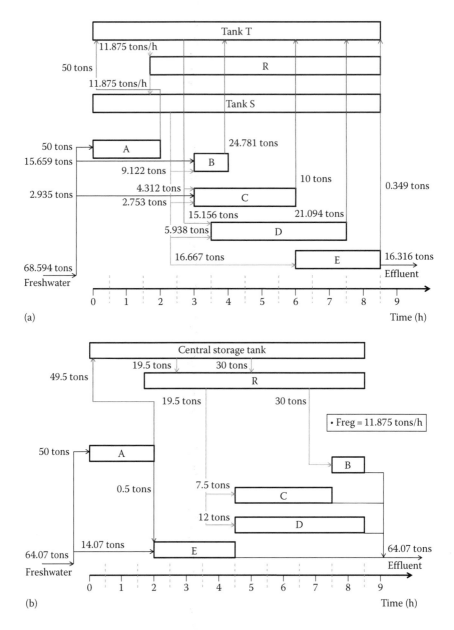

FIGURE 7.1 (a) Schedule obtained from the case study by Liu et al. (2009). (b) Schedule obtained when time is treated as a variable.

For example, process E occurs at 2 h in Figure 7.1b as opposed to 6 h in Figure 7.1a, thus providing an opportunity for the direct reuse of water from process A. In addition, process B occurs at 7.5 h which allows for the use of only regenerated water as opposed to 3 h in Figure 7.1a whereby an insufficient amount of regenerated water is available. It is important to note, however, that in certain situations the order of processes is specified beforehand due to the given recipe. In such situations, a fixed/predefined schedule is unavoidable.

The formulation presented in this chapter incorporates regeneration such that the batch production schedule and water network that correspond to minimum water requirement are determined simultaneously within the same optimization framework as opposed to a two-step procedure which will inherently rely on a given schedule for wastewater minimization. The proposed method is especially significant in multipurpose batch facilities where units must be washed before subsequent usage. Without proper integration between production scheduling and water management, useful water may be discharged as effluent instead of being reused.

7.3   PROBLEM STATEMENT

The problem addressed in this chapter can be stated as follows:

Given:

1. The production recipe for each product, including mean processing times in each unit operation
2. The available units and their capacities
3. The necessary costs and stoichiometric data
4. The contaminant mass load of each contaminant
5. Water requirement and the cleaning duration for each unit to achieve the required cleanliness
6. Maximum inlet and outlet concentrations of each contaminant
7. The maximum storage available for water reuse
8. The performance of the regenerator
9. The time horizon of interest

Determine:

The production schedule that achieves the minimum amount of wastewater generated by exploring recycle and reuse opportunities in the presence of a central storage vessel and a wastewater regenerator.

7.4   PROBLEM SUPERSTRUCTURE

The problem superstructure on which the mathematical formulation is based is depicted in Figure 7.2. In the figure, only the water using operations which are part of a complete batch process are depicted. Unit j represents a water using operation in which the water used consists of freshwater, regenerated water, stored water or recycle/reuse water. Water from unit j can be recycled into the same unit, reused elsewhere or sent to storage. The storage vessel can send water to a unit or to the wastewater regenerator. The philosophy of operation of the regenerator is as follows:

1. Water is sent to the regenerator provided there exists an available unit requiring that water. Whenever the regenerator is active, it operates in a continuous mode with steady inlet and outlet streams until the sink process is satisfied.

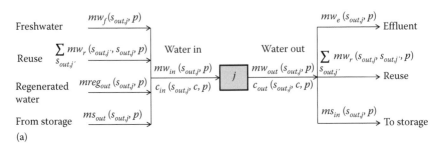

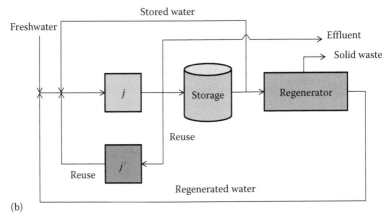

(b)

FIGURE 7.2 Superstructure for the mathematical formulation for processing unit (a) and storage (b).

2. Transfer of water to regenerator occurs to purify contaminated water because the contaminant concentration of water in storage is too high to be utilized in a unit.
3. The regenerator operates intermittently in the sense that it is only active if there is a unit that requires water; otherwise it remains in the passive mode.
4. The regenerator performance is given in terms of a constant removal ratio of contaminants.

7.5 MATHEMATICAL FORMULATION

The mathematical formulation involves the following sets, variables and parameters.

7.5.1 SETS

P $\{p|p = \text{time point}\}$

J $\{j|j = \text{unit}\}$

C $\{c|c = \text{contaminant}\}$

S_{in} $\{s_{in}|s_{in} = \text{input state into any unit}\}$

S_{out} $\{s_{out}|s_{out} = \text{output state from any unit}\}$

$S_{in,j}^{*}$ $\{s_{in,j}^{*}|s_{in,j}^{*} = \text{input state into unit}\} \subseteq S_{in}$

$S_{out,j}$ $\{s_{out,j}|s_{out,j} = \text{output state from unit } j\} \subseteq S_{out}$

7.5.2 Variables

$mw_{in}\left(s_{out,j},p\right)$	Mass of water into unit j for cleaning state s_{out} at time point p
$mw_{out}\left(s_{out,j},p\right)$	Mass of water produced at time point p from unit j
$mw_f\left(s_{out,j},p\right)$	Mass of freshwater into unit j at time point p
$mw_e\left(s_{out,j},p\right)$	Mass of effluent water from unit j at time point p
$mw_r\left(s_{out,j},s_{out,j'},p\right)$	Mass of water recycled to unit j' from j at time point p
$ms_{in}\left(s_{out,j},p\right)$	Mass of water transferred from unit j to storage at time point p
$ms_{out}\left(s_{out,j},p\right)$	Mass of water transferred from storage to unit j at time point p
$mreg_{in}\left(s_{out,j},p\right)$	Mass of water transferred from storage to regenerator at time point p
$mreg_{out}\left(s_{out,j},p\right)$	Mass of water transferred from regenerator to unit j at time point p
$m_{dirt}\left(c,p\right)$	Mass of contaminant c removed from water by regenerator at time point p amount of state delivered to customers at time point p
$c_{in}\left(s_{out,j},c,p\right)$	Inlet concentration of contaminant c, to unit j at time point p
$c_{out}\left(s_{out,j},c,p\right)$	Outlet concentration of contaminant c, from unit j at time point p
$cs_{in}\left(c,p\right)$	Inlet concentration of contaminant c, to storage at time point p
$cs_{out}\left(c,p\right)$	Outlet concentration of contaminant c, from storage at time point p
$cr_{out}\left(c,p\right)$	Outlet concentration of contaminant c, from regenerator at time point p
$qw_s\left(p\right)$	Amount of water stored in storage at time point p
$t_{out}\left(s_{out,j},p\right)$	Time at which a state is produced from unit j at time point p
$tw_{in}\left(s_{out,j},p\right)$	Time that the water is used at time point p in unit j
$tw_{out}\left(s_{out,j},p\right)$	Time at which water is produced at time point p from unit j
$tw_r\left(s_{out,j},s_{out,j'},p\right)$	Time at which water is recycled from unit j to unit j' at time point p
$ts_{in}\left(s_{out,j},p\right)$	Time at which water is transferred from unit j to storage at time point p
$ts_{out}\left(s_{out,j},p\right)$	Time at which water is transferred from storage to unit j at time point p
$treg_{in}\left(s_{out,j},p\right)$	Time at which water is transferred from storage to regenerator at time point p
$treg_{out}\left(s_{out,j},p\right)$	Time at which water is transferred from regenerator to unit j at time point p
$y\left(s^{*}_{in,j},p\right)$	Binary variable associated with usage of state s at time point p
$yw_r\left(s_{out,j},s_{out,j'},p\right)$	Binary variable showing usage of recycle from unit j to unit j' at time point p
$yw\left(s_{out,j},p\right)$	Binary variable showing usage of water in unit j at time point p

$ys_{in}\left(s_{out,j},p\right)$	Binary variable showing transfer of water from unit j to storage at time point p
$ys_{out}\left(s_{out,j},p\right)$	Binary variable showing transfer of water from storage to unit j at time point p
$yreg_{in}\left(s_{out,j},p\right)$	Binary variable showing transfer of water from storage to regenerator at time point p
$yreg_{out}\left(s_{out,j},p\right)$	Binary variable showing transfer of water from regenerator to unit j at time point p

7.5.3 PARAMETERS

CE	Cost of effluent water treatment (c.u./kg water)
CF	Cost of freshwater (c.u./kg water)
$CP(s)$	Selling price of product s, s = product
$M\left(s_{out,j},c\right)$	Mass load of contaminant c added from unit j to the water stream
$Mw^{U}\left(s_{out,j}\right)$	Maximum inlet water mass of unit j
$C_{in}^{U}\left(s_{out,j},c\right)$	Maximum inlet concentration of contaminant c in unit j
$C_{out}^{U}\left(s_{out,j},c\right)$	Maximum outlet concentration of contaminant c from unit j
$CS_{out}^{U}\left(c\right)$	Maximum outlet concentration of contaminant c from storage
$CR_{out}^{U}\left(c\right)$	Maximum outlet concentration of contaminant c from regenerator
$CS_{out}^{o}\left(c\right)$	Initial concentration of contaminant in storage
$RR\left(c\right)$	Removal ratio of contaminant c from regenerator
Qw_{s}^{o}	Initial amount of water in storage
Qw_{s}^{U}	Maximum capacity of storage
$\tau w\left(s_{out,j}\right)$	Mean processing time of unit j
$freg$	Regenerator flowrate
H	Time horizon of interest

In the mathematical formulation, wastewater minimization is embedded in the production scheduling framework. The production scheduling framework adopted here, which is based on the state sequence network (SSN) and the uneven discretization of the time horizon, has been explained in detail in Majozi and Gouws (2009). It will not be discussed in this chapter. However, critical constraints have been provided again to facilitate understanding.

7.5.4 MASS BALANCE CONSTRAINTS

It is necessary to define the mass balances around each processing unit: the central storage vessel and the regenerator.

7.5.4.1 Mass Balance around a Unit j

Constraint (7.1) is the water balance over the inlet to a unit. Water entering the unit is a combination of reuse/recycle streams from other units j', freshwater, water from storage and water from the regenerator. Constraint (7.2) states that the water leaving a unit could be recycled/reused, discarded as effluent or sent to storage. Constraint (7.3) states that the amount of water exiting a unit must equal the amount of water entering the unit at the previous time point. This constraint captures the fact that water is not produced in the unit during the washing operation.

$$
mw_{in}\left(s_{out,j}, p\right) = \sum_{s_{out,j'}} mwr\left(s_{out,j'}, s_{out,j}, p\right) + mw_f\left(s_{out,j}, p\right) + ms_{out}\left(s_{out,j}, p\right)
$$

$$
+ mreg_{out}\left(s_{out,j}, p\right) \quad \forall j, j' \in J, \quad s_{out,j} \in S_{out,j}, \quad p \in P \qquad (7.1)
$$

$$
mw_{out}\left(s_{out,j}, p\right) = \sum_{s_{out,j'}} mwr\left(s_{out,j}, s_{out,j'}, p\right) + mw_e\left(s_{out,j}, p\right) + ms_{in}\left(s_{out,j}, p\right)
$$

$$
\forall j, j' \in J, \quad s_{out,j} \in S_{out,j}, \quad p \in P \qquad (7.2)
$$

$$
mw_{in}\left(s_{out,j}, p-1\right) = mw_{out}\left(s_{out,j}, p\right) \quad \forall j \in J, \quad s_{out,j} \in S_{out,j}, \quad p \in P, \quad p > p_1 \quad (7.3)
$$

Constraint (7.4) defines the inlet contaminant mass balance. The contaminant mass load in the inlet stream consists of contaminant mass load in recycle/reuse water, the contaminant mass load in water from storage and the contaminant mass load in water from the regenerator. Constraint (7.5) defines the outlet contaminant mass as the mass of contaminant that entered the unit at the previous time point and the mass load of contaminant picked up in the unit during its operation.

$$
mw_{in}\left(s_{out,j}, p\right)c_{in}\left(s_{out,j}, c, p\right) = \sum_{s_{out,j'}} mw_r\left(s_{out,j'}, s_{out,j}, p\right)c_{out}\left(s_{out,j}, c, p\right)
$$

$$
+ ms_{out}\left(s_{out,j}, p\right)cs_{out}\left(c, p\right)
$$

$$
+ mreg_{out}\left(s_{out,j}, p\right)cr_{out}\left(c, p\right)
$$

$$
\forall j, j' \in J, \quad s_{out,j} \in S_{out,j}, \quad p \in P, \quad c \in C \qquad (7.4)
$$

$$
mw_{out}\left(s_{out,j}, p\right)c_{out}\left(s_{out,j}, c, p\right) = M\left(s_{out,j}, c\right)yw\left(s_{out,j}, p-1\right)
$$

$$
+ mw_{in}\left(s_{out,j}, p-1\right)c_{in}\left(s_{out,j}, c, p-1\right)
$$

$$
\forall j \in J, \quad s_{out,j} \in S_{out,j}, \quad p \in P, \quad p > p_1, \quad c \in C
$$

$$
(7.5)
$$

Constraints (7.6) and (7.7) ensure that the inlet and outlet contaminant concentrations do not exceed the allowed maximum. Similarly, the maximum allowable water in a unit must not be exceeded. This is governed by Constraint (7.8). Constraints (7.9) through (7.11) restrict the mass of water entering the unit from the various sources to the maximum allowable for the unit. Constraint (7.9) is with respect to recycle/reuse, Constraint (7.10) is with respect to water from storage, while Constraint (7.11) is with respect to regenerated water.

$$c_{in}\left(s_{out,j},c,p\right) \leq C_{in}^{U}\left(s_{out,j},c\right)yw\left(s_{out,j},p\right) \quad \forall j \in J, \quad s_{out,j} \in S_{out,j}, \quad p \in P, \quad c \in C \tag{7.6}$$

$$C_{out}\left(s_{out,j},c,p\right) \leq C_{out}^{U}\left(s_{out,j},c\right)yw\left(s_{out,j},p-1\right)$$
$$\forall j \in J, \quad s_{out,j} \in S_{out,j}, \quad p \in P, \quad p > p_1, \quad c \in C \tag{7.7}$$

$$mw_{in}\left(s_{out,j},p\right) \leq Mw^{U}\left(s_{out,j}\right)yw\left(s_{out,j},p\right) \quad \forall j \in J, \quad s_{out,j} \in S_{out,j}, \quad p \in P \tag{7.8}$$

$$mw_r\left(s_{out,j'},s_{out,j},p\right) \leq Mw^{U}\left(s_{out,j}\right)yw_r\left(s_{out,j'},s_{out,j},p\right)$$
$$\forall j,j' \in J, \quad s_{out,j} \in S_{out,j}, \quad p \in P \tag{7.9}$$

$$ms_{out}\left(s_{out,j},p\right) \leq Mw^{U}\left(s_{out,j}\right)ys_{out}\left(s_{out,j},p\right) \quad \forall j \in J, \quad s_{out,j} \in S_{out,j}, \quad p \in P \tag{7.10}$$

$$mreg_{out}\left(s_{out,j},p\right) \leq Mw^{U}\left(s_{out,j}\right)yreg_{out}\left(s_{out,j},p\right) \quad \forall j \in J, \quad s_{out,j} \in S_{out,j}, \quad p \in P \tag{7.11}$$

The maximum quantity of water into a unit is represented by Equation 7.12. It is important to note that for multi-contaminant wastewater the outlet concentration of the individual components cannot all be set to the maximum, since the contaminants are not limiting simultaneously. The limiting contaminant(s) will always be at the maximum outlet concentration and the non-limiting contaminants will be below their respective maximum outlet concentrations.

$$Mw^{U}\left(s_{out,j}\right) = \max_{c \in C}\left\{ \frac{M\left(s_{out,j},c\right)}{C_{out}^{U}\left(s_{out,j},c\right)-C_{in}^{U}\left(s_{out,j},c\right)} \right\}, \quad \forall j \in J, \quad s_{out,j} \in S_{out,j} \tag{7.12}$$

7.5.4.2 Mass Balance around Central Storage

Constraint (7.13) is the water mass balance around the storage tank. The amount of water stored in the storage tank consists of water stored at the previous time point and the difference between water entering the storage tank from processing unit and water leaving the storage tank either to processing unit or regenerator. Constraint (7.14) defines the initial amount of water in the tank.

$$qw_s\left(p\right) = qw_s\left(p-1\right) + \sum_{s_{out,j}} ms_{in}\left(s_{out,j},p\right) - \sum_{s_{out,j}} ms_{out}\left(s_{out,j},p\right)$$

$$- \sum_{s_{out,j}} mreg_{in}\left(s_{out,j},p\right) \quad \forall j \in J, \quad s_{out,j} \in S_{out,j}, \quad p \in P, \quad p > p_1 \tag{7.13}$$

$$qw_s(p_1) = Qw_s^o - \sum_{S_{out,j}} ms_{out}(s_{out,j}, p_1) - \sum_{S_{out,j}} mreg_{in}(s_{out,j}, p_1)$$

$$\forall j \in J, \quad s_{out,j} \in S_{out,j} \tag{7.14}$$

The definition of the inlet contaminant concentration to the storage tank is given in Constraint (7.15). The concentration of water exiting the storage tank is assumed to be equal to the concentration of water in the tank as given in Constraint (7.16). This condition is true in the case of perfect mixing. The initial concentration in the storage tank is given in Constraint (7.17). Constraints (7.18) and (7.19) ensure that the maximum capacity of the tank is not exceeded.

$$cs_{in}(c,p) = \frac{\displaystyle\sum_{S_{out,j}} \left(ms_{in}(s_{out,j},p)c_{out}(s_{out,j},c_p)\right)}{\displaystyle\sum_{S_{out,j}} ms_{in}(s_{out,j},p)} \quad \forall j \in J, \quad s_{out,j} \in S_{out,j}, \quad p \in P \quad c \in C$$

$$\tag{7.15}$$

$$cs_{out}(c,p) = \frac{qw_s(p-1)cs_{out}(c,p-1) + \left(\displaystyle\sum_{S_{out,j}} ms_{in}(s_{out,j},p)\right)cs_{in}(c,p)}{qw_s(p-1) + \displaystyle\sum_{S_{out,j}} ms_{in}(s_{out,j},p)}$$

$$\forall j \in J, \quad s_{out,j} \in S_{out,j}, \quad p \in P, \quad p > p_1, \quad c \in C \tag{7.16}$$

$$cs_{out}(c,p_1) = CS_{out}^o(c) \quad \forall c \in C \tag{7.17}$$

$$qw_s(p) \leq Qw_s^U \quad \forall p \in P \tag{7.18}$$

$$ms_{in}(s_{out,j}, p) \leq Qw_s^U ys_{in}(s_{out,j}, p) \quad \forall j \in J, \quad s_{out,j} \in S_{out,j}, \quad p \in P \tag{7.19}$$

Constraint (7.20) ensures that no water is stored in the storage vessel at the end of the time horizon in order to give a true optimum. Otherwise, the minimum target could be misleading.

$$qw_s(p) = 0 \quad \forall p = |P| \tag{7.20}$$

7.5.4.3 Mass Balance around the Regenerator

Constraint (7.21) states that the total amount of outlet water from the regenerator at time point p consists of the total amount of inlet water to the regenerator at the previous time point. This constraint captures the assumption that during the operation of the regenerator, no water is produced or lost.

$$\sum_{s_{out,j}} mreg_{in}\left(s_{out,j}, p-1\right) = \sum_{s_{out,j}} mreg_{out}\left(s_{out,j}, p\right)$$

$$\forall j \in J, \quad s_{out,j} \in S_{out,j}, \quad p \in P, \quad p > p_1 \tag{7.21}$$

The performance of the regenerator is given in terms of the removal ratio of contaminant, which specifies a constant ratio between the inlet and outlet mass of contaminant. This is stated in Constraint (7.22) provided the flowrate of the regenerator is constant.

$$cr_{out}\left(c, p\right) = cs_{out}\left(c, p-1\right)\left(1 - RR(c)\right) \quad \forall c \in C, \quad p \in P, \quad p > p_1 \tag{7.22}$$

The contaminant mass balance around the regenerator is expressed in Constraint (7.23). The contaminant mass load exiting the regenerator to a unit is the difference between the contaminant mass load entering the regenerator and the contaminant mass removed from the water during regeneration, $m_{dirt}(c, p)$.

$$cs_{out}\left(c, p-1\right)\sum_{s_{out,j}} mreg_{in}\left(s_{out,j}, p-1\right) = cr_{out}\left(c, p\right)\sum_{s_{out,j}} mreg_{out}\left(s_{out,j}, p\right) + m_{dirt}\left(c, p\right)$$

$$\forall j \in J, \quad s_{out,j} \in S_{out,j}, \quad p \in P, \quad p > p_1 \tag{7.23}$$

7.5.5 SCHEDULING CONSTRAINTS

The scheduling constraints take care of the time dimension associated with batch processes. The scheduling aspects of the model are divided into four groups as follows:

1. Those related to task scheduling
2. Those that cater for recycle/reuse
3. Those that account for the central storage vessel
4. Those that account for the regenerator

7.5.5.1 Task Scheduling

These constraints ensure that each water-using operation is integrated with production scheduling. Constraints (7.24) and (7.25) together ensure that unit j is washed immediately after a task that produced $s_{out,j}$. Constraints assuming the form of Constraints (7.24) or (7.25) are called big-M constraints. If water is used in the unit, $yw\left(s_{out,j}, p\right)$ has a value of 1 causing Constraints (7.24) and (7.25) to become active, and the start time of washing is forced to coincide with the end time of production. Otherwise, when water is not used in the unit, that is $yw\left(s_{out,j}, p\right)$ has a value of zero, the two constraints become relaxed. Constraint (7.26) defines the duration of the washing operation performed in unit j. Constraint (7.27) stipulates that the washing operation can only commence at time point p if the task producing $s_{out,j}$ was active at the previous time point.

$$tw_{in}\left(s_{out,j},p\right) \geq t_{out}\left(s_{out,j},p\right) - H\left(1 - yw\left(s_{out,j},p\right)\right) \quad \forall j \in J, \quad s_{out,j} \in S_{out,j}, \quad p \in P \tag{7.24}$$

$$tw_{in}\left(s_{out,j},p\right) \leq t_{out}\left(s_{out,j},p\right) + H\left(1 - yw\left(s_{out,j},p\right)\right) \quad \forall j \in J, \quad s_{out,j} \in S_{out,j}, \quad p \in P \tag{7.25}$$

$$tw_{out}\left(s_{out,j},p\right) = tw_{in}\left(s_{out,j},p-1\right) + \tau w\left(s_{out,j}\right) yw\left(s_{out,j},p-1\right)$$
$$\forall j \in J, \quad s_{out,j} \in S_{out,j}, \quad p \in P, \quad p > p_1 \tag{7.26}$$

$$yw\left(s_{out,j},p\right) = y\left(s^*_{in,j},p-1\right)$$
$$\forall j \in J, \quad s_{out,j} \in S_{out,j}, \quad s^*_{in,j} \in S^*_{in,j} \rightarrow s_{out,j}, \quad p \in P, \quad p > p_1 \tag{7.27}$$

7.5.5.2 Recycle/Reuse Scheduling

Wastewater can only be directly recycled/reused if the unit producing wastewater and the unit receiving wastewater finish operating and begin operating at the same time, respectively. Constraint (7.28) describes the relationship between usage of water in a unit and the opportunity for recycle and reuse. The constraint states that for a unit j to transfer water to unit j', unit j' should require water at that time point. It does not, however, mean that unit j' must use water from unit j; it could still obtain water from other sources. Constraints (7.29) and (7.30) state that the time at which water recycle/reuse takes place coincides with the time at which the water is produced. Constraints (7.31) and (7.32) ensure that the time at which water recycle/reuse takes place coincides with the starting time of the unit receiving the water.

$$yw_r\left(s_{out,j},s_{out,j'},p\right) \leq yw\left(s_{out,j'},p\right) \quad \forall j,j' \in J, \quad s_{out,j} \in S_{out,j}, \quad p \in P \tag{7.28}$$

$$tw_r\left(s_{out,j},s_{out,j'},p\right) \leq tw_{out}\left(s_{out,j},p\right) - H\left(1 - yw_r\left(s_{out,j},s_{out,j'},p\right)\right)$$
$$\forall j,j' \in J, \quad s_{out,j} \in S_{out,j}, \quad p \in P \tag{7.29}$$

$$tw_r\left(s_{out,j},s_{out,j'},p\right) \geq tw_{out}\left(s_{out,j},p\right) - H\left(1 - yw_r\left(s_{out,j},s_{out,j'},p\right)\right)$$
$$\forall j,j' \in J, \quad s_{out,j} \in S_{out,j}, \quad p \in P \tag{7.30}$$

$$tw_r\left(s_{out,j}, s_{out,j'}, p\right) \leq tw_{in}\left(s_{out,j}, p\right) + H\left(1 - yw_r\left(s_{out,j}, s_{out,j'}, p\right)\right)$$

$$\forall j, j' \in J, \quad s_{out,j} \in S_{out,j}, \quad p \in P \tag{7.31}$$

$$tw_r\left(s_{out,j}, s_{out,j'}, p\right) \geq tw_{in}\left(s_{out,j}, p\right) - H\left(1 - yw_r\left(s_{out,j}, s_{out,j'}, p\right)\right)$$

$$\forall j, j' \in J, \quad s_{out,j} \in S_{out,j}, \quad p \in P \tag{7.32}$$

7.5.5.3   Central Storage Scheduling

Constraint (7.33) relates water usage in a unit and water transfer from storage. It states that water can only be transferred to a unit if it uses water at the same time point. However, it is not a prerequisite for the unit to use stored water, the water could be provided from other sources. Constraints (7.34) and (7.35) ensure that the time at which water is sent from storage to a unit coincides with the start time of washing of the same unit.

$$ys_{out}\left(s_{out,j}, p\right) \leq yw\left(s_{out,j}, p\right) \quad \forall j \in J, \quad s_{out,j} \in S_{out,j}, \quad p \in P \tag{7.33}$$

$$ts_{out}\left(s_{out,j}, p\right) \geq tw_{in}\left(s_{out,j}, p\right) - H\left(2 - ys_{out}\left(s_{out,j}, p\right) - yw\left(s_{out,j}, p\right)\right)$$

$$\forall j \in J, \quad s_{out,j} \in S_{out,j}, \quad p \in P \tag{7.34}$$

$$ts_{out}\left(s_{out,j}, p\right) \leq tw_{in}\left(s_{out,j}, p\right) + H\left(2 - ys_{out}\left(s_{out,j}, p\right) - yw\left(s_{out,j}, p\right)\right)$$

$$\forall j \in J, \quad s_{out,j} \in S_{out,j}, \quad p \in P \tag{7.35}$$

Constraint (7.36) relates water usage in a unit and water transfer to storage. It states that water can only be transferred from a unit to storage if the unit used water at the previous time point. However, washing can take place in the unit without discharging water to the storage tank. The water could be discharged to other sinks. Constraints (7.37) and (7.38) ensure that the time at which water is sent to storage from a unit must coincide with the finishing time of washing of the same unit.

$$ys_{in}\left(s_{out,j}, p\right) \leq yw\left(s_{out,j}, p-1\right) \quad \forall j \in J, \quad s_{out,j} \in S_{out,j}, \quad p \in P, \quad p > p_1 \tag{7.36}$$

$$ts_{in}\left(s_{out,j}, p\right) \geq tw_{out}\left(s_{out,j}, p\right) - H\left(2 - ys_{out}\left(s_{out,j}, p\right) - yw\left(s_{out,j}, p-1\right)\right)$$

$$\forall j \in J, \quad s_{out,j} \in S_{out,j}, \quad p \in P, \quad p > p_1 \tag{7.37}$$

$$ts_{in}\left(s_{out,j}, p\right) \leq tw_{out}\left(s_{out,j}, p\right) + H\left(2 - ys_{out}\left(s_{out,j}, p\right) - y\left(s_{out,j}, p-1\right)\right)$$

$$\forall j \in J, \quad s_{out,j} \in S_{out,j}, \quad p \in P, \quad p > p_1 \tag{7.38}$$

If water is transferred from storage to a unit at a later time point, the time at which this happens must correspond to a later time in the time horizon. This is specified in Constraint (7.39). Constraint (7.40) ensures that if water is transferred from a unit to storage at a later time point, the time at which this happens corresponds to a later time in the time horizon.

$$ts_{out}\left(s_{out,j},p\right) \geq ts_{out}\left(s_{out,j'},p'\right) - H\left(2 - ys_{out}\left(s_{out,j},p\right) - ys_{out}\left(s_{out,j'},p'\right)\right)$$

$$\forall j, j' \in J, \quad s_{out,j} \in S_{out,j}, p, \quad p' \in P, \quad p \geq p' \tag{7.39}$$

$$ts_{in}\left(s_{out,j},p\right) \geq ts_{in}\left(s_{out,j'},p'\right) - H\left(2 - ys_{in}\left(s_{out,j},p\right) - ys_{in}\left(s_{out,j'},p'\right)\right)$$

$$\forall j, j' \in J, \quad s_{out,j} \in S_{out,j}, p, \quad p' \in P, \quad p \geq p' \tag{7.40}$$

Constraints (7.41) and (7.42) state that if water is transferred to storage from more than one unit at the same time point, the time at which they do so must coincide. Constraints (7.43) and (7.44) state that if water is discharged from storage to more than one unit at the same time point, the time at which the water is discharged must coincide.

$$ts_{in}\left(s_{out,j},p\right) \geq ts_{in}\left(s_{out,j'},p\right) - H\left(2 - ys_{in}\left(s_{out,j},p\right) - ys_{in}\left(s_{out,j'},p\right)\right)$$

$$\forall j, j' \in J, \quad s_{out,j} \in S_{out,j}, \quad p \in P \tag{7.41}$$

$$ts_{in}\left(s_{out,j},p\right) \leq ts_{in}\left(s_{out,j'},p\right) + H\left(2 - ys_{in}\left(s_{out,j},p\right) - ys_{in}\left(s_{out,j'},p\right)\right)$$

$$\forall j, j' \in J, \quad s_{out,j} \in S_{out,j}, \quad p \in P \tag{7.42}$$

$$ts_{out}\left(s_{out,j},p\right) \geq ts_{out}\left(s_{out,j'},p\right) - H\left(2 - ys_{out}\left(s_{out,j},p\right) - ys_{out}\left(s_{out,j'},p\right)\right)$$

$$\forall j, j' \in J, \quad s_{out,j} \in S_{out,j}, \quad p \in P \tag{7.43}$$

$$ts_{out}\left(s_{out,j},p\right) \leq ts_{out}\left(s_{out,j'},p\right) + H\left(2 - ys_{out}\left(s_{out,j},p\right) - ys_{out}\left(s_{out,j'},p\right)\right)$$

$$\forall j, j' \in J, \quad s_{out,j} \in S_{out,j}, \quad p \in P \tag{7.44}$$

If water is simultaneously being transferred to and discharged from storage, the time at which this happens should coincide. This is given in Constraints (7.45) and (7.46)

$$ts_{in}\left(s_{out,j},p\right) \geq ts_{out}\left(s_{out,j'},p\right) - H\left(2 - ys_{in}\left(s_{out,j},p\right) - ys_{out}\left(s_{out,j'},p\right)\right)$$

$$\forall j, j' \in J, \quad s_{out,j} \in S_{out,j}, \quad p \in P \tag{7.45}$$

$$ts_{in}\left(s_{out,j}, p\right) \le ts_{out}\left(s_{out,j'}, p\right) - H\left(2 - ys_{in}\left(s_{out,j}, p\right) - ys_{out}\left(s_{out,j'}, p\right)\right)$$

$$\forall j, j' \in J, \quad s_{out,j} \in S_{out,j}, \quad p \in P \tag{7.46}$$

Constraint (7.47) ensures that if water leaves storage at a later time point compared to water entering the storage, the time at which water leaves the storage must correspond to a later time in the time horizon.

$$ts_{out}\left(s_{out,j}, p\right) \ge ts_{in}\left(s_{out,j'}, p'\right) - H\left(2 - ys_{out}\left(s_{out,j}, p\right) - ys_{in}\left(s_{out,j'}, p'\right)\right)$$

$$\forall j, j' \in J, \quad s_{out,j} \in S_{out,j}, p, \quad p' \in P, \quad p \ge p' \tag{7.47}$$

Worthy of mention is the fact that most of the constraints in this chapter were presented in Majozi and Gouws (2009) but are given again here to facilitate understanding.

7.5.5.4 Regenerator Scheduling

Constraint (7.48) defines the relationship between the transfer of water to the regenerator and the discharge of water from the regenerator. It should be re-emphasized at this point that water is only transferred to the regenerator provided that there is a unit available downstream requiring that water and concentration in the storage unit must exceed the maximum allowed inlet concentration in the receiving unit. Hence, the water transferred to the regenerator at any time is in terms of an available unit. Constraints (7.49) and (7.50) ensure that the time at which water is discharged from the regenerator to a unit coincides with the time at which the unit starts to use water. Constraint (7.51) gives the duration of regeneration.

$$yreg_{out}\left(s_{out,j}, p\right) = yreg_{in}\left(s_{out,j}, p-1\right) \quad \forall j \in J, \quad s_{out,j} \in S_{out,j}, \quad p \in P, \quad p > p_1 \tag{7.48}$$

$$treg_{out}\left(s_{out,j}, p\right) \le tw_{in}\left(s_{out,j}, p\right) + H\left(2 - yreg_{out}\left(s_{out,j}, p\right) - yw\left(s_{out,j}, p\right)\right)$$

$$\forall j \in J, \quad s_{out,j} \in S_{out,j}, \quad p \in P \tag{7.49}$$

$$treg_{out}\left(s_{out,j}, p\right) \ge tw_{in}\left(s_{out,j}, p\right) - H\left(2 - yreg_{out}\left(s_{out,j}, p\right) - yw\left(s_{out,j}, p\right)\right)$$

$$\forall j \in J, \quad s_{out,j} \in S_{out,j}, \quad p \in P \tag{7.50}$$

$$treg_{out}\left(s_{out,j}, p\right) = treg_{in}\left(s_{out,j}, p-1\right) + \left[\frac{\sum\limits_{s_{out,j}} mreg_{in}\left(s_{out,j}, p-1\right)}{freg}\right] yreg_{in}\left(s_{out,j}, p-1\right)$$

$$\forall j \in J, \quad s_{out,j} \in S_{out,j}, \quad p \in P, \quad p > p_1 \tag{7.51}$$

Constraint (7.52) ensures that storage does not supply a unit with water through the regenerator and also supply water directly to the unit at the same time point. If the unit is directly supplied with water from storage, the need to regenerate water no longer exists. Constraint (7.53) is included to ensure that the full potential for waste-water regeneration is realized. It states that a unit does not receive water from both the storage tank and the regenerator at the same time point. This is to avoid unnecessary mixing.

$$yreg_{in}\left(s_{out,j},p\right)+ys_{out}\left(s_{out,j},p\right)\le 1 \quad \forall j \in J, \quad s_{out,j} \in S_{out,j}, \quad p \in P \quad (7.52)$$

$$yreg_{out}\left(s_{out,j},p\right)+ys_{out}\left(s_{out,j},p\right)\le 1 \quad \forall j \in J, \quad s_{out,j} \in S_{out,j}, \quad p \in P \quad (7.53)$$

Constraint (7.54) together with Constraint (7.48) defines the relationship between the usage of water in a unit and the regeneration of water for the unit. Constraint (7.54) states that for water to be charged to the regenerator from storage, the unit requiring the water should begin operation at the next time point. This, however, does not mean that the unit cannot begin operation without regenerated water, since water required could be provided from other sources.

$$yw\left(s_{out,j},p\right)\ge yreg_{in}\left(s_{out,j},p-1\right) \quad \forall j \in J, \quad s_{out,j} \in S_{out,j}, \quad p \in P, \quad p > p_1 \quad (7.54)$$

Constraint (7.55) ensures that if water is charged to the regenerator at a later time point, the time at which it does so is at a later time in the time horizon. Constraint (7.56) is similar, but it applies to the time at which water is discharged from the regenerator.

$$treg_{in}\left(s_{out,j},p\right)\ge treg_{in}\left(s_{out,j'},p'\right)+H\left(2-yreg_{in}\left(s_{out,j},p\right)-yreg_{in}\left(s_{out,j'},p'\right)\right)$$
$$\forall j,j' \in J, \quad s_{out,j} \in S_{out,j}, p, \quad p' \in P, \quad p \ge p' \quad\quad (7.55)$$

$$treg_{out}\left(s_{out,j},p\right)\le treg_{out}\left(s_{out,j'},p'\right)+H\left(2-yreg_{out}\left(s_{out,j},p\right)-yreg_{out}\left(s_{out,j'},p'\right)\right)$$
$$\forall j,j' \in J, \quad s_{out,j} \in S_{out,j}, p, \quad p' \in P, \quad p \ge p' \quad\quad (7.56)$$

7.5.6 FEASIBILITY AND TIME HORIZON CONSTRAINTS

Constraint (7.57) ensures that if a processing unit j is reusing water from unit j' at time point p, then unit j' cannot reuse water from unit j at the same time point. Constraint (7.58) ensures that at a certain time point both the charging of the regenerator and the discharging from the regenerator on behalf of a particular unit cannot take place at the same time point.

$$yw_r\left(s_{out,j},s_{out,j'},p\right)+yw_r\left(s_{out,j'},s_{out,j},p\right)\le 1 \quad \forall j,j' \in J, \quad s_{out,j} \in S_{out,j}, \quad p \in P$$
$$(7.57)$$

$$yreg_{in}\left(s_{out,j},p\right)+yreg_{out}\left(s_{out,j},p\right)\le 1 \quad \forall j\in J, \quad s_{out,j}\in S_{out,j}, \quad p\in P \quad (7.58)$$

Constraints (7.59) through (7.65) ensure that each event occurs within the time horizon of time of interest.

$$tw_{in}\left(s_{out,j},p\right)\le H \quad \forall j\in J, \quad s_{out,j}\in S_{out,j}, \quad p\in P \quad\quad\quad (7.59)$$

$$tw_{out}\left(s_{out,j},p\right)\le H \quad \forall j\in J, \quad s_{out,j}\in S_{out,j}, \quad p\in P \quad\quad\quad (7.60)$$

$$tw_{r}\left(s_{out,j},s_{out,j'},p\right)\le H \quad \forall j,j'\in J, \quad s_{out,j}\in S_{out,j}, \quad p\in P \quad\quad\quad (7.61)$$

$$ts_{in}\left(s_{out,j},p\right)\le H \quad \forall j\in J, \quad s_{out,j}\in S_{out,j}, \quad p\in P \quad\quad\quad (7.62)$$

$$ts_{out}\left(s_{out,j},p\right)\le H \quad \forall j\in J, \quad s_{out,j}\in S_{out,j}, \quad p\in P \quad\quad\quad (7.63)$$

$$treg_{in}\left(s_{out,j},p\right)\le H \quad \forall j\in J, \quad s_{out,j}\in S_{out,j}, \quad p\in P \quad\quad\quad (7.64)$$

$$treg_{out}\left(s_{out,j},p\right)\le H \quad \forall j\in J, \quad s_{out,j}\in S_{out,j}, \quad p\in P \quad\quad\quad (7.65)$$

7.5.7 Objective Function

The objective of the formulation depends on whether the production is given or not. In the situation where production is not given, the objective function is expressed as Constraint (7.66). This objective here is the maximization of profit while taking into account freshwater and effluent treatment costs.

$$max\sum_{s}\sum_{p}CP(s)d(s,p)-CF\sum_{s_{out,j}}\sum_{p}mw_{f}\left(s_{out,j},p\right)-CE\sum_{s_{out,j}}\sum_{p}mw_{e}\left(s_{out,j},p\right) \quad (7.66)$$

Constraint (7.67) is the objective function when the production is given. It is the minimization of effluent.

$$min\sum_{p}\left(\sum_{s_{out,j}}mw_{e}\left(s_{out,j},p\right)\right) \quad\quad\quad (7.67)$$

Due to Constraints (7.4), (7.5), (7.15), (7.16) and (7.23) which involve bilinear terms, the presented mathematical formulation is a mixed integer nonlinear program, (MINLP). The method proposed by Majozi and Gouws (2009) was to solve the MINLP using a two-step procedure in which the MINLP problem was linearized and solved as a mixed integer linear program (MILP) to provide a starting point for the exact MINLP problem. The method of linearization is discussed in the Appendix.

7.6 CASE STUDIES

This section presents the application of this method to two problems: one from literature and the other from a practical production facility.

7.6.1 CASE STUDY I

This case study is commonly known as BATCH 1 in literature. The SSN representation of the problem is illustrated in Figure 7.3. The BATCH 1 facility consists of a heater which is used to heat Feed A before reaction. Three different chemical reactions can be performed in two common reactors. A separator exists to purify product from reaction 3. The data required for the production aspects are given in Table 7.1.

Similar to the work by Majozi and Gouws (2009), the philosophy is that after a reaction has taken place in a reactor, the reactor must be washed before another reaction can take place. This ensures that any product residue is removed to retain the integrity of subsequent product. Data pertaining to the washing operations are given in Table 7.2. In addition, the cost of freshwater is given as 2 c.u./kg water, while the effluent treatment cost is 3 c.u./kg.

A central storage tank for wastewater is available with a capacity of 200 kg. The flowrate of the regenerator is given to be 100 kg/h. The removal ratios of the various contaminants are provided in Table 7.3. The objective function is the maximization of profit as given in Constraint (7.66). The time horizon of interest is 10 h.

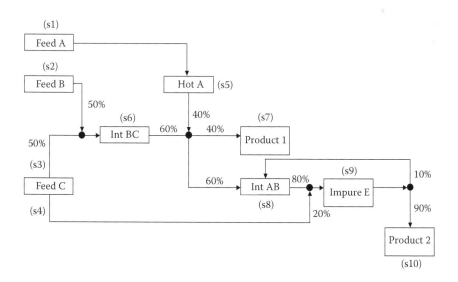

FIGURE 7.3   SSN for literature example.

TABLE 7.1

Production Data for Literature Example

Units	Capacity	Suitability	Mean Processing Time (τ)
Heater	100	Heating	1
Reactor 1	50	Reaction 1,2,3	2,2,1
Reactor 2	80	Reaction 1,2,3	2,2,1
Still	200	Separation	1 for product 2, 2 for IntAB

States	Storage Capacity	Initial Amount	Price
Feed A	Unlimited	Unlimited	0
Feed B	Unlimited	Unlimited	0
Feed C	Unlimited	Unlimited	0
Hot A	100	0	0
IntAB	200	0	0
IntBC	150	0	0
Impure E	200	0	0
Product 1	Unlimited	0	100
Product 2	Unlimited	0	100

7.6.1.1 Results and Discussion

The resulting model for Case Study I was formulated in GAMS 22.0 and solved using the DICOPT algorithm, with CPLEX 9.1.2 as the MIP solver and CONOPT3 as the NLP solver. The model had 532 binary variables. The optimum number of time points was 10. The value of the objective function, the maximization of profit in Constraint (7.67) was 21,129 c.u.

The optimum amount of freshwater used was 401.7 kg. If only recycle/reuse of water and the use of a central storage tank had been considered, the objective function would have been 20,180 c.u. with an optimum freshwater usage of 479 kg. This amounts to a 4.7% increase in profit and a 19.24% decrease in effluent production. The solution time was 46.3 s using a Pentium 4, 3.4 GHz processor. The schedule corresponding to the MILP was infeasible. Hence, the resulting solution of 21,129 c.u. is a local optimum.

The schedule obtained from the optimization is shown in Figure 7.4. The clear blocks represent the production operation in the unit, while the dark blocks represent the washing operation in the unit. The numbers above the clear blocks represent the amount of material used in the unit for production, the numbers in brackets represent freshwater. The sum of all these amounts to 401.7 kg, hence the target. Water transfer to and from storage as well as water transfer from regenerator have been clearly labelled.

From Figure 7.4, it can be seen that reaction 1 occurs twice and reactions 2 and 3 occur thrice. The water used for washing Reactors 1 and 2 (R1 and R2) consists of freshwater only. 150 kg of water produced in R2 after washing is sent to storage at 2.25 h. This provides opportunity for indirect water reuse or wastewater regeneration. The storage tank discharges 145.5 kg of water to the regenerator with respect to R1 at 2.80 h. The regenerated water is discharged to R1 at 4.25 h for washing.

TABLE 7.2
Wastewater Minimization Data for Literature Example

		Maximum Concentration (g Contaminant/kg Water)		
		Contaminant 1	Contaminant 2	Contaminant 3
Reaction 1 (Reactor 1)	Max. inlet	0.5	0.5	2.3
	Max. outlet	1	0.9	3
Reaction 2 (Reactor 1)	Max. inlet	0.01	0.05	0.3
	Max. outlet	0.2	0.1	1.2
Reaction 3 (Reactor 1)	Max. inlet	0.15	0.2	0.35
	Max. outlet	0.3	1	1.2
Reaction 1 (Reactor 2)	Max. inlet	0.05	0.2	0.05
	Max. outlet	0.1	1	12
Reaction 2 (Reactor 2)	Max. inlet	0.03	0.1	0.2
	Max. outlet	0.075	0.2	1
Reaction 3 (Reactor 2)	Max. inlet	0.3	0.6	1.5
	Max. outlet	2	1.5	2.5
		Mass Load (g)		
Reaction 1	Reactor 1	4	80	10
	Reactor 2	15	24	358
Reaction 2	Reactor 1	28.5	7.5	135
	Reactor 2	9	2	16
Reaction 3	Reactor 1	15	80	85
	Reactor 2	22.5	45	36.5
		Duration of Washing (h)		
		Reaction 1	Reaction 2	Reaction 3
Reactor 1		0.25	0.5	0.25
Reactor 2		0.3	0.25	0.25

TABLE 7.3
Removal Ratio for Contaminants

Contaminant	Removal Ratio
Contaminant 1	0.98
Contaminant 2	0.97
Contaminant 3	0.96

The storage tank later discharges 7.5 kg of water to the regenerator with respect to R2 at 4.43 h. The regenerated water is discharged to R2 at 4.5 h for washing. A similar scenario is encountered with the water sent to storage from R2 and R1 at 4.75 h. The regenerated water is utilized in the later washing operations that follow. It can be observed from this solution that direct water reuse between units did not occur.

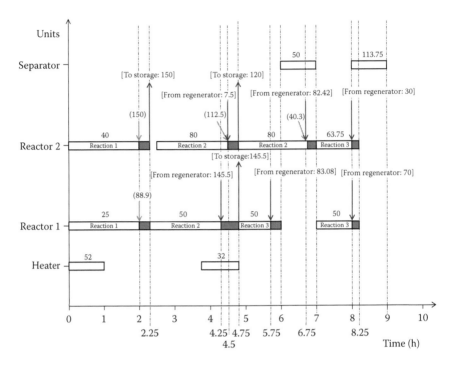

FIGURE 7.4   Resulting production schedule for literature example.

7.6.2 CASE STUDY II

This case study is based on a section of a pharmaceuticals production plant which produces four types of products, namely shampoos, deodorants, lotions and creams. Four mixers for product mixing were available in this section. Each product was produced by a specific mixer. Mixer 1 was dedicated to the mixing of shampoos, mixer 2 was dedicated to the mixing of deodorants, mixer 3 was dedicated to the mixing of lotions and mixer 4 was dedicated to the mixing of creams. The general production procedure is as follows. Raw material is charged to a mixer. The raw material is then mixed until the required physical characteristics are obtained. Once a product is mixed, it is removed and stored. The mixers are then washed. There is sufficient storage available for product. The typical production requirement over a 24-h time horizon is given in Table 7.4. In Table 7.4, the average production time for each product is also given.

The data pertaining to wastewater minimization are given in Table 7.5 together with the residue mass left in each mixer after product has been removed. These are average values taken from actual process data. The maximum outlet concentrations from each mixer are also given. It is important to note that there is maximum outlet concentration defined for only one component from each mixer. This is because each mixer will at any point only contain residue of a specific product. Also given is the maximum contaminant load that can enter with the water for a cleaning operation.

TABLE 7.4
Production Data for Case Study II

Mixer	Product	Number of Batches	Duration (h)
1	Shampoo	2	7
2	Deodorant	3	5.5
3	Lotion	1	11
4	Cream	2	11

TABLE 7.5
Wastewater Minimization Data for Case Study II

Mixer	Contaminant	Residue Mass (kg)	Limiting Water (kg)	C^{max} Outlet (kg/kg)
1	Shampoos	15	576.9	0.04
2	Deodorants	15	361.4	0.045
3	Lotions	30	697.6	0.05
4	Creams	70	1238.9	0.06

In this problem, the residue left from the specific product in a mixer accounts for the contaminant mass added to water. This then makes the limiting component in each mixer the component that leaves residue in the mixer. For example, mixer 1 has shampoo as the limiting contaminant, since this is the only component which leaves a residue in the mixer. The limiting amount of water for each mixer is calculated using Equation 7.12 and is given for each mixer in Table 7.5.

A central storage vessel with a capacity of 10 tons is available for storing wastewater used for washing. The washing of each mixer takes 30 min.

The maximum inlet concentration of each mixer is given in Table 7.6. In this case, the maximum inlet concentration of the deodorant in mixers 1, 3 and 4 is set to zero. This is because the reuse of wastewater containing deodorant in a mixer with

TABLE 7.6
Maximum Inlet Concentrations for Cleaning Operation in the Case Study II

Mixer	Shampoo (kg Product/kg Water)	Deodorant (kg Product/kg Water)	Lotion (kg Product/kg Water)	Creams (kg Product/kg Water)
1	0.014	0	0.007	0.0035
2	0.014	0.0035	0.007	0.007
3	0.014	0	0.007	0.0035
4	0.014	0	0.007	0.0035

TABLE 7.7

Removal Ratio for Contaminants

Contaminant	Removal Ratio
Shampoo	0.95
Deodorant	0.99
Lotion	0.96
Cream	0.98

any other residue is undesirable. However, wastewater containing the deodorant can be reused in a cleaning operation in mixer 2, containing another deodorant residue.

A regenerator with a flowrate of 466 kg/h is available. The removal ratio of the various contaminants is given in Table 7.7.

7.6.2.1 Results and Discussion

Case Study II was formulated using the presented methodology in GAMS 22.0 and solved using the DICOPT2 algorithm, with CPLEX 9.1.2 as the MIP and CONOPT3 as the NLP solvers. The resulting model had 576 binary variables. The optimum number of time points was 16. The value of the objective function, which was the minimization of effluent as given in Constraint (7.67), was 2653 kg of water. If regeneration was not considered, the objective function was 3587 kg of water. The incorporation of the regenerator resulted in a 26% decrease in effluent production. The solution time was 464 s using a Pentium 4, 3.4 GHz processor. The schedule corresponding to the MILP was infeasible. Hence, the resulting solution of 2653 kg is a local optimum.

The resulting schedule of the solution is given in Figure 7.5. The clear block represents the production operation in the unit, while the dark blocks represent the washing operation in the unit. The numbers in normal brackets represent freshwater. The sum of these amounts to 2653 kg, hence the target. The numbers in square brackets represent direct reuse water. Water transfer to and from storage as well as water transfer from regenerator have been clearly labelled.

From the Gantt chart in Figure 7.5, it can be observed that the schedule meets the production requirement, that is, two batches of shampoo, three batches of deodorant, one batch of lotion and two batches of cream. Mixer 3 uses freshwater in addition to direct reuse water from mixer 1. Mixer 3 sends 600 kg of water to storage at 11.5 h. Of this water, 201.7 kg is indirectly reused in mixer 4 at 11.5 h. At 20.2 h, the storage tank discharges 1166.7 kg of water to the regenerator on behalf of mixer 4. At 23 h, the regenerated water is discharged to mixer 4. The rest of the production schedule can be observed from Figure 7.5. As mentioned earlier, it is not acceptable to use water from mixer 2 containing deodorant. This was ensured in the model by setting the binary variable associated with water transfer from mixer 2 to any other mixer to zero and also by setting the binary variable associated with mixer 2 transferring water to the storage tank to zero. These conditions were met, as can be seen from the Gantt chart.

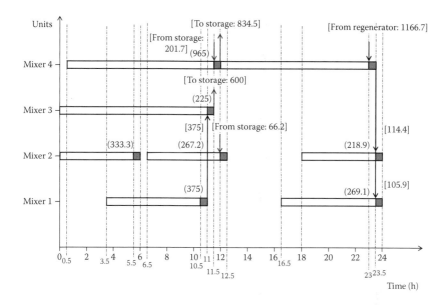

FIGURE 7.5 Production schedule for the Case Study II.

7.7 CONCLUSIONS

The regeneration methodology presented in this chapter deals with wastewater minimization where wastewater can contain multiple contaminants. The methodology does not require the production schedule to be given beforehand. The methodology is able to determine the minimum wastewater generated and the corresponding schedule that achieves this minimum. The methodology was applied to two case studies. In Case Study I where the production was not given, a 19.24% reduction in wastewater was achieved. In Case Study II where the production requirement was given, a 26% reduction in wastewater was achieved.

The model has a considerable number of bilinear terms. Upon linearization, additional variables and constraints are introduced in the model, which results in the model being very large. The amount of wastewater minimized depends on the given removal ratios of contaminants in the regenerator.

7A APPENDIX

The mathematical formulation is nonlinear due to the bilinear terms in Constraints (7.4), (7.5), (7.15), (7.16) and (7.23). Consider Constraint (7.4) containing the following bilinear terms:

1. $mw_{in}\left(s_{out,j},p\right)c_{in}\left(s_{out,j},c,p\right)$
2. $mw_r\left(s_{out,j'},s_{out,j},p\right)c_{out}\left(s_{out,j},c,p\right)$

3. $ms_{out}\left(s_{out,j},p\right)cs_{out}\left(c,p\right)$

4. $mreg_{out}\left(s_{out,j},p\right)cr_{out}\left(c,p\right)$

These terms are linearized using the method by Sherali and Alameddine (1992), as discussed by Quesada and Grossmann (1995).

Let

$$mw_{in}\left(s_{out,j},p\right)c_{in}\left(s_{out,j},c,p\right)=\Gamma_1\left(s_{out,j},c,p\right)$$

$$mw_r\left(s_{out,j'},s_{out,j},p\right)c_{out}\left(s_{out,j},c,p\right)=\Gamma_2\left(s_{out,j'},s_{out,j},c,p\right)$$

$$ms_{out}\left(s_{out,j},p\right)cs_{out}\left(c,p\right)=\Gamma_3\left(s_{out,j},c,p\right)$$

$$mreg_{out}\left(s_{out,j},p\right)cr_{out}\left(c,p\right)=\Gamma_4\left(s_{out,j},c,p\right)$$

with each variable having the following bounds:

$$0\leq mw_{in}\left(s_{out,j},p\right)\leq Mw^U\left(s_{out,j}\right)$$

$$0\leq c_{in}\left(s_{out,j},c,p\right)\leq C_{in}^U\left(s_{out,j},c\right)$$

$$0\leq mw_r\left(s_{out,j'},s_{out,j},p\right)\leq Mw^U\left(s_{out,j}\right)$$

$$0\leq c_{out}\left(s_{out,j},c,p\right)\leq C_{out}^U\left(s_{out,j},c\right)$$

$$0\leq ms_{out}\left(s_{out,j},p\right)\leq Mw^U\left(s_{out,j}\right)$$

$$0\leq cs_{out}\left(c,p\right)\leq CS_{out}^U\left(c\right)$$

$$0\leq mreg_{out}\left(s_{out,j},p\right)\leq Mw^U\left(s_{out,j}\right)$$

$$0\leq cr_{out}\left(c,p\right)\leq CR_{out}^U\left(c\right)$$

The following are true for Γ_1:

$$\Gamma_1\left(s_{out,j},c,p\right)\geq 0,\quad \forall j\in J,\quad s_{out,j}\in S_{out,j},\quad c\in C,\quad p\in P \tag{7A.1}$$

$$\Gamma_1\left(s_{out,j},c,p\right) \geq Mw^U\left(s_{out,j}\right)c_{in}\left(s_{out,j},c,p\right) + mw_{in}\left(s_{out,j},p\right)C_{in}^U\left(s_{out,j},c\right)$$

$$-Mw^U\left(s_{out,j}\right)C_{in}^U\left(s_{out,j},c\right), \quad \forall j \in J, \quad s_{out,j} \in S_{out,j}, \quad c \in C, \quad p \in P \quad (7A.2)$$

$$\Gamma_1 \leq Mw^U\left(s_{out,j}\right)c_{in}\left(s_{out,j},c,p\right), \quad \forall j \in J, \quad s_{out,j} \in S_{out,j}, \quad c \in C, \quad p \in P \quad (7A.3)$$

$$\Gamma_1 \leq mw_{in}\left(s_{out,j},p\right)C_{in}^U\left(s_{out,j},c\right), \quad \forall j \in J, \quad s_{out,j} \in S_{out,j}, \quad c \in C, \quad p \in P \quad (7A.4)$$

The following are true for Γ_2:

$$\Gamma_2\left(s_{out,j'},s_{out,j},c,p\right) \geq 0, \quad \forall j,j' \in J, \quad s_{out,j} \in S_{out,j}, \quad c \in C, \quad p \in P \quad (7A.5)$$

$$\Gamma_2\left(s_{out,j'},s_{out,j},c,p\right) \geq Mw^U\left(s_{out,j}\right)c_{out}\left(s_{out,j'},c,p\right) + mw_r\left(s_{out,j'},s_{out,j},p\right)C_{out}^U\left(s_{out,j'},c\right)$$

$$-Mw^U\left(s_{out,j}\right)C_{out}^U\left(s_{out,j'},c\right), \forall j,j' \in J, \quad s_{out,j} \in S_{out,j}, \quad c \in C, \quad p \in P \quad (7A.6)$$

$$\Gamma_2\left(s_{out,j'},s_{out,j},c,p\right) \leq Mw^U\left(s_{out,j}\right)c_{out}\left(s_{out,j'},c,p\right)$$

$$\forall j,j' \in J, \quad s_{out,j} \in S_{out,j}, \quad c \in C, \quad p \in P \quad (7A.7)$$

$$\Gamma_2\left(s_{out,j'},s_{out,j},c,p\right) \leq mw_r\left(s_{out,j'},s_{out,j},p\right)C_{out}^U\left(s_{out,j'},c\right),$$

$$\forall j,j' \in J, \quad s_{out,j} \in S_{out,j}, \quad c \in C, \quad p \in P \quad (7A.8)$$

The following are true for Γ_3:

$$\Gamma_3\left(s_{out,j},c,p\right) \geq 0, \quad \forall j \in J, \quad s_{out,j} \in S_{out,j}, \quad c \in C, \quad p \in P \quad (7A.9)$$

$$\Gamma_3\left(s_{out,j},c,p\right) \geq Mw^U\left(s_{out,j}\right)cs_{out}\left(c,p\right) + ms_{out}\left(s_{out,j},p\right)CS_{out}^U\left(c\right)$$

$$-Mw^U\left(s_{out,j}\right)CS_{out}^U\left(c\right), \quad \forall j \in J, \quad s_{out,j} \in S_{out,j}, \quad c \in C, \quad p \in P \quad (7A.10)$$

$$\Gamma_3\left(s_{out,j},c,p\right) \leq Mw^U\left(s_{out,j}\right)cs_{out}\left(c,p\right), \quad \forall j \in J, \quad s_{out,j} \in S_{out,j}, \quad c \in C, \quad p \in P$$

$$(7A.11)$$

$$\Gamma_3\left(s_{out,j},c,p\right) \leq ms_{out}\left(s_{out,j},p\right)CS_{out}^U\left(c\right), \quad \forall j \in J, \quad s_{out,j} \in S_{out,j}, \quad c \in C, \quad p \in P$$

$$(7A.12)$$

The following are true for Γ_4:

$$\Gamma_4\left(s_{out,j},c,p\right) \geq 0, \quad \forall j \in J, \quad s_{out,j} \in S_{out,j}, \quad c \in C, \quad p \in P \quad (7A.13)$$

$$\Gamma_4\left(s_{out,j}, c, p\right) \geq Mw^U\left(s_{out,j}\right) cr_{out}\left(c, p\right) + mreg_{out}\left(s_{out,j}, p\right) CR_{out}^U\left(c\right)$$

$$- Mw^U\left(s_{out,j}\right) CR_{out}^U\left(c\right), \quad \forall j \in J, \quad s_{out,j} \in S_{out,j}, \quad c \in C, \quad p \in P \quad\quad (7A.14)$$

$$\Gamma_4\left(s_{out,j}, c, p\right) \leq Mw^U\left(s_{out,j}\right) cr_{out}\left(c, p\right), \quad \forall j \in J, \quad s_{out,j} \in S_{out,j}, \quad c \in C, \quad p \in P$$
$$(7A.15)$$

$$\Gamma_4\left(s_{out,j}, c, p\right) \leq mreg_{out}\left(s_{out,j}, p\right) CR_{out}^U\left(c\right), \quad \forall j \in J, \quad s_{out,j} \in S_{out,j}, \quad c \in C, \quad p \in P$$
$$(7A.16)$$

Substituting these linearized variables into Constraint (7.4) gives Constraint (7A.17):

$$\Gamma_1\left(s_{out,j}, c, p\right) = \sum_{s_{out,j'}} \Gamma_2\left(s_{out,j'}, s_{out,j}, c, p\right) + \Gamma_3\left(s_{out,j}, c, p\right) + \Gamma_4\left(s_{out,j}, c, p\right)$$

$$\forall j, j' \in J, \quad s_{out,j} \in S_{out,j}, \quad p \in P, \quad c \in C \quad\quad (7A.17)$$

The bilinear terms in Constraints (7.5), (7.15), (7.16) and (7.23) are linearized in a similar manner and substituted in the original constraints to yield new constraints similar to Constraint (7A.17).

7A.1   SOLUTION PROCEDURE

The relaxed MILP problem consists of Constraints (7.1) through (7.65), (7.66) or (7.67) with the new substituted constraints instead of Constraints (7.4), (7.5), (7.15), (7.16) and (7.23). Also included are the respective forms of Constraints (7A.1) through (7A.4) for each linearized variable. The relaxed MILP problem is solved to provide a starting point for the exact MINLP problem. The solution procedure is illustrated in Figure 7A.1.

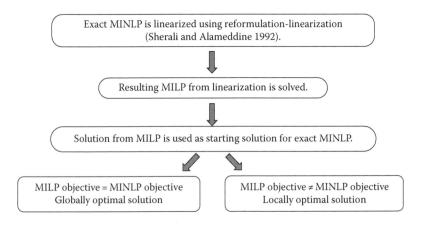

FIGURE 7A.1   Solution procedure.

REFERENCES

Liu, Y., Li, G., Wang, L., Zhang, J., Shams, K., 2009. Optimal design of an integrated discontinuous water-using network coordinating with a central regeneration unit. *Industrial Engineering and Chemical Research*. 48, 10924–10940.

Majozi, T., Gouws, J.F., 2009. A mathematical optimisation approach for wastewater minimisation in multipurpose batch plants: Multiple contaminants. *Computers and Chemical Engineering*. 33, 1826–1840.

Quesada, I., Grossmann, I.E., 1995. Global optimization of bilinear process networks with multicomponent flows. *Computers and Chemical Engineering*. 19(12), 1219–1242.

Sherali, H.D., Alameddine, A., 1992. A new reformulation–linearization technique for bilinear programming problems. *Journal of Global Optimum*. 2, 379–410.

8 Optimization of Energy and Water Use in Multipurpose Batch Plants Using an Improved Mathematical Formulation

8.1 INTRODUCTION

Presented in this chapter is a formulation that addresses optimization of both water and energy, while simultaneously optimizing the batch process schedule. The scheduling framework used in this study is based on a recent and efficient formulation. This formulation has been shown to result in a significant reduction of computational time and an improvement of the objective function and lead to fewer time points. The objective is to improve the profitability of the plant by minimizing wastewater generation and utility usage. From a case study, it was found that through applying only water integration the cost is reduced by 11.6%, by applying only energy integration the cost is reduced by 29.1% and by applying both energy and water integration the cost is reduced by 34.6%. This indicates that optimizing water and energy integration in the same scheduling framework will reduce the operating cost and environmental impact significantly.

8.2 NECESSARY BACKGROUND

In recent years, batch processes have been getting more attention due to their suitability for the production of small volume, high value–added products. The flexibility of batch plants allows the production of different products within the same facility. Batch manufacturing is typically used in the pharmaceutical, polymer, food and specialty chemical industries as demand for such products are highly seasonal and are influenced by changing markets. A common feature of many batch plants is that they utilize fossil fuels as the energy source and use water for process equipment cleaning, due to inherent sharing of equipment by different tasks. Despite the advantage of batch plants being flexible, they also pose a challenging task to operate in a sustainable way. In the past, batch industries could tolerate high inefficiencies in energy and water consumption due to the high value of final products which outstrips

197

the production costs. However, greater public awareness of the impact of industrial pollution, more stringent environmental regulations and escalating raw materials, energy, and waste treatment costs have now motivated energy and water saving measures for more sustainable operations (Halim and Srinivasan, 2011). Since scheduling, energy and wastewater minimization for multipurpose batch plants go hand in hand, published works in those areas are reviewed.

8.2.1 SCHEDULING OF BATCH PLANTS

Much research has been done on developing mathematical models to improve batch plant efficiency. The substantial advancement in modern computers allows the possibility of handling large and more complex problems by using optimization techniques. Excellent reviews of current scheduling techniques based on different time representations and associated challenges have been conducted (Floudas and Lin, 2004; Méndez et al., 2006; Shaik et al., 2006). In the reviews, with regard to time representation, the models are classified as slot based, event based and precedence based (sequence based). In slot-based models (Pinto and Grossmann, 1994; Lim and Karimi, 2003; Liu and Karimi, 2008), the time horizon is divided into 'non-uniform unknown slots' and tasks start and finish in the same slot. On the other hand, slot models exist that use non-uniform unknown slots where tasks are allowed to continue to the next slots (Schilling and Pantelides, 1996; Karimi and McDonald, 1997; Reddy et al., 2004; Sundaramoorthy and Karimi, 2005; Erdirik-Dogan and Grossmann, 2008; Susarla et al., 2010). The event-based models can also be categorized into those that use uniform unknown events, where the time associated with the events is common across all units (Maravelias and Grossmann, 2003; Castro et al., 2004), and those that use unit-specific events where the time associated with the events can be different across the units (Ierapetritou and Floudas, 1998; Majozi and Zhu, 2001; Shaik et al., 2006; Janak and Floudas, 2008; Shaik and Floudas, 2009; Li et al., 2010). The heterogeneous location of events across the units gives fewer event points as compared to both the global event–based and slot-based models. As a result, unit-specific event-based models are computationally superior. The sequence-based or precedence-based representation uses either direct precedence (Hui and Gupta, 2000; Liu and Karimi, 2007) or indirect precedence sequencing of pairs of tasks in units (Méndez and Cerdá, 2000, 2003; Méndez et al., 2001; Ferrer-Nadal et al., 2008). The models do not require pre-postulation of events and slots. Seid and Majozi (2012) presented a mixed integer linear programming (MILP) formulation based on the state sequence network (SSN) and unit-specific time points, which can handle proper sequencing of tasks and fixed intermediate storage (FIS) policy. The model results in a reduction of event or time points required and as a result gives better performance in terms of objective value and CPU time required when compared to previous literature models.

8.2.2 ENERGY INTEGRATION IN BATCH PLANTS

Many heat integration techniques are applied to predefined schedules which are inherently suboptimal. Vaklieva-Bancheva et al. (1996) considered direct heat integration with the objective of minimizing total costs. The resulting overall formulation was an MILP problem, solved to global optimality, although only specific pairs of units were allowed to undergo heat integration. Uhlenbruck et al. (2000) improved OMNIUM, which is a tool developed for heat exchanger network synthesis by Hellwig and Thöne (1994). The improved OMNIUM tool increased the energy recovery by 20%. Bozan et al. (2001) developed a single-step, interactive computer program (BatcHEN) used for the determination of the campaigns, that is the set of products which can be produced simultaneously, the heat exchange areas of all possible heat exchangers in the campaigns and the heat exchanger network. This work addressed the limitation of the graph theory method for the determination of the campaign by Bancheva et al. (1996). Krummenacher and Favrat (2001) proposed a new systematic procedure, supported by graphics, which made it possible to determine the minimum number of heat storage units. Chew et al. (2005) applied cascade analysis proposed by Kemp and Macdonald (1987) to reduce the utility requirement for the production of oleic acid from palm olein using immobilized lipase. The result obtained showed savings of 71.4% and 62.5% for hot and cold utilities respectively. Pires et al. (2005) developed the BatchHeat software, whose aim was to highlight the energy inefficiencies in the process and thereby enabling the scope for possible heat recovery to be established through direct heat exchange or storage through implementation of cascade analysis.

Boer et al. (2006) evaluated the technical and economic feasibility of an industrial heat storage system for an existing production facility of organic surfactants. Fritzson and Berntsson (2006) applied process integration methods to investigate the potential to decrease the energy usage in the slaughtering and meat processing industry. The result obtained illustrates that 30% of the external heat demand and more than 10% of the shaftwork used can be saved. Morrison et al. (2007) developed a user-friendly software package known as optimal batch integration (OBI). Chen and Ciou (2008) formulated a method to design an optimization of indirect energy storage systems for batch process. Their work aimed at simultaneously solving the problem of indirect heat exchange network synthesis and its associated thermal storage policy for recirculated hot/cold heat storage medium (HEN). Most of the previous works solved this sequentially. Foo et al. (2008) extended the minimum units targeting and network evolution techniques that were developed for batch mass exchange network (MEN) into batch HEN. They applied the technique for energy integration of oleic acid production from palm olein using immobilized lipase. Halim and Srinivasan (2009) discussed a sequential method using direct heat integration. A number of optimal schedules with minimum makespan were found, and heat integration analysis was performed on each. The schedule with minimum utility requirement was chosen as the best. Later, Halim and Srinivasan (2011) extended their technique to carry out water reuse network synthesis simultaneously. One key feature of this method is its ability to find the heat integration and water reuse solution without sacrificing the quality of the scheduling solution.

Atkins et al. (2010) applied indirect heat integration using heat storage for a milk powder plant in New Zealand. The traditional composite curves have been used to estimate the maximum heat recovery and to determine the optimal temperatures of the stratified tank. Tokos et al. (2010) applied a batch heat integration technique to a large beverage plant. The opportunities of heat integration between batch operations were analyzed by a MILP model, which was slightly modified by considering specific industrial circumstances. Muster-Slawitsch et al. (2011) came up with the Green Brewery concept to demonstrate the potential for reducing thermal energy consumption in breweries. Three detailed case studies were investigated. The 'Green Brewery' concept has shown a saving potential of more than 5000 t/year fossil CO_2 emissions from thermal energy supply for the three breweries that were closely considered. Becker et al. (2012) applied time average energy integration approach to a real case study of a cheese factory with non-simultaneous process operations. Their work addressed appropriate heat pump integration. A cost saving of more than 40% was reported.

For a more optimal solution, scheduling and heat integration may be combined into an overall problem. Papageorgiou et al. (1994) embedded a heat integration model within the scheduling formulation of Kondili et al. (1993). Opportunities for both direct and indirect heat integration were considered as well as possible heat losses from a heat storage tank. The operating policy, in terms of heat integrated or standalone, was predefined for tasks. Adonyi et al. (2003) used the 'S-Graph' scheduling approach and incorporated one to one direct heat integration. Barbosa-Póvoa et al. (2001) presented a mathematical formulation for the detailed design of multipurpose batch process facilities with heat integration. Pinto et al. (2003) extended the work of Barbosa-Póvoa et al. (2001) with the consideration of the economic savings in utility requirements, while considering the cost of both the auxiliary structures that is heatexchanger through their transfer area and the design of the utility circuits and associated piping costs. Majozi (2006) presented a direct heat integration formulation based on the state sequence network of Majozi and Zhu (2001) which uses an unevenly discretized time horizon. The direct heat integration model developed by Majozi (2006) was extended to incorporate heat storage for more flexible schedules and utility savings in the later work by Majozi (2009). However, the storage size is a parameter in his formulation which is addressed later by Stamp and Majozi (2011), where the storage size is determined by the optimization exercise. Chen and Chang (2009) extended the work of Majozi (2006) to periodic scheduling, based on the resource task network (RTN) scheduling framework. The reader can get a more comprehensive and detailed review on energy recovery for batch processes in the chapter by Fernández et al. (2012).

8.2.3 WASTEWATER MINIMIZATION IN BATCH PLANTS

Wastewater is generated in batch plants during the cleaning of multipurpose equipment and when water is used as a solvent. Tight environmental regulations and increased public awareness demand that batch plants consider rational use of water during their operation. Many researchers have developed methodologies for the efficient use of water through direct reuse, indirect reuse and regeneration of wastewater. Direct reuse consists of recycle and reuse. Recycle refers to the reuse of an outlet wastewater stream from a processing unit in the same unit, while reuse refers to the use of an outlet wastewater stream from a processing unit in another processing unit.

Indirect reuse is when wastewater is temporarily stored in a storage vessel and later reused in a processing unit requiring water.

Based on the analogy of heat and mass transfer, several methodologies for synthesizing water reuse network in batch processes have also been developed. Gouws et al. (2010) reviewed these techniques based on graphical-based pinch analysis and mathematical optimization approach. The seminal work on pinch analysis application to batch water network was reported by Wang and Smith (1994). Foo et al. (2005) proposed a time-dependent water cascade analysis to obtain minimum required water flows in a process. While these graphical-based techniques are useful, they share a common drawback that their application is limited to single contaminant cases. The mathematical optimization-based techniques, which are capable of solving multiple contaminant problems, can be differentiated into two groups, namely, those based on the schedule being known a priori, that is sequential approach and those that simultaneously determine the process schedule and minimize the freshwater usage.

Almató et al. (1997) addressed the problem of water reuse through storage tank allocation based on the optimal schedule being known a priori. Kim and Smith (2004) proposed a more generalized method for optimal design of discontinuous water reuse network. In their approach, a production schedule was fixed and direct reuse of water between operations within the same time interval was allowed without passing through storage tanks. Most of the mathematically based models are based on a superstructure approach. Majozi and Gouws (2009) proposed a continuous-time scheduling framework to simultaneously optimize the schedule and water reuse while addressing both single and multiple contaminants. Cheng and Chang (2007) considered the optimization of the batch production schedule, water reuse schedule and wastewater treatment schedule in a single problem based on discrete time scheduling framework. At the end of optimization, the production schedule, the number and sizes of buffer tanks and the physical configuration of the pipeline network were obtained. Adekola and Majozi (2011) extended the work of Majozi and Gouws (2009) by incorporating wastewater regenerator for further improvement of water utilization.

From the review, it can be seen that wastewater minimization and heat integration in batch plants are addressed separately. To the knowledge of the authors, only the work presented by Halim and Srinivasan (2011) and Adekola et al. (2013) addressed this literature gap. In the work of Halim and Srinivasan (2011), the overall problem is decomposed into three parts, namely scheduling, heat integration and water reuse optimization and solved sequentially. Batch scheduling is solved first to meet an economic objective function. Next, alternate schedules are generated through a stochastic search-based integer cut procedure. For each resulting schedule, minimum energy and water reuse targets are established and networks identified which might lead to suboptimal results. Adekola et al. (2013) also addressed this problem by developing a model that simultaneously optimize energy, water and production throughput. They demonstrated that the unified approach where all resources are optimized simultaneously gives a better economic performance compared to the common sequential techniques for wastewater and energy integration techniques developed for multipurpose batch plants. However, the model has two basic limitations. The first drawback is the model is not based on time average model (TAM) and treats the temperature driving force based on initial and target temperatures of cold and hot streams. This assumption makes the model impossible to apply for a case where the starting and finishing time of the heat integrated units to be anywhere between the starting and finishing time of the processing tasks since it is required to calculate the intermediate temperatures to ensure for the minimum thermal driving force. The second limitation is it forces the heat integrating units to start simultaneously which results suboptimal because of restricting the flexibility of the schedule.

In this chapter, a section is made to close the literature gap by simultaneously solving energy integration and wastewater minimization problem in the same scheduling framework. The model is based on TAM and time slice model (TSM) where the time slice is a variable determined by optimization to keep the flexibility of the schedule as compared to previous models based on fixed schedule and fixed time slice for heat integration. The model also addressed the two basic limitations as discussed in the model of Adekola et al. (2013). Additionally, the proposed model used the resent robust scheduling framework of Seid and Majozi (2012) as a platform since the model gave better objective value as compared to previous literature models. The rest of the chapter is organized as follows. Section 8.3 defines the problem statement. Section 8.4 describes the detail mathematical formulation. Section 8.5 describes the application of the mathematical model to literature problems. Finally, conclusions are drawn from this work in Section 8.6.

8.3 PROBLEM STATEMENT

Given:

1. The production recipe (STN or SSN representation)
2. The capacity of units and the type of tasks each unit can perform
3. The maximum storage capacity for each material
4. The task processing times
5. Hot duties for tasks require heating and cold duties for tasks that require cooling
6. Operating temperatures of heat sources and heat sinks
7. Minimum allowable temperature differences
8. The material heat capacities
9. The units' washing time
10. The mass load of each contaminant
11. The concentration limits of each contaminant
12. The costs of raw materials, products and utilities
13. The scheduling time horizon (for profit maximization problem)
14. Production demand (for makespan minimization problem)

Determine:

1. The optimum production schedule, that is allocation of tasks to units, timing of all tasks, and batch sizes
2. Optimum energy requirement and associated heat exchange configuration
3. Optimum water requirement and associated water-reuse network.

8.4   MATHEMATICAL FORMULATION

The scheduling model by Seid and Majozi (2012) was adopted as a scheduling plat-form since it has proven to result in better CPU time and optimal objective value compared to other scheduling models. Uneven discretization of the time horizon so-called continuous time was used.

8.4.1   HEAT INTEGRATION MODEL

The mathematical model is based on the superstructure in Figure 8.1. Each task may operate using either direct or standalone mode by using only external utilities. If direct integration is not sufficient to satisfy the required duty, external utilities may make up for any deficit.

Constraints (8.1) and (8.2) are active simultaneously and ensure that one hot unit will be integrated with one cold unit when direct heat integration takes place in order to simplify operation of the process. It is worth noting that mathematically it is also possible for one unit to integrate with more than one unit at a given time point when the summation notation is not used. However, this is practically very dif-ficult to implement. Also, if two units are to be heat integrated at a given time point, they must both be active at that time point. For better understanding, the difference between time point p and extended time point $pp.$ is explained using Figure 8.2. If a unit j that is active at time point p is integrated with more than one unit in different temperature and time intervals, an extended time point $pp.$ must be defined. Unit $j1$ active at time point p can be integrated with units $j2$ and $j3$ in different time and tem-perature intervals. At the beginning, unit $j1$ is integrated with unit $j2$ at time point p and the extended time point $pp.$ is the same as time point $p.$ Later, $j1$ is integrated with unit $j3$ in another time interval where extended time point $pp.$ equals to $p + 1.$ $pp.$ is equal to or greater than time point p and less than or equal to $n + p$, where n is a parameter which is greater than or equal to zero. If n equals 2, then a unit that is active at time point p can be integrated in three different intervals. The model should

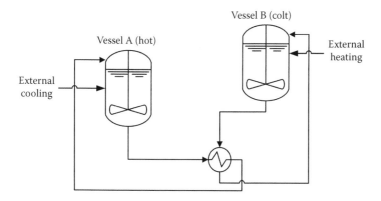

FIGURE 8.1   Superstructure for the energy integration.

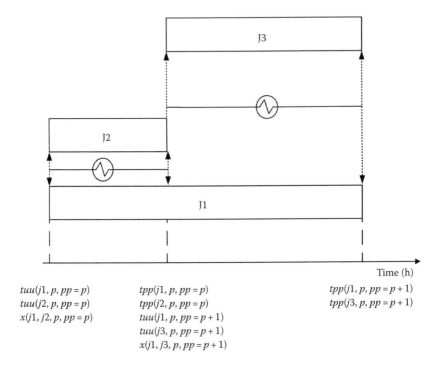

FIGURE 8.2   Differentiating time point p and extended time point pp.

be solved starting from n equals zero and adding one at a time until no better objective value is achieved.

$$\sum_{s_{inj_c}} x\left(s_{inj_c}, s_{inj_h}, p, pp\right) \leq y\left(s_{inj_h}, p\right), \quad \forall p, pp \in P, \quad s_{inj_h} \in S_{inJ_h}, \quad s_{inj_c} \in S_{inJ_c} \quad (8.1)$$

$$\sum_{s_{inj_h}} x\left(s_{inj_c}, s_{inj_h}, p, pp\right) \leq y\left(s_{inj_c}, p\right), \quad \forall p, pp \in P, \quad s_{inj_h} \in S_{inJ_h}, \quad s_{inj_c} \in S_{inJ_c} \quad (8.2)$$

Constraint (8.3) describes the amount of cooling load required by the hot unit from its initial temperature to its target temperature. In a situation where the temperature in the reactor unit is fixed during exothermic reaction, the heat load becomes the product of the amount of mass that undergoes reaction and the heat of reaction.

$$cl\left(s_{inj_h}, p\right) = mu\left(s_{inj_h}, p\right) cp\left(s_{inj_h}\right)\left(T_{s_{inj_h}}^{in} - T_{s_{inj_h}}^{out}\right), \quad \forall p \in P, \quad s_{inj_h} \in S_{inJ_h} \quad (8.3)$$

Constraint (8.4) describes the amount of heating load required by the cold unit from its initial temperature to its target temperature. In a situation where the temperature in the reactor unit is fixed during endothermic reaction, the heat load becomes the product of the amount of mass that undergoes reaction and the heat of reaction.

$$hl\left(s_{inj_c},p\right)=mu\left(s_{inj_c},p\right)cp\left(s_{inj_c}\right)\left(T_{s_{inj_c}}^{out}-T_{s_{inj_c}}^{in}\right),\quad \forall p\in P,\quad s_{inj_c}\in S_{inJ_c}\quad (8.4)$$

Constraints (8.5) and (8.6) describe the average heat flow for the hot and cold unit, respectively, during the processing time which is the same as TAM model to address the energy balance during heat integration properly.

$$cl\left(s_{inj_h},p\right)=avcl\left(s_{inj_h},p\right)\left(tp\left(s_{inj_h},p\right)-tu\left(s_{inj_h},p\right)\right),\quad \forall p\in P,\quad s_{inj_h}\in S_{inJ_h}\quad (8.5)$$

$$hl\left(s_{inj_c},p\right)=avhl\left(s_{inj_c},p\right)\left(tp\left(s_{inj_c},p\right)-tu\left(s_{inj_c},p\right)\right),\quad \forall p\in P,\quad s_{inj_c}\in S_{inJ_c}\quad (8.6)$$

Constraints (8.7) and (8.8) define the heat load at time point p and extended time point pp. for the hot and cold unit.

$$hlp\left(s_{inj_c},p,pp\right)=avhl\left(s_{inj_c},p\right)\left(tpp\left(s_{inj_c},p,pp\right)-tuu\left(s_{inj_c},p,pp\right)\right),$$
$$\forall p,pp\in P,\quad s_{inj_c}\in S_{inJ_c}\quad (8.7)$$

$$clp\left(s_{inj_h},p,pp\right)=avcl\left(s_{inj_h},p\right)\left(tpp\left(s_{inj_h},p,pp\right)-tuu\left(s_{inj_h},p,pp\right)\right),$$
$$\forall p,pp\in P,\quad s_{inj_h}\in S_{inJ_h}\quad (8.8)$$

Constraints (8.9) and (8.10) are used to calculate the temperature of the hot and cold unit at the intervals.

$$clp\left(s_{inj_h},p,pp\right)=mu\left(s_{inj_h},p\right)cp\left(s_{inj_h}\right)\left(T^{in}\left(s_{inj_h},p,pp\right)-T^{out}\left(s_{inj_h},p,pp\right)\right),$$
$$\forall p,pp\in P,\quad s_{inj_h}\in S_{inJ_h}\quad (8.9)$$

$$hlp\left(s_{inj_c},p,pp\right)=mu\left(s_{inj_c},p\right)cp\left(s_{inj_c}\right)\left(T^{out}\left(s_{inj_c},p,pp\right)-T^{in}\left(s_{inj_c},p,pp\right)\right),$$
$$\forall p,pp\in P,\quad s_{inj_c}\in S_{inJ_c}\quad (8.10)$$

Constraint (8.11) states that the amount of heat exchanged by the hot unit with the cold units should be less than the cooling load required by the hot unit during the interval.

$$\sum_{s_{inj_c}}qe\left(s_{inj_c},s_{inj_h},p,pp\right)\leq clp\left(s_{inj_h},p,pp\right),\quad \forall p,pp\in P,\quad s_{inj_h}\in S_{inJ_h},\quad s_{inj_c}\in S_{inJ_c}$$
$$(8.11)$$

Constraint (8.12) states that the amount of heat exchanged by the cold unit with the hot units should be less than the heat load required by the cold unit during the interval.

$$\sum_{s_{inj_h}}qe\left(s_{inj_c},s_{inj_h},p,pp\right)\leq hlp\left(s_{inj_c},p,pp\right),\quad \forall p,pp\in P,\quad s_{inj_h}\in S_{inJ_h},\quad s_{inj_c}\in S_{inJ_c}$$
$$(8.12)$$

Constraints (8.13) and (8.14) state that the temperature of the unit at the start of an interval should be equal to the temperature at the end of the previous interval.

$$T^{in}\left(s_{inj_h}, p, pp\right) = T^{out}\left(s_{inj_h}, p, pp-1\right), \quad \forall p, pp \in P, \quad s_{inj_h} \in S_{inJ_h} \qquad (8.13)$$

$$T^{in}\left(s_{inj_c}, p, pp\right) = T^{out}\left(s_{inj_c}, p, pp-1\right), \quad \forall p, pp \in P, \quad s_{inj_c} \in S_{inJ_c} \qquad (8.14)$$

Constraints (8.15) and (8.16) state that the temperature at the start of the first interval, which is time point p, which is also pp, should be equal to the initial temperature of the task.

$$T^{in}\left(s_{inj_h}, p, pp\right) = T^{in}_{s_{inj_h}}, \quad \forall p, pp \in P, \quad p = pp, \quad s_{inj_h} \in S_{inJ_h} \qquad (8.15)$$

$$T^{in}\left(s_{inj_c}, p, pp\right) = T^{in}_{s_{inj_c}}, \quad \forall p, pp \in P, \quad p = pp, \quad s_{inj_c} \in S_{inJ_c} \qquad (8.16)$$

Constraints (8.17) and (8.18) ensure that the minimum thermal driving forces are obeyed when there is direct heat integration between a hot and a cold unit.

$$T^{in}\left(s_{inj_h}, p, pp\right) - T^{out}\left(s_{inj_c}, p, pp\right) \geq \Delta T - \Delta T^U\left(1 - x\left(s_{inj_c}, s_{inj_h}, p, pp\right)\right),$$
$$\forall p, pp \in P, \quad s_{inj_h} \in S_{inJ_h}, \quad s_{inj_c} \in S_{inJ_c} \qquad (8.17)$$

$$T^{out}\left(s_{inj_h}, p, pp\right) - T^{in}\left(s_{inj_c}, p, pp\right) \geq \Delta T - \Delta T^U\left(1 - x\left(s_{inj_c}, s_{inj_h}, p, pp\right)\right),$$
$$\forall p, pp \in P, \quad s_{inj_h} \in S_{inJ_h}, \quad s_{inj_c} \in S_{inJ_c} \qquad (8.18)$$

Constraints (8.19) through (8.22) ensure that the times at which units are active are synchronized when direct heat integration takes place.

$$tuu\left(s_{inj_h}, p, pp\right) \geq tuu\left(s_{inj_c}, p, pp\right) - M\left(1 - x\left(s_{inj_c}, s_{inj_h}, p, pp\right)\right),$$
$$\forall p, pp \in P, \quad s_{inj_h} \in S_{inJ_h}, \quad s_{inj_c} \in S_{inJ_c} \qquad (8.19)$$

$$tuu\left(s_{inj_h}, p, pp\right) \leq tuu\left(s_{inj_c}, p, pp\right) + M\left(1 - x\left(s_{inj_c}, s_{inj_h}, p, pp\right)\right),$$
$$\forall p, pp \in P, \quad s_{inj_h} \in S_{inJ_h}, \quad s_{inj_c} \in S_{inJ_c} \qquad (8.20)$$

$$tpp\left(s_{inj_h}, p, pp\right) \geq tpp\left(s_{inj_c}, p, pp\right) - M\left(1 - x\left(s_{inj_c}, s_{inj_h}, p, pp\right)\right),$$
$$\forall p, pp \in P, \quad s_{inj_h} \in S_{inJ_h}, \quad s_{inj_c} \in S_{inJ_c} \qquad (8.21)$$

$$tpp\left(s_{inj_h}, p, pp\right) \leq tpp\left(s_{inj_c}, p, pp\right) + M\left(1 - x\left(s_{inj_c}, s_{inj_h}, p, pp\right)\right),$$
$$\forall p, pp \in P, \quad s_{inj_h} \in S_{inJ_h}, \quad s_{inj_c} \in S_{inJ_c} \qquad (8.22)$$

Constraints (8.23) and (8.24) stipulate that the starting time of the heating load required for the cold unit and cooling load required for the hot unit at the first interval should be equal to the starting time of the hot and cold unit.

$$tuu\left(s_{inj_h}, p, pp\right) = tu\left(s_{inj_h}, p\right), \quad \forall p, pp \in P, \quad p = pp, \quad s_{inj_h} \in S_{inJ_h} \quad (8.23)$$

$$tuu\left(s_{inj_c}, p, pp\right) = tu\left(s_{inj_c}, p\right), \quad \forall p, pp \in P, \quad p = pp, \quad s_{inj_c} \in S_{inJ_c} \quad (8.24)$$

Constraints (8.25) and (8.26) state that the starting time of heating and cooling in an interval should be equal to the finishing time at the previous interval.

$$tuu\left(s_{inj_h}, p, pp\right) = tpp\left(s_{inj_h}, p, pp - 1\right), \quad \forall p, pp \in P, \quad s_{inj_h} \in S_{inJ_h} \quad (8.25)$$

$$tuu\left(s_{inj_c}, p, pp\right) = tpp\left(s_{inj_c}, p, pp - 1\right), \quad \forall p, pp \in P, \quad s_{inj_c} \in S_{inJ_c} \quad (8.26)$$

Constraint (8.27) ensures that if heat integration occurs, the heat load should have a value that is less than the maximum amount of heat exchangeable. When the binary variable associated with heat integration takes a value of zero, no heat integration occurs and the associated heat load is zero.

$$qe\left(s_{inj_c}, s_{inj_h}, p, pp\right) \le Q^U x\left(s_{inj_c}, s_{inj_h}, p, pp\right), \quad \forall p, pp \in P, \quad s_{inj_h} \in S_{inJ_h}, \quad s_{inj_c} \in S_{inJ_c}$$
$$(8.27)$$

Constraints (8.28) and (8.29) state that if the binary variable associated with heat integration is active, then the binary variable associated with heating and cooling must be active at that interval.

$$x\left(s_{inj_c}, s_{inj_h}, p, pp\right) \le y_{int}\left(s_{inj_h}, p, pp\right), \quad \forall p, pp \in P, \quad s_{inj_h} \in S_{inJ_h}, \quad s_{inj_c} \in S_{inJ_c}$$
$$(8.28)$$

$$x\left(s_{inj_c}, s_{inj_h}, p, pp\right) \le y_{int}\left(s_{inj_c}, p, pp\right), \quad \forall p, pp \in P, \quad s_{inj_h} \in S_{inJ_h}, \quad s_{inj_c} \in S_{inJ_c}$$
$$(8.29)$$

Constraints (8.30) and (8.31) state that the heating and cooling loads take on a value for a certain duration when the binary variables associated with heating and cooling are active.

$$tpp\left(s_{inj_h}, p, pp\right) - tuu\left(s_{inj_h}, p, pp\right) \le Hy_{int}\left(s_{inj_h}, p, pp\right), \quad \forall p, pp \in P, \quad s_{inj_h} \in S_{inJ_h}$$
$$(8.30)$$

$$tpp\left(s_{inj_c}, p, pp\right) - tuu\left(s_{inj_c}, p, pp\right) \le Hy_{int}\left(s_{inj_c}, p, pp\right), \quad \forall p, pp \in P, \quad s_{inj_c} \in S_{inJ_c}$$
$$(8.31)$$

Constraints (8.32) and (8.33) state that temperatures change in the heating and cooling unit when the binary variables associated with heating and cooling are active.

$$T^{in}\left(s_{inj_h}, p, pp\right) - T^{out}\left(s_{inj_h}, p, pp\right) \le \Delta T^{U}\left(s_{inj_h}\right) y_{int}\left(s_{inj_h}, p, pp\right),$$

$$\forall p, pp \in P, \quad s_{inj_h} \in S_{inJ_h} \tag{8.32}$$

$$T^{out}\left(s_{inj_c}, p, pp\right) - T^{in}\left(s_{inj_c}, p, pp\right) \le \Delta T^{U}\left(s_{inj_c}\right) y_{int}\left(s_{inj_c}, p, pp\right),$$

$$\forall p, pp \in P, \quad s_{inj_c} \in S_{inJ_c} \tag{8.33}$$

Constraint (8.34) states that the cooling of a hot unit will be satisfied by direct heat integration and external cooling utility if required.

$$cl\left(s_{inj_h}, p\right) = cw\left(s_{inj_h}, p\right) + \sum_{s_{inj_c}} qe\left(s_{inj_c}, s_{inj_h}, p, pp\right),$$

$$\forall p, pp \in P, \quad s_{inj_h} \in S_{inJ_h}, \quad s_{inj_c} \in S_{inJ_c} \tag{8.34}$$

Constraint (8.35) states that the heating of a cold unit will be satisfied by direct heat integration and external heating utility if required.

$$hl\left(s_{inj_c}, p\right) = st\left(s_{inj_h}, p\right) + \sum_{s_{inj_h}} qe\left(s_{inj_c}, s_{inj_h}, p, pp\right),$$

$$\forall p, pp \in P, \quad s_{inj_h} \in S_{inJ_h}, \quad s_{inj_c} \in S_{inJ_c} \tag{8.35}$$

8.4.2 WASTEWATER MINIMIZATION MODEL

The superstructure on which the wastewater minimization model is based is depicted in Figure 8.3. Only the water-using operations which are part of a complete batch process are depicted. Unit j represents a water-using operation in which the water

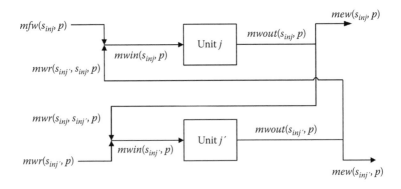

FIGURE 8.3 Superstructure for water usage.

used can consist of freshwater, reuse water or reuse and freshwater. Water from unit j can be reused elsewhere or sent to effluent treatment.

Constraint (8.36) defines the amount of water entering the unit as the sum of freshwater and reuse water from other units.

$$mwin\left(s_{inj}, p\right) = mfw\left(s_{inj}, p\right) + \sum_{s_{inj'}} mrw\left(s_{inj'}, s_{inj}, p\right), \quad \forall p \in P, \quad s_{inj}, s_{inj'} \in S_{inJ} \quad (8.36)$$

Constraint (8.37) states that the amount of water leaving the unit is equal to the sum of reuse water sent to other units and water sent to effluent treatment.

$$mwout\left(s_{inj}, p\right) = \sum_{s_{inj'}} mrw\left(s_{inj}, s_{inj'}, p\right) + mew\left(s_{inj}, p\right), \quad \forall p \in P, \quad s_{inj}, s_{inj'} \in S_{inJ} \quad (8.37)$$

Constraint (8.38) is the water balance around the unit and states that the amount of water entering the unit equals the amount of water leaving the unit.

$$mwin\left(s_{inj}, p\right) = mwout\left(s_{inj}, p\right), \quad \forall p \in P, \quad s_{inj} \in S_{inJ} \quad (8.38)$$

Constraint (8.39) defines the inlet contaminant load as the mass of contaminant, entering with reuse water.

$$cin\left(s_{inj}, c, p\right) mwin\left(s_{inj}, p\right) = \sum_{s_{inj'}} cout\left(s_{inj'}, c, p\right) mrw\left(s_{inj'}, s_{inj}, p\right),$$

$$\forall p \in P, \quad s_{inj}, s_{inj'} \in S_{inJ}, \quad c \in C \quad (8.39)$$

Constraint (8.40) states that the amount of contaminant leaving the unit equals the sum of the contaminant entering into the unit and the contaminant removed from the process.

$$mwout\left(s_{inj}, p\right) cout\left(s_{inj}, c, p\right) = SMC\left(s_{inj}\right) mu\left(s_{inj}, p\right) + cin\left(s_{inj}, c, p\right) mwin\left(s_{inj}, p\right),$$

$$\forall p \in P, \quad s_{inj} \in S_{inJ}, \quad c \in C \quad (8.40)$$

Constraint (8.41) ensures that the amount of reused water from unit j to other units does not exceed the maximum allowable water in the receiving units. It also indicates whether water from unit j is reused or not.

$$mrw\left(s_{inj}, s_{inj'}, p\right) \leq W_{in}^{U}\left(s_{inj'}\right) y_{re}\left(s_{inj}, s_{inj'}, p\right), \quad \forall p \in P, \quad s_{inj}, s_{inj'} \in S_{inJ} \quad (8.41)$$

Constraint (8.42) ensures that the reuse of water from unit j in other units can occur only if the units are active.

$$y_{re}\left(s_{inj}, s_{inj'}, p\right) \leq y\left(s_{inj'}, p\right), \quad \forall p \in P, \quad s_{inj}, s_{inj'} \in S_{inJ} \quad (8.42)$$

Constraint (8.43) gives the upper bound on the water entering into unit j. It also ensures that water enters into the unit only if it is active.

$$mwin\left(s_{inj}, p\right) \leq W_{in}^{U}\left(s_{inj}\right) y\left(s_{inj}, p\right), \quad \forall p \in P, \quad s_{inj} \in S_{inJ} \tag{8.43}$$

In Constraints (8.44) and (8.45), wastewater can only be directly reused if the finishing time of the unit producing wastewater and the starting time of the unit receiving wastewater coincide.

$$tuw\left(s_{inj}, p\right) \geq tpw\left(s_{inj'}, p\right) - M * y_{re}\left(s_{inj}, s_{inj'}, p\right), \quad \forall p \in P, \quad s_{inj}, s_{inj'} \in S_{inJ} \tag{8.44}$$

$$tuw\left(s_{inj}, p\right) \leq tpw\left(s_{inj'}, p\right) + M * y_{re}\left(s_{inj}, s_{inj'}, p\right), \quad \forall p \in P, \quad s_{inj}, s_{inj'} \in S_{inJ} \tag{8.45}$$

Constraint (8.46) defines the finishing time of the washing operation as the starting time of the washing operation added to the duration of washing.

$$tpw\left(s_{inj}, p\right) \geq tuw\left(s_{inj}, p\right) + \tau w\left(s_{inj}\right) y\left(s_{inj}, p\right), \quad \forall p \in P, \quad s_{inj} \in S_{inJ} \tag{8.46}$$

Constraint (8.47) ensures that the starting time of a task in a unit is greater than the finishing time of the washing operations.

$$tu\left(s_{inj}, p\right) \geq tpw\left(s'_{inj}, p-1\right), \quad \forall p \in P, \quad s_{inj}, s'_{inj} \in S_{inJ}, S_{inj}^{*} \tag{8.47}$$

Constraint (8.48) stipulates that the starting time of the washing operation in a unit occurs after the completion of the task in the unit.

$$tuw\left(s_{inj}, p\right) \geq tp\left(s_{inj}, p\right), \quad \forall p \in P, \quad s_{inj} \in S_{inJ} \tag{8.48}$$

Constraints (8.49) and (8.50) ensure that the inlet and outlet concentrations do not exceed the maximum allowable concentration.

$$cin\left(s_{inj}, c, p\right) \leq cin^{U}\left(s_{inj}, c\right), \quad \forall p \in P, \quad s_{inj} \in S_{inJ}, \quad c \in C \tag{8.49}$$

$$cout\left(s_{inj}, c, p\right) \leq cout^{U}\left(s_{inj}, c\right), \quad \forall p \in P, \quad s_{inj} \in S_{inJ}, \quad c \in C \tag{8.50}$$

Constraint (8.51) is the objective function in terms of profit maximization, with profit defined as the difference between revenue from product, cost of utility, raw material cost, freshwater cost and effluent treatment cost.

$$\max \left(\begin{array}{c} \sum_{s^P} price\left(s^P\right) d\left(s^P\right) - \sum_{p}\sum_{s_{injh}} costcw * cw\left(s_{injh}, p\right) \\[2mm] - \sum_{p}\sum_{s_{injc}} costst * st\left(s_{injc}, p\right) - \sum_{p}\sum_{s_{inj}} costfw * mfw\left(s_{inj}, p\right) \\[2mm] - \sum_{p}\sum_{s_{inj}} costew * mew\left(s_{inj}, p\right) \end{array} \right),$$

$$\forall p, \in P, \quad s_{injh} \in S_{inJh}, \quad s_{injc} \in S_{inJc}, \quad s_{inj} \in S_{inJ} \tag{8.51}$$

Constraint (8.52) defines minimization of energy and wastewater if the product demand is known.

$$
\min \left(
\begin{array}{c}
\sum_{p}\sum_{s_{inj_h}} costcw * cw\left(s_{inj_h}, p\right) + \sum_{p}\sum_{s_{inj_c}} costst * st\left(s_{inj_c}, p\right) \\
+ \sum_{p}\sum_{s_{inj}} costfw * mfw\left(s_{inj}, p\right) + \sum_{p}\sum_{s_{inj}} costew * mew\left(s_{inj}, p\right)
\end{array}
\right),
$$

$$\forall p, pp \in P, \quad s_{inj_h} \in S_{inJ_h}, \quad s_{inj_c} \in S_{inJ_c}, \quad s_{inj} \in S_{inJ} \tag{8.52}$$

8.5 CASE STUDIES

Case studies from published literature were selected to demonstrate the application of the proposed model. The results from the proposed models were obtained using CPLEX 9 as MILP solver and CONOPT 3 as NLP solver in DICOPT interface of GAMS 22.0 and were solved using a 2.4 GHz, 4 GB of RAM, Acer TravelMate 5740G computer.

8.5.1 CASE STUDY I

This case study has been investigated extensively in published literature (Halim and Srinivasan, 2011). It is a simple batch plant requiring only one raw material to yield a product as depicted in the state task network (STN) representation in Figure 8.4. The plant comprises of five units and two intermediate storage units. The conversion of the raw material into product is achieved through three sequential processes. The first task can be performed in two units ($j1$ and $j2$), the second task can be performed only in unit $j3$ and the third task can be performed in units $j4$ and $j5$. Tasks 1 and 2 require cooling during their operation, while task 3 requires heating. The cooling and heating demands are satisfied by external utilities and heat integration. The operational philosophy requires that the units are cleaned before the next batch is processed. Both freshwater and reuse water can be used as cleaning agents. Table 8.1 gives the capacities of the units, durations of processing and washing tasks, initial availability of states, storage capacities and selling prices and costs for the states. Table 8.2 gives data pertaining to initial and target temperatures for the tasks, specific heat capacities for the states, maximum inlet and outlet contaminant concentrations which are unit dependent and the specific contaminant loads.

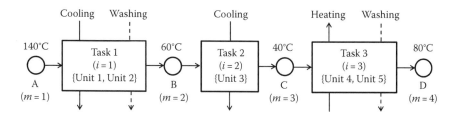

FIGURE 8.4 STN representation of a simple batch plant producing one product.

TABLE 8.1

Scheduling Data for Case Study I

Task (i)	Unit (j)	Max Batch Size (kg)	Total Operation Time (h)	Washing Time (h)	Material State (m)	Initial Inventory (kg)	Max Storage (kg)	Revenue or Cost ($/kg or $/MJ)
Task 1	Unit 1	100	1.5	0.25	A	1000	1000	0
	Unit 2	150	2	0.3	B	0	200	0
Task2	Unit 3	200	1.5	0	C	0	250	0
Task3	Unit 4	100	1	0.25	D	0	1000	5
	Unit 5	150	1.5	0.3	Wash water			0.1
					Wastewater			0.05
					Cooling water			0.02
					Steam			1

Note: Total operation time includes processing time and washing time.

TABLE 8.2

Energy and Cleaning Requirements for Case Study I

Task (i)	T^{in}(°C)	T^{out}(°C)	Unit (j)	C_p(kJ/kg°C)	Max Inlet Concentration (ppm)	Max Outlet Concentration (ppm)	Contaminant Loading (g Contaminant/kg Batch)
Task 1	140	60	Unit 1	4	500	1000	0.2
			Unit 2	4	50	100	0.2
Task 2	60	40	Unit 3	3.5	—	—	0.2
Task 3	40	80	Unit 4	3	150	300	0.2
			Unit 5	3	300	2000	0.2
Cooling water	20	30					
Steam	170	160					

8.5.1.1 Results and Discussion

The computational results for Case Study I using the proposed model for the different scenarios and results obtained from literature are presented in Table 8.3. For the scenario without energy and water integration, the total cost of utilities was $293.5. Applying only water integration, the total cost obtained was $259.5, which is an 11.6% reduction, compared to the standalone operation without energy and water integration. For the scenario with energy integration only, a total cost of $208 was obtained, which is a 29.1% reduction compared to the standalone operation. The fifth column shows the results obtained with combined energy and water integration solved simultaneously giving a total cost of $191.8, which is a 34.6% saving compared to the standalone operation. These results show that in order to achieve the best economic performance, the scheduling problem has to be solved simultaneously considering both water and energy integration.

The performance of the proposed model was compared to the sequential optimization technique by Halim and Srinivasan (2011) which resulted in an overall cost of $239.5, which is an 18.4% saving, much less than the 34.6% saving obtained by the proposed model. This work also gives much better result compared to the recent simultaneous optimization technique of Adekola et al. (2013) with a cost saving of 23.4% compared to 34.6% saving obtained by the proposed model. The suboptimality results of the method by Adekola et al. (2013) are attributed to two basic drawbacks. The first drawback is due to restricting the flexibility of the schedule by forcing the heat integrated units to start at the same time. The second drawback is the model is not based on TAM and the possibility of heat integration between pairs of tasks as well as possible ΔT violations was investigated for each pair of hot and cold tasks before optimization using the initial and target temperatures of the heat integrated tasks. This limits the chance of a unit to be integrated in multiple intervals with different intermediate temperatures with other units. Using the proposed model, we keep the schedule flexibility by allowing the heat integrated units to start anywhere between the starting and finishing time of the heat integrated tasks. This benefit can be demonstrated in Figure 8.5, for example, Unit 2 during processing a task from 3.2 to 4.9 h is integrated to exchange heat with Unit 4 during processing a task from 4.25 to 5 h. These two units are integrated from 4.25 to 4.9 h to exchange heat which is not possible by the method of Adekola et al. (2013). Consequently, this work reduces the steam requirement by 40.9% compared to technique by Adekola et al. (2013). The efficiency of the proposed model can be attributed to solving the scheduling problem while incorporating water and energy integration in the same framework and also using the recent efficient scheduling technique by Seid and Majozi (2012). Figure 8.5 details the possible amount of energy integration between the cold and hot units and the time intervals during which energy integration occurred.

The energy requirements of unit $j2$ and unit $j4$ during the interval 3.2–5 h are emphasized to elaborate on the application of the proposed model. The cooling load of unit $j2$ between 3.2 and 4.9 h was 32 MJ. This is partly satisfied through energy integration with unit $j4$ in the same time interval, resulting in an external cooling requirement of 26.8 MJ rather than 32 MJ if operated in standalone mode. At the

TABLE 8.3
Computational Results for Case Study I

	Proposed Formulation without Water and Energy Integration	Proposed Formulation with Water Integration	Proposed Formulation with Energy Integration	Proposed Formulation with Water and Energy Integration	Halim and Srinivasan (2011) with Water and Energy Integration	Adekola et al. (2013) with Water and Energy Integration
Profit ($)	4706.5	4740.5	4791.5	4808.2	4764.1	4777.3
Steam (MJ)	120	120	36.63	39	43.9	66
Cooling water (MJ)	390	390	281.2	309	313.9	336
Total freshwater (kg)	1105	878.2	1105	977.7	1238.4	1013.3
Revenue from product ($)	5000	5000	5000	5000	5000	5000
Cost of steam ($)	120	120	36.63	39	43.9	66
Cost of cooling water ($)	7.8	7.8	5.623	6.2	6.3	6.72
Cost of freshwater ($)	110.5	87.8	110.5	97.7	123.8	101.3
Cost of wastewater ($)	55.25	43.9	55.2	48.9	61.9	50.7
Total cost ($)	293.5	259.5	208	191.8	235.9	224.72
CPU time (s)	2.3	5000	5000	5000	Not reported	28797

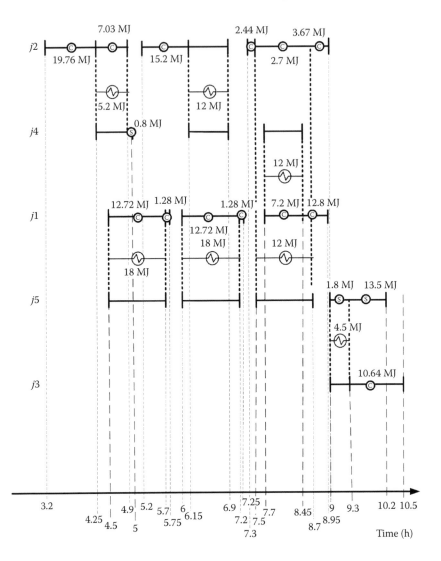

FIGURE 8.5 Possible energy integration within the time horizon of 12 h for Case Study I.

beginning of the operation of unit $j2$ from 3.2 to 4.25 h, the cooling requirement was 19.76 MJ. This value was obtained using the time average model by multiplying the duration (4.25–3.2 h) and the energy demand per hour (32 MJ/1.7 h (total duration of the task) = 18.823 MJ) where the cooling requirement is fully satisfied by external cooling. For the rest of its operation between 4.25 and 4.9 h, the cooling requirement was 12.24 MJ, satisfied partly with energy integration (5.2 MJ) and the difference by external cooling. The heating requirement of unit $j4$ when it is operated during the interval 4.25–5 h was 6 MJ. From 4.25 to 4.9 h, the steam requirement was 5.2 MJ obtained from the time average model. This heating requirement was fully satisfied

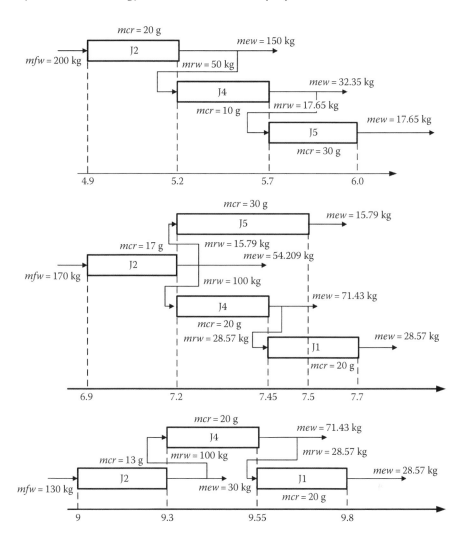

FIGURE 8.6 Water network with water integration within the time horizon of 12 h for Case Study I. *Note*: mfw, freshwater; mcr, contaminant removed; mrw, recycled water; mew, water sent to effluent.

during the interval, by integrating with the hot unit *j*2. The rest of the heating, 0.8 MJ, required during its operation between 4.9 and 5 h was satisfied by external steam.

Figure 8.6 shows the amount of contaminant removed, freshwater usage, amount of reused water and wastewater produced from washing the necessary units. The washing operation of unit *j*2 between 4.9 and 5.2 h required 200 kg of freshwater to remove a contaminant load of 20 g, producing water with a contaminant concentration of 100 ppm. Part of this water produced from unit *j*2, 50 kg, was used for cleaning unit *j*4 to remove a contaminant load of 10 g. This was possible because the outlet concentration from unit *j*2 (100 ppm) was lower than the maximum inlet contaminant

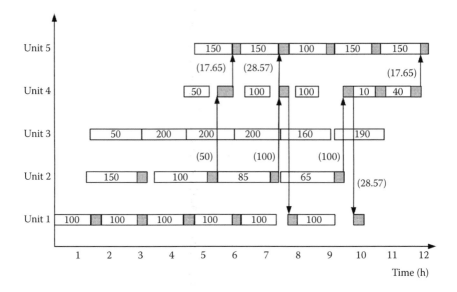

FIGURE 8.7   Gantt chart for the time horizon of 12 h incorporating energy and water integration for Case Study I.

concentration (150 ppm) for unit $j4$. From Figure 8.6, the total amount of reused water was 358.23 kg, thereby reducing the water usage from 1105 kg (without water integration) to 977.7 kg (with water integration). This resulted in a saving of 11.5% freshwater usage and wastewater produced.

The amount of material produced, the starting and finishing times of the processes and washing tasks are shown in Figure 8.7 in the form of a Gantt chart.

8.5.2 Case Study II

This case study obtained from Kondili et al. (1993) has become one of the most commonly used examples in literature. However, this case study has been adapted by Halim and Srinivasan (2011) to include energy and water integration. The batch plant produces two different products sharing the same processing units, where Figure 8.8 shows the plant flowsheet. The unit operations consist of preheating, three different reactions and separation. The plant accommodates many common features of multipurpose batch plants such as units performing multiple tasks, multiple units suitable for a task and dedicated units for specific tasks. The STN and SSN representations of the flowsheet are shown in Figure 8.9. Tables 8.4 and 8.5 give the required data to solve the scheduling problem. The production recipe is as follows:

1. Raw material, Feed A, is heated from 50°C to 70°C to form HotA used in reaction 2.
2. Reactant materials, 50% Feed B and 50% Feed C are used in reaction 1 to produce IntBC. During the reaction, the material has to be cooled from 100°C to 70°C.

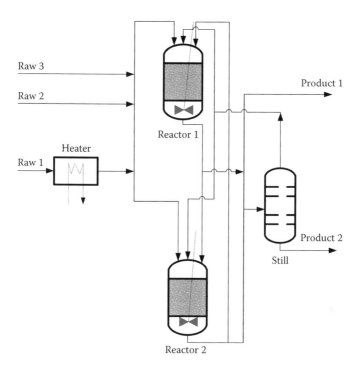

Raw 3

Raw 2

Product 1

Reactor 1

Heater

Raw 1

Product 2

Still

Reactor 2

FIGURE 8.8   Flowsheet for Case Study II.

3. 60% of the intermediate material, IntBC, and 40% of HotA are used in reaction 2 to produce product 1 and IntAB. The process needs to be heated from 70°C to 100°C during its operation.

4. 20% of the reactant, Feed C, and 80% of intermediate, IntAB, from reaction 2 are used in reaction 3 to produce ImpureE. The reaction needs its temperature to be raised from 100°C to 130°C during its operation.

5. The separation process produces 90% product 2 and 10% IntAB from Impure E. Cooling water is used to lower its temperature from 130°C to 100°C.

The processing time of a task i in unit j is assumed to be linearly dependent on its batch size B, that is, $\alpha_i + \beta_i B$. Where α_i is a constant term of the processing time of task i and β_i is a coefficient of variable processing time of task i. The batch-dependent processing time makes this case study more complex. Table 8.4 gives the relevant data on coefficients of processing times, the capacity of the processing units, duration of washing, initial inventory of raw materials, storage capacity and relevant costs. Four contaminants are considered in the case study. The maximum inlet and outlet concentrations are given in Table 8.5. The production demand is given as 200 kg for both Prod1 and Prod2. The objective here is to optimize with respect to makespan, energy and water consumption.

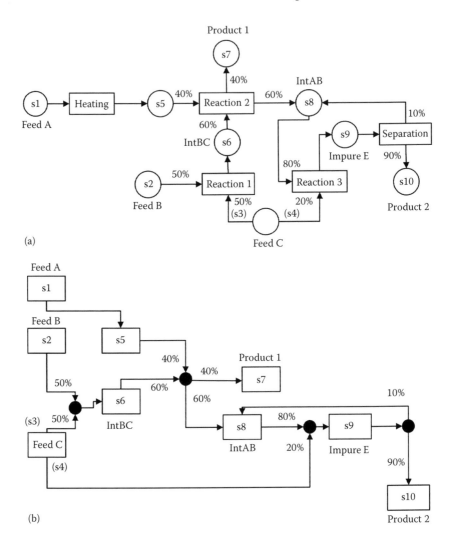

FIGURE 8.9 STN (a) and (b) SSN representations for Case Study II.

8.5.2.1 Results and Discussion

The computational statistics for this case study using the proposed model and results obtained from literature are presented in Table 8.6. For makespan minimization, an objective value of 19.5 h was obtained using the proposed model, which is better than 19.96 h obtained by Halim and Srinivasan (2011) and 19.93 obtained by Adekola et al. (2013). Using the makespan obtained, the case study was solved using the different scenarios for water minimization, energy minimization and the simultaneous minimization of energy and water by setting customer requirement for Product 1 and Product 2. The total energy and freshwater required for the standalone operation were 125.5 MJ and 357.94 kg, respectively.

TABLE 8.4

Scheduling Data for Case Study II

Task (i)	Unit (j)	Max Batch Size (kg)	$\alpha(S_{inj})$	$\beta(S_{inj})$	Washing Time	Material State (s)	Initial Inventory	Max Storage (kg)	Revenue or Cost ($/kg or $/MJ)
Heating (H)	HR	100	0.667	0.007	0	Feed A	1000	1000	0
Reaction-1 (R1)	RR1	50	1.334	0.027	0.25	Feed B	1000	1000	0
	RR2	80	1.334	0.017	0.3	Feed C	1000	1000	0
Reaction-2 (R2)	RR1	50	1.334	0.027	0.25	HotA	0	100	0
	RR2	80	1.334	0.017	0.3	IntAB	0	200	0
Reaction-3 (R3)	RR1	50	0.667	0.013	0.25	IntBC	0	150	0
	RR2	80	0.667	0.008	0.3	ImpureE	0	200	0
Separation (S)	SR	200	1.334	0.007	0	Prod 1	0	1000	20
						Prod2	0	1000	20
						Wash water			0.1
						Wastewater			0.05
						Cooling water			0.02
						Steam			1

TABLE 8.5
Data Required for Energy and Water Integration

Task (i)	T^{in}_{Sinj} (°C)	T^{in}_{Sinj} (°C)	Unit (j)	C_p (kJ/kg °C)	Max Inlet Concentration (ppm)				Max Outlet Concentration (ppm)				Contaminants (ar, br, cp and dw) Loading (g Contaminant/kg Batch)
					ar	br	cp	dw	ar	br	cp	dw	
Heating (H)	50	70	HR	2.5									
Reaction-1	100	70	RR1	3.5	300	500	800	400	700	800	1200	900	0.2
			RR2	3.5	300	500	800	400	700	800	1200	900	0.2
Reaction-2	70	100	RR1	3.2	700	600	300	400	1200	1000	600	800	0.2
			RR2	3.2	700	600	300	400	1200	1000	600	800	0.2
Reaction-3	100	130	RR1	2.6	500	200	400	300	800	500	700	900	0.2
			RR2	2.6	500	200	400	300	800	500	700	900	0.2
Separation	130	100	SR	2.8									
Cooling water	20	30											
Steam	170	160											

TABLE 8.6

Computational Results for Case Study II

	Proposed Formulation without Water and Energy Integration	Proposed Formulation with Water Integration	Proposed Formulation with Energy Integration	Proposed Formulation with Water and Energy Integration	Halim and Srinivasan (2011) with Water and Energy Integration	Adekola et al. (2013) with Water and Energy Integration
Makespan (h)	19.5	19.5	19.5	19.5	19.96	19.93
Objective ($)	127.5	112	96.4	94.3	103.3	96.5
Steam (MJ)	75.3	75.3	44.9	43.3	61.4	44.88
Cooling water (MJ)	50.2	50.2	19.7	18.1	35.4	19.72
Total freshwater (kg)	357.94	238.1	341.3	337.7	275.1	341.2
Revenue from product ($)	8000	8000	8000	8000	8000	8000
Cost of steam ($)	75.3	75.3	44.9	43.3	61.4	44.88
Cost of cooling water ($)	1	1	0.4	0.36	0.7	0.3994
Cost of freshwater ($)	35.8	23.8	34.1	33.8	27.5	34.12
Cost of wastewater ($)	17.9	11.9	17.1	16.9	13.8	17.06
Number of time points/ slots	11	11	11	11	N/A	17
CPU time (s)	5000	5000	5000	6074	Not reported	24,532

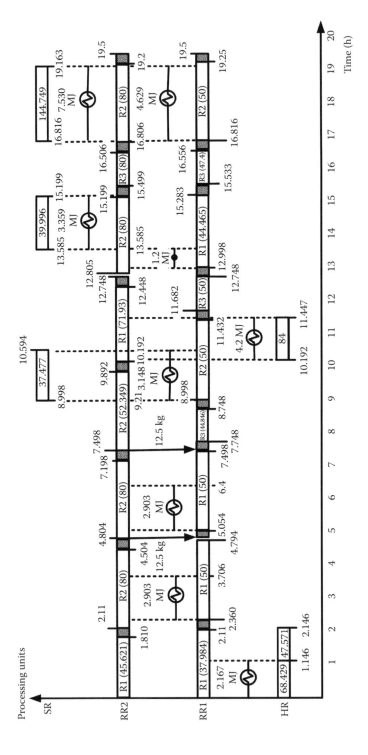

FIGURE 8.10 Resulting production schedule for Case Study II with direct heat integration and direct water reuse.

For the scenario of water integration only allowing the use of reuse water, the total cost was $112, resulting in 12.2% saving when compared to the standalone operation which had a total cost of $127.52. By using only energy integration, the total energy requirement was reduced from 125.5 MJ in standalone operation to 64.56 MJ, resulting in a 48.6% energy saving and a total cost saving of 24.4%. For the case of simultaneous optimization of energy and water, a significant total cost saving was obtained compared to energy integration alone and water integration alone. A total cost saving of 29.4% was obtained, compared to the standalone operation. The performance of the proposed model was also compared to the technique by Halim and Srinivasan (2011), a total cost of $103 was found using their technique which is significantly higher than $94.3 obtained using the proposed model. Furthermore, the proposed technique is very easy to adopt as opposed to their approach which required to solve 3500 MILP scheduling problem to find the best schedule compared to only 3 MILP major iterations of the MINLP problem. Each MILP problem is solved in a specified CPU time of 2000 s. This complex case study was solved in a reasonable CPU time of 6074 s, which is less than 2 h, using the proposed model. When this work is compared to the model by Adekola et al. (2013), the number of event points required reduced considerably from 17 to 11 which have a direct effect on reducing CPU time required. Additionally, the usage of hot and cold utilities, freshwater and wastewater is also improved.

Figure 8.10 shows the Gantt chart related to the optimal usage of energy and water. It also indicates the types of tasks performed in each equipment, the starting and finishing times of the processes and washing tasks and the amount of material processed in each batch.

8.6 CONCLUSIONS

In the presented method, wastewater minimization and heat integration are both embedded within the scheduling framework and solved simultaneously, thus leading to a truly flexible process schedule. Results from case studies show that addressing profit maximization together with heat integration and wastewater minimization gives much better overall economic performance. From the case studies, a better objective value was achieved using the proposed model compared to previous literature models. Forthcoming communications will address the usage of heat storage, wastewater storage and wastewater regenerator with the consideration of capital investment to investigate further improvement in energy and water usage. Although this invariably complicates the model formulation. Additionally, this work only addressed short-term scheduling problem. Extending this work to medium-term scheduling problem using a cyclic approach will be reported in future communication.

NOMENCLATURE

SETS

S_{inJ_h} $\{s_{inj_h}|s_{inj_h}$ task which needs cooling$\}$
S_{inJ_c} $\{s_{inj_c}|s_{inj_c}$ task which needs heating$\}$

S_{inJ} $\{s_{inj}|s_{inj}$ any task$\}$
P $\{p|p$ time point$\}$
S_{inJ_w} $\{s_{inj_w}|s_{inj_w}$ task which needs washing afterwards$\}$
C $\{c|c$ contaminant$\}$

PARAMETERS

$cp\left(s_{inj_h}\right)$	Specific heat capacity for the heating task
$cp\left(s_{inj_c}\right)$	Specific heat capacity for the cooling task
$T^{in}_{s_{inj_h}}$	Inlet temperature of the heating task
$T^{out}_{s_{inj_h}}$	Outlet temperature of the heating task
$T^{in}_{s_{inj_c}}$	Inlet temperature of the cooling task
$T^{out}_{s_{inj_c}}$	Outlet temperature of the cooling task
ΔT^U	Maximum thermal driving force
$\Delta T^U\left(s_{inj}\right)$	Maximum temperature change for a task
ΔT	Minimum thermal driving force
M	Big-M mostly equivalent to the time horizon
Q^U	Maximum heat requirement from the heating and cooling task
$SMC\left(s_{inj}\right)$	Specific contaminant load produced by a task
$W^U_{in}\left(s_{inj}\right)$	Maximum water inlet to a processing task
$\tau w\left(s_{inj}\right)$	Minimum duration required for a washing task
$cin^U\left(s_{inj},c\right)$	Maximum inlet contaminant concentration allowed for contaminant c
$cout^U\left(s_{inj},c\right)$	Maximum outlet contaminant concentration allowed for contaminant c
$price\left(s^p\right)$	Price of a product
$d\left(s^p\right)$	Amount of product produced at the end of the time horizon
H	Time horizon of interest
$costfw$	Cost of freshwater
$costew$	Cost of effluent water
$costst$	Cost of steam
$costcw$	Cost of cooling water

VARIABLES

$x\left(s_{inj_c}, s_{inj_h}, p, pp\right)$	Binary variable signifying whether heat integration occurs between the hot and cold unit
$y\left(s_{inj_h}, p\right)$	Binary variable associated to whether the hot state is active at time point p or not
$y\left(s_{inj_c}, p\right)$	Binary variable associated to whether the cold state is active at time point p or not

$y_{\text{int}}\left(s_{inj}, p, pp\right)$	Binary variable associated to whether the hot and cold states are active at time point p and extended time point pp
$y_{re}\left(s_{inj}, s_{inj'}, p\right)$	Binary variable associated with reuse of water from unit j to j' at time point p
$cl\left(s_{inj_h}, p\right)$	Cooling load required by the hot task at time point p
$hl\left(s_{inj_c}, p\right)$	Heating load required by the cold task at time point p
$avcl\left(s_{inj_h}, p\right)$	Average cooling load required by the hot task at time point p using time average model
$avhl\left(s_{inj_c}, p\right)$	Average heating load required by the cold task at time point p using time average model
$mu\left(s_{inj_h}, p\right)$	Amount of material processed by the hot task
$mu\left(s_{inj_c}, p\right)$	Amount of material processed by the cold task
$tp\left(s_{inj}, p\right)$	End time of a heat flow for a task
$tu\left(s_{inj}, p\right)$	Starting time of a heat flow for a task
$clp\left(s_{inj_h}, p, pp\right)$	Cooling load required by the hot task active at time point p and extended time point pp
$hlp\left(s_{inj_c}, p, pp\right)$	Heating load required by the cold task active at time point p and extended time point pp
$tuu\left(s_{inj}, p, pp\right)$	Starting time of a heat flow for a task active at time point p and extended time point pp
$tpp\left(s_{inj}, p, pp\right)$	Finishing time of a heat flow for a task active at time point p and extended time point pp
$T^{in}\left(s_{inj}, p, pp\right)$	Inlet temperature of a task active at time point p and extended time point pp
$T^{out}\left(s_{inj}, p, pp\right)$	Outlet temperature of a task active at time point p and extended time point pp
$qe\left(s_{inj_c}, s_{inj_h}, p, pp\right)$	Amount of heat load exchanged by the hot and cold unit active at time point p and extended time point pp
$cw\left(s_{inj_h}, p\right)$	External cooling water used by the hot task
$st\left(s_{inj_c}, p\right)$	External heating used by the cold task
$mwin\left(s_{inj}, p\right)$	Mass of water entering to wash a unit after a task is performed
$mwout\left(s_{inj}, p\right)$	Mass of water leaving after washing
$mfw\left(s_{inj}, p\right)$	Mass of freshwater entering to a unit
$mrw\left(s_{inj}, s_{inj'}, p\right)$	Mass of water recycled from unit j to another unit j'
$mew\left(s_{inj}, p\right)$	Mass of water entering to effluent treatment produced from washing
$cin\left(s_{inj}, c, p\right)$	Inlet contaminant concentration at time point p
$cout\left(s_{inj}, c, p\right)$	Outlet contaminant concentration at time point p
$tuw\left(s_{inj}, p\right)$	Starting time of washing operation for unit j
$tpw\left(s_{inj}, p\right)$	Finishing time of washing operation for unit j

REFERENCES

Adekola, O., Majozi, T., 2011. Wastewater minimization in multipurpose batch plants with a regeneration unit: Multiple contaminants. *Computers and Chemical Engineering.* 35, 2824–2836.

Adekola, O., Stamp, J., Majozi, T., Garg, A., Bandyopadhyay, S., 2013. Unified approach for the optimization of energy and water in multipurpose batch plants using a flexible scheduling framework. *Industrial and Engineering Chemistry Research.* 52, 8488–8506.

Adonyi, R., Romero, J., Puigjaner, L., Friedler, F., 2003. Incorporating heat integration in batch process scheduling. *Applied Thermal Engineering.* 23, 1743–1762.

Almató, M., Sanmartí, E., Espuña, A., Puigjaner, L., 1997. Rationalizing the water use in the batch process industry. *Computers and Chemical Engineering.* 21, S971–S976.

Atkins, M.J., Walmsley, M.R.W., Neale, J.R., 2010. The challenge of integrating noncontinuous processes—Milk powder plant case study. *Journal of Cleaner Production.* 34, 9276–9296.

Bancheva, N., Ivanov, B., Shah, N., Pantelides, C.C., 1996. Heat exchanger network design for multipurpose batch plants. *Computers and Chemical Engineering.* 20, 989–1001.

Barbosa-Póvoa, A.P.F.D., Pinto, T., Novais, A.Q., 2001. Optimal design of heat-integrated multipurpose batch facilities: A mixed-integer mathematical formulation. *Computers and Chemical Engineering.* 25, 547–559.

Becker, H., Vuillermoz, A., Maréchal, F., 2012. Heat pump integration in a cheese factory. *Applied Thermal Engineering.* 43, 118–127.

Boer, R., Smeding, S.F., Bach, P.W., 2006. Heat storage systems for use in an industrial batch process (Results of) a case study. In *10th International Conference on Thermal Energy Storage ECOSTOCK*, Stockton, CA.

Bozan, M., Borak, F., Or, I., 2001. A computerized and integrated approach for heat exchanger network design in multipurpose batch plants. *Chemical Engineering Process.* 40, 511–524.

Castro, P.M., Barbosa-Póvoa, A.P., Matos, H.A., Novais, A.Q., 2004. Simple continuous-time formulation for short-term scheduling of batch and continuous processes. *Industrial and Engineering Chemistry Research.* 43, 105–118.

Chen, C.L., Chang, C.Y., 2009. A resource-task network approach for optimal short-term/periodic scheduling and heat integration in multipurpose batch plants. *Applied Thermal Engineering.* 29, 1195–1208.

Chen, C.L., Ciou, Y.J., 2008. Design and optimization of indirect energy storage systems for batch process plants. *Industrial and Engineering Chemistry Research.* 47, 4817–4829.

Cheng, K.F., Chang, C.T., 2007. Integrated water network designs for batch processes. *Industrial and Engineering Chemistry Research.* 46, 1241–1253.

Chew, Y.H., Lee, C.T., Foo, C.Y., 2005. Evaluating heat integration scheme for batch production of oleic acid. In *Malaysian Science and Technology Congress (MSTC)*, (pp. 18–20).

Erdirik-Dogan, M., Grossmann, I.E., 2008. Slot-based formulation for the short-term scheduling of multistage, multiproduct batch plants with sequence-dependent changeovers. *Industrial and Engineering Chemistry Research.* 47, 1159–1163.

Fernández, I., Renedo, C.J., Pérez, S.F., Ortiz, A., Mañana, M., 2012. A review: Energy recovery in batch processes. *Renewable and Sustainable Energy Reviews.* 16, 2260–2277.

Ferrer-Nadal, S., Capón-Garćia, E., Méndez, C.A., Puigjaner, L., 2008. Material transfer operations in batch scheduling A critical modeling issue. *Industrial and Engineering Chemistry Research.* 47, 7721–7732.

Floudas, C.A., Lin, X., 2004. Continuous-time versus discrete-time approaches for scheduling of chemical processes: A review. *Computers and Chemical Engineering.* 28, 2109–2129.

Foo, D.C.Y., Chew, Y.H., Lee, C.T., 2008. Minimum units targeting and network evolution for batch heat exchanger network. *Applied Thermal Engineering.* 28, 2089–2099.

Foo, D.C.Y., Manan, Z.A., Tan, Y.L., 2005. Synthesis of maximum water recovery network for batch process systems. *Journal of Cleaner Production.* 13, 1381–1394.

Fritzson, A., Berntsson, T., 2006. Efficient energy use in a slaughter and meat processing plant—Opportunities for process integration. *Journal of Food Engineering.* 76, 594–604.

Gouws, J.F., Majozi, T., Foo., D.C.Y., Chen., C.L., Lee, J Y., 2010. Water minimization techniques for batch processes. *Industrial and Engineering Chemistry Research.* 48(19), 8877–8893.

Halim, I., Srinivasan, R., 2009. Sequential methodology for scheduling of heat-integrated batch plants. *Industrial and Engineering Chemistry Research.* 48, 8551–8565.

Halim, I., Srinivasan, R., 2011. Sequential methodology for integrated optimization of energy and water use during batch process scheduling. *Computers and Chemical Engineering.* 35, 1575–1597.

Hellwig, T., Thöne, E., 1994. Omnium: ein verfahren zur optimierung der abwarmenutzung. *BWK (Brennstoff, Warme, Kraft).* 46, 393–397 [in German].

Hui, C.W., Gupta, A., 2000. A novel MILP formulation for short-term scheduling of multi-stage multi-product batch plants. *Computers and Chemical Engineering.* 24, 2705–2717.

Ierapetritou, M.G., Floudas, C.A., 1998. Effective continuous-time formulation for short-term scheduling: 1 Multipurpose batch processes. *Industrial and Engineering Chemistry Research.* 37, 4341–4359.

Janak, S.L., Floudas, C.A., 2008. Improving unit-specific event based continuous time approaches for batch processes: Integrality gap and task splitting. *Computers and Chemical Engineering.* 32, 913–955.

Karimi, I.A., McDonald, C.M., 1997. Planning and scheduling of parallel semicontinuous processes II. Short-term scheduling. *Industrial and Engineering Chemistry Research.* 36, 2701–2714.

Kemp, I.C., Macdonald, E.K., 1987. Energy and process integration in continuous and batch processes Innovation in process energy utilization. *IChemE Symposium Series.* 105, 185–200.

Kim, J.K., Smith, R., 2004. Automated design of discontinuous water systems. *Processing Safety Environment.* 82, 238–248.

Kondili, E., Pantelides, C.C., Sargent, R.W.H., 1993. A general algorithm for short-term scheduling of batch operations I. MILP formulation. *Computers and Chemical Engineering.* 17, 211–227.

Krummenacher, P., Favrat, D., 2001. Indirect and mixed direct–indirect heat integration of batch processes based on Pinch Analysis. *International Journal of Applied Thermodynamics.* 4, 135–143.

Li, J., Susarla, N., Karimi, I.A., Shaik, M., Floudas, C.A., 2010. An analysis of some unit-specific event-based models for the short-term scheduling of non-continuous processes II. Short-term scheduling. *Industrial and Engineering Chemistry Research.* 49, 633–647.

Lim, M.F., Karimi, I.A., 2003. Resource-constrained scheduling of parallel production lines using asynchronous slots. *Industrial and Engineering Chemistry Research.* 42, 6832–6842.

Liu, Y., Karimi, I.A., 2007. Scheduling multistage, multiproduct batch plants with non identical parallel units and unlimited intermediate storage. *Chemical Engineering Science.* 62, 1549–1566.

Liu, Y., Karimi, I.A., 2008. Scheduling multistage batch plants with parallel units and no interstage storage. *Computers and Chemical Engineering.* 32, 671–693.

Majozi, T., 2006. Heat integration of multipurpose batch plants using a continuous-time framework. *Applied Thermal Engineering.* 26, 1369–1377.

Majozi, T., 2009. Minimization of energy use in multipurpose batch plants using heat storage: An aspect of cleaner production. *Journal of Cleaner Production.* 17, 945–950.

Majozi, T., Gouws, J.F., 2009. A mathematical optimisation approach for wastewater minimization in multipurpose batch plants: Multiple contaminants. *Computers and Chemical Engineering.* 33, 1826–1840.

Majozi, T., Zhu, X.X., 2001. A novel continuous-time MILP Formulation for multipurpose bach plants. *Industrial and Engineering Chemistry Research.* 40, 5935–5949.

Maravelias, C.T., Grossmann, I.E., 2003. New general continuous-time state-task network formulation for short-term scheduling of multipurpose batch plants. *Industrial and Engineering Chemistry Research.* 42, 3056–3074.

Méndez, C.A., Cerdá, J., 2000. Optimal scheduling of a resource-constrained multiproduct batch plant supplying intermediates to nearby end product facilities. *Computers and Chemical Engineering.* 24, 369–376.

Méndez, C.A., Cerdá, J., 2003. An MILP continuous-time framework for short-term scheduling of multipurpose batch processes under different operation strategies. *Optical Engineering.* 4, 7–22.

Méndez, C.A., Cerdá, J., Grossmann, I.E., Harjunkoski, I., Fahl, M., 2006. State of-the-art review of optimization methods for short-term scheduling of batch processes. *Computers and Chemical Engineering.* 30, 913–946.

Méndez, C.A., Henning, G.P., Cerdá, J., 2001. An MILP continuous-time approach to short-term scheduling of resource-constrained multistage flowshop batch facilities. *Computers and Chemical Engineering.* 25, 701–711.

Morrison, A.S., Walmsley, M.R.W., Neale, J.R., Burrell, C.P., Kamp, P.J.J., 2007. Non-continuous and variable rate processes: Optimisation for energy use. *Asia-Pacific Journal of Chemical Engineering.* 5, 380–387.

Muster-Slawitsch, B., Weiss, W., Schnitzer, H., Brunner, C., 2011. The green brewery concept Energy efficiency and the use of renewable energy sources in breweries. *Applied Thermal Engineering.* 31, 2123–2134.

Papageorgiou, L.G., Shah, N., Pantelides, C.C., 1994. Optimal scheduling of heat-integrated multipurpose plants. *Industrial and Engineering Chemistry Research.* 33, 3168–3186.

Pinto, J.M., Grossmann, I.E., 1994. Optimal cyclic scheduling of multistage continuous multiproduct plants. *Computers and Chemical Engineering.* 1994, 18, 797–816.

Pinto, T., Novais, A.Q., Barbosa-Póvoa, A.P.F.D., 2003. Optimal design of heat-integrated multipurpose batch facilities with economic savings in utilities: A mixed integer mathematical formulation. *Annals of Operation Research.* 120, 201–230.

Pires, A.C., Fernandes, C.M., Nunes, C.P., 2005. An energy integration tool for batch process, sustainable development of energy, water and environment systems. In *Proceedings of the third Dubrovnik Conference*, pp. 5–10.

Reddy, P.C.P., Karimi, I.A., Srinivasan, R., 2004. A new continuous-time formulation for scheduling crude oil operations. *Chemical Engineering Science.* 59, 1325–1341.

Schilling, G., Pantelides, C., 1996. A simple continuous-time process scheduling formulation and a novel solution algorithm. *Computers and Chemical Engineering.* 20, 1221–1226.

Seid, R., Majozi, T., 2012. A robust mathematical formulation for multipurpose batch plants. *Chemical Engineering Science.* 68, 36–53.

Shaik, M., Floudas, C., 2009. Novel unified modeling approach for short term scheduling. *Industrial and Engineering Chemistry Research.* 48, 2947–2964.

Shaik, M.A., Janak, S.L., Floudas, C.A., 2006. Continuous-time models for short-term scheduling of multipurpose batch plants: A comparative study. *Industrial and Engineering Chemistry Research.* 45, 6190–4209.

Stamp, J., Majozi, T., 2011. Optimal heat storage design for heat integrated multipurpose batch plants. *Energy*. 36(8), 1–13.

Sundaramoorthy, A., Karimi, I.A., 2005. A simpler better slot-based continuous-time formulation for short-term scheduling in multipurpose batch plants. *Chemical Engineering Science*. 60, 2679–2702.

Susarla, N., Li, J., Karimi, I.A., 2010. A novel approach to scheduling of multipurpose batch plants using unit slots. *AICHE Journal*. 56, 1859–1879.

Tokos, H., Pintarič, Z.N., Glavič, P., 2010. Energy saving opportunities in heat integrated plant retrofit. *Applied Thermal Engineering*. 30, 36–44.

Uhlenbruck, S., Vogel, R., Lucas, K., 2000. Heat integration of batch processes. *Chemical Engineering and Technology*. 23, 226–229.

Vaklieva-Bancheva, N., Ivanov, B.B., Shah, N., Pantelides, C.C., 1996. Heat exchanger network design for multipurpose batch plants. *Computers and Chemical Engineering*. 20, 989–1001.

Wang, Y.P., Smith, R., 1994. Wastewater minimization. *Chemical Engineering Science*. 49(7), 981–1006.

9 Targeting for Long-Term Time Horizons
Water Optimization

9.1 INTRODUCTION

Most of the methodologies published in literature on wastewater minimization for batch processes are based on short-term scheduling techniques. When these methods are applied to longer time horizons, the computational time becomes intractable, hence the focus of this chapter. This chapter presents a methodology for simultaneous optimization of production schedule and wastewater minimization in a multipurpose batch facility. The key feature of the presented methodology is the adaption of cyclic scheduling concepts to wastewater minimization. The methodology is developed based on continuous-time formulation and the state sequence network (SSN) representation. The methodology is successfully applied to two common literature examples and an industrial case study to demonstrate its effectiveness. None of the currently published wastewater minimization techniques could solve the case study for a time horizon of 168 h. However, through the application of the presented methodology, a time horizon of 168 h for the case study was reduced to eight cycles with the cycle length of 23 h, for which the CPU time for the optimum cycle is 64.53 s.

9.2 PROBLEM STATEMENT

The problem considered in this chapter can be formally stated as follows:

For each water-using operation

Given:

1. The production recipe for each product
2. The available units and their capacities
3. The mass load, maximum inlet and outlet concentration for each contaminant
4. Water requirement and the cleaning duration for each unit to achieve the required cleanliness
5. The maximum storage available for water reuse
6. Time horizon of interest

Determine:

The optimal production schedule which will generate the minimum amount of wastewater through reuse and recycle opportunities. *Reuse* refers to the use of an outlet water stream from operation *j* in another operation *j'*, whereas *recycle* refers to the use of an outlet water stream from operation *j* in the same operation *j*. It is worthy of note that the minimization of wastewater generation is concomitant with reduction in freshwater consumption. The absolute minimum wastewater generation can be achieved through a flexible rather than fixed schedule of the operations. Thus, time is treated as a variable instead of a parameter in this chapter.

9.3   MATHEMATICAL FORMULATION

The concept of cyclic scheduling and the proposed mathematical formulation are presented in this section.

9.3.1   CONCEPTS OF CYCLIC SCHEDULING

The concept of cyclic scheduling is based on the following axiom, which was originally proposed by Shah et al. (1993):

> *Consider a case where the time horizon (H) is much longer when compared to the duration of the individual tasks, a sub-schedule exists with a much smaller time horizon (T), periodic execution of which achieves production very close to the optimal production of the original longer time horizon (H).*

The axiom is depicted in Figure 9.1, the optimal cycle length being *T* on the diagram.

From Figure 9.1, it is clear that some of the tasks cross the boundary of the optimal cycle length *T*. A task that has such an effect can be viewed as a task extended past the cycle of interest notionally *wrapping around* to the beginning of the cycle. This notion was introduced in the work by Shah et al. (1995) and it is depicted in Figure 9.2 by the tasks in units 2 and 3.

The key feature of the presented methodology is to adapt these concepts of cyclic scheduling into wastewater minimization for batch processes.

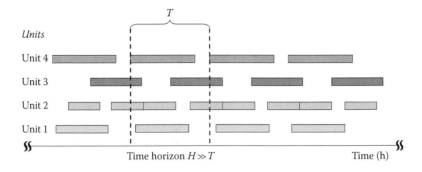

FIGURE 9.1   The original time horizon *H* is much greater than the optimal cycle length *T*.

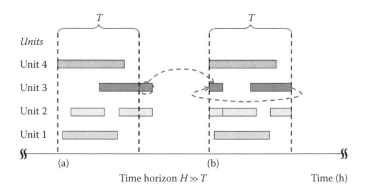

FIGURE 9.2   (a) Task 2 in unit 2 overlaps over the cycle. (b) Task 2 is wrapped around the beginning of the cycle.

9.3.2 Nomenclature

All the sets, variables and parameters present in the formulation are listed here.

9.3.2.1   Sets

P $\{p|p = \text{time point}\}$

J $\{j|j = \text{unit}\}$

C $\{c|c = \text{contaminant}\}$

S_{in} $\{s_{in}|s_{in} = \text{input state into any unit}\}$

S_{out} $\{s_{out}|s_{out} = \text{output state from any unit}\}$

S $\{s|s = \text{any state}\} = S_{in} \cup S_{out}$

$S_{in,j}$ $\{s_{in,j}|s_{in,j} = \text{input state into unit } j\} \subseteq S_{in}$

$S_{in,j}^{*}$ $\{s_{in,j}^{*}|s_{in,j}^{*} = \text{input state into unit } j\} \subseteq S_{in}$

$S_{out,j}$ $\{s_{out,j}|s_{out,j} = \text{output state from unit } j\} \subseteq S_{out}$

9.3.2.2   Variables Associated with Wastewater Minimization

$mw_{in}\left(s_{out,j}, p\right)$	Mass of water into unit j for cleaning state s_{out} at time point p
$mw_{out}\left(s_{out,j}, p\right)$	Mass of water produced at time point p from unit j
$mw_{f}\left(s_{out,j}, p\right)$	Mass of freshwater into unit j at time point p
$mw_{e}\left(s_{out,j}, p\right)$	Mass of effluent water from unit j at time point p
$mw_{r}\left(s_{out,j}, s_{out,j'}, p\right)$	Mass of water recycled to unit j' from j at time point p
$ms_{in}\left(s_{out,j}, p\right)$	Mass of water transferred from unit j to storage at time point p
$ms_{out}\left(s_{out,j}, p\right)$	Mass of water transferred from storage to unit j at time point p
$c_{in}\left(s_{out,j}, c, p\right)$	Inlet concentration of contaminant c, to unit j at time point p
$c_{out}\left(s_{out,j}, c, p\right)$	Outlet concentration of contaminant c, from unit j at time point p
$cs_{in}\left(s_{out,j}, c, p\right)$	Inlet concentration of contaminant c, to storage at time point p

$cs_{out}\left(s_{out,j}, c, p\right)$	Outlet concentration of contaminant c, from storage at time point p
$qw_s\left(p\right)$	Amount of water stored in storage at time point p
$ts_{in}\left(s_{out,j}, p\right)$	Time at which water is transferred from unit j to storage at time point p
$ts_{out}\left(s_{out,j}, p\right)$	Time at which water is transferred from storage to unit j at time point p
$tw_{in}\left(s_{out,j}, p\right)$	Time that the water is used at time point p in unit j
$tw_{out}\left(s_{out,j}, p\right)$	Time at which water is produced at time point p from unit j
$tw_r\left(s_{out,j}, s_{out,j'}, p\right)$	Time at which water is recycled from unit j to unit j' at time point p
$yw_r\left(s_{out,j}, s_{out,j'}, p\right)$	Binary variable showing usage of recycle from unit j to unit j' at time point p
$yw\left(s_{out,j}, p\right)$	Binary variable showing the usage of unit j at time point p
$ys_{in}\left(s_{out,j}, p\right)$	Binary variable showing transfer of water from unit j to storage at time point p
$ys_{out}\left(s_{out,j}, p\right)$	Binary variable showing transfer of water from storage to unit j at time point p

9.3.2.3 Variables Associated with Production Scheduling

$t_{out}\left(s_{out,j}, p\right)$	Time at which a state is produced from unit j at time point p
$t_{in}\left(s_{in,j}, p\right)$	Time at which a state is used in or enters unit j at time point p
$q_s\left(s, p\right)$	Amount of state s stored at time point p
$m_{out}\left(s_{out,j}, p\right)$	Amount of state produced from unit j at time point p
$m_{in}\left(s_{in,j}, p\right)$	Amount of state used in or enters unit j at time point p
$y\left(s_{in,j}^*, p\right)$	Binary variable associated with usage of state s at time point p
$d\left(s_{out}, p\right)$	Amount of state delivered to customers at time point p
H	Time horizon for a single cycle

9.3.2.4 Parameters Associated with Wastewater Minimization

CE	Cost of effluent water treatment (c.u./kg water)
CF	Cost of freshwater (c.u./kg water)
$M\left(s_{out,j}, c\right)$	Mass load of contaminant c added from unit j to the water stream
$Mw^U\left(s_{out,j}\right)$	Maximum inlet water mass of unit j
$C_{in}^U\left(s_{out,j}, c\right)$	Maximum inlet concentration of contaminant c in unit j
$C_{out}^U\left(s_{out,j}, c\right)$	Maximum outlet concentration of contaminant c from unit j
$\tau w\left(s_{out,j}\right)$	Mean processing time of unit j
Qw_s^0	Initial amount of water in storage
Qw_s^U	Maximum capacity of storage

9.3.2.5 Parameters Associated with Production Scheduling

V_j^U Maximum design capacity of a particular unit j

V_j^L Minimum design capacity of a particular unit j

H^u Upper bound of the cycle time length

$\tau\left(s_{in,j}^*\right)$ Mean processing time for a state

$Q_s^0(s)$ Initial amount of state s stored

$Q_s^U(s)$ Maximum amount of state s stored within the time horizon

CF Interest selling price of product s,s = product

9.3.3 Mathematical Model

As mentioned earlier, the presented mathematical formulation is, in essence, an extension of the published work by Majozi and Gouws (2009). The constraints considered in the mathematical formulation are divided into three modules. The first module deals with the water mass balance constraints, for the case where a central storage vessel for reusable water is available and a case where it is absent. The second module deals with the sequencing and scheduling constraints, that is time, for direct and indirect reuse/recycle of water. The third module, which is presented in detail in the work by Majozi and Zhu (2001), deals with the necessary scheduling of operations for production purposes. More emphasis will be on the new constraints added to cater for cyclic scheduling.

9.3.3.1 Mass Balance Constraints

In this sub-section, all the essential water and contaminant mass balances are for a washing operation presented. It is assumed throughout the formulation that none of the water using operations produces or consumes water, that is, water is conserved during a washing operation.

9.3.3.1.1 Mass Balance Constraints without Storage

The mathematical formulation for multiple contaminant system without storage is based on the superstructure given in Figure 9.3, adopted from the work by Majozi and Gouws (2009). For each water-using operation, water into the unit is a combination of freshwater and water recycled from other units to that unit. Water leaving the unit can either be discarded as effluent or be recycled/reused in other units. It is highly imperative to realize that the water-use and reuse variables shown in Figure 9.3 correspond to the washing operation that follows immediately after the production of a particular state in unit j, that is, $s_{out,j}$. Consequently, these variables are task-specific, which allows scheduling to be readily embedded within the water reuse and recycle framework.

The first constraints considered are the mass balance constraints around unit j. Constraint (9.1) is a water mass balance over the inlet of unit j. The total water into a unit is the sum of all the recycles to the unit and the freshwater into the unit j at time point p. As mentioned earlier, it is assumed that any operation that takes place in unit j does not produce water. Consequently, water is conserved as captured in

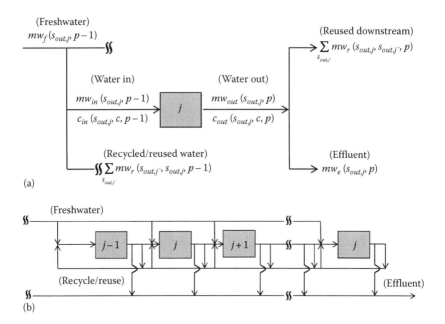

FIGURE 9.3 Superstructure for the mathematical formulation with no reusable water storage (Majozi and Gouws, 2009).

Constraint (9.2). Constraint (9.3) is the outlet water balance from unit j. Here, the total water out of a unit is the sum of all the recycle streams to other units and the water discarded as effluent. Constraint (9.4) is a contaminant balance over unit j. The mass of the contaminants out of the unit j is the sum of the contaminant into the unit j and the mass of the contaminant added during the washing operation of the unit. As there are more than one contaminant present in the system, the balance has to be done for each contaminant c. Constraint (9.5) is the definition of the inlet concentration of contaminant c to unit j.

$$mw_{in}\left(s_{out,j}, p\right) = \sum_{s_{out,j'}} mw_r\left(s_{out,j'}, s_{out,j}, p\right) + mw_f\left(s_{out,j}, p\right),$$
$$\forall j, j' \in J, \quad s_{out,j} \in S_{out,j}, \quad p \in P \tag{9.1}$$

$$mw_{in}\left(s_{out,j}, p-1\right) = mw_{out}\left(s_{out,j}, p\right), \quad \forall j \in J, \quad s_{out,j} \in S_{out,j}, \quad p \in P, \quad p > p_1 \tag{9.2}$$

$$mw_{out}\left(s_{out,j}, p\right) = \sum_{s_{out,j'}} mw_r\left(s_{out,j}, s_{out,j'}, p\right) + mw_e\left(s_{out,j}, p\right),$$
$$\forall j, j' \in J, \quad s_{out,j} \in S_{out,j}, \quad p \in P \tag{9.3}$$

$$mw_{out}\left(s_{out,j},p\right)c_{out}\left(s_{out,j},c,p\right)=mw_{in}\left(s_{out,j},p-1\right)c_{in}\left(s_{out,j},c,p-1\right)$$
$$+M\left(s_{out},c\right)yw\left(s_{out,j},p-1\right) \quad \forall j \in J, \quad s_{out,j} \in S_{out,j}, \quad p \in P, \quad p > p_1, \quad c \in C \quad (9.4)$$

$$c_{in}\left(s_{out,j},c,p\right)=\frac{\displaystyle\sum_{s_{out,j'}} mw_r\left(s_{out,j'},s_{out,j},p\right)c_{out}\left(s_{out,j'},c,p\right)}{mw_{in}\left(s_{out,j},p\right)},$$

$$\forall j,j' \in J, \quad s_{out,j} \in S_{out,j}, \quad p \in P, \quad c \in C \quad (9.5)$$

The outlet concentration of each contaminant c in unit j cannot exceed its maximum limit as stated in Constraint (9.6). Constraint (9.7) ensures that the total water into a unit j does not exceed the maximum allowable water for the operation in unit j. Constraint (9.8) restricts mass of water recycled into the unit j to the maximum allowable water for operation in unit j. Constraint (9.9) stipulates that the inlet concentration for contaminant c into unit j cannot exceed its upper limit.

$$c_{out}\left(s_{out,j},c,p\right)\leq C_{out}^{U}\left(s_{out,j},c\right)yw\left(s_{out,j},p-1\right),$$
$$\forall j \in J, \quad s_{out,j} \in S_{out,j}, \quad p \in P, \quad p > p_1, \quad c \in C \quad (9.6)$$

$$mw_{in}\left(s_{out,j},p\right)\leq Mw^{U}\left(s_{out,j}\right)yw\left(s_{out,j},p\right), \quad \forall j \in J, \quad s_{out,j} \in S_{out,j}, \quad p \in P \quad (9.7)$$

$$mw_r\left(s_{out,j},s_{out,j},p\right)\leq Mw^{U}\left(s_{out,j}\right)yw_r\left(s_{out,j'},s_{out,j},p\right),$$
$$\forall j,j' \in J, \quad s_{out,j} \in S_{out,j}, \quad p \in P \quad (9.8)$$

$$c_{in}\left(s_{out,j},c,p\right)\leq C_{in}^{U}\left(s_{out,j},c\right)yw\left(s_{out,j},p\right),$$
$$\forall j \in J, \quad s_{out,j} \in S_{out,j}, \quad p \in P, \quad p > p_1, \quad c \in C \quad (9.9)$$

The maximum water quantity into a unit is represented by Equation 9.10. It is important to note that for multi-contaminant wastewater the outlet concentration of the individual components cannot all be set to the maximum, since the contaminants are not limiting simultaneously. The limiting contaminant(s) will always be at the maximum outlet concentration and the non-limiting contaminants will be below their respective maximum outlet concentrations.

$$Mw^{U}\left(s_{out,j}\right)=\max_{c \in C}\left\{\frac{M\left(s_{out,j},c\right)}{C_{out}^{U}\left(s_{out,j},c\right)-C_{in}^{U}\left(s_{out,j},c\right)}\right\}, \quad \forall j \in J, \quad s_{out,j} \in S_{out,j} \quad (9.10)$$

9.3.3.1.2 Linearization

Constraints (9.4) and (9.5) contain bilinear terms. This makes the model thus far nonlinear. These two constraints are not the only source of nonlinearity in the model, as this will be more apparent as the rest of the formulation unfolds. These nonlinear terms can be linearized according to the linearization of bilinear terms proposed by Quesada and Grossmann (1995). The detailed linearization procedure of these terms is detailed in the work by Majozi and Gouws (2009) and appears in the Appendix of this chapter.

9.3.3.1.3 Mass Balance Constraints with Storage

The mathematical formulation for the case where there is reusable water storage available is based on the superstructure given in Figure 9.4. This superstructure is similar to the previous one. However, there is a central storage vessel available. The total water flowing into the unit in this case is the sum of the freshwater, the water recycled/reused from the other unit and water from the central storage. The water leaving the unit is the sum of the water going to effluent, water going to storage and the water being recycled/reused.

Constraints (9.1), (9.3) and (9.5) need to be modified to cater for the water from storage, thus yielding (9.11), (9.12) and (9.13). The difference in the new constraints is that water into and out of the unit includes water from and to storage as well. Moreover, the inlet concentration is not only affected by the concentration from the previous recycle as in the previous case, but also the concentration in the water from the storage tank as stated in Constraints (9.13). Constraint (9.13) contains additional

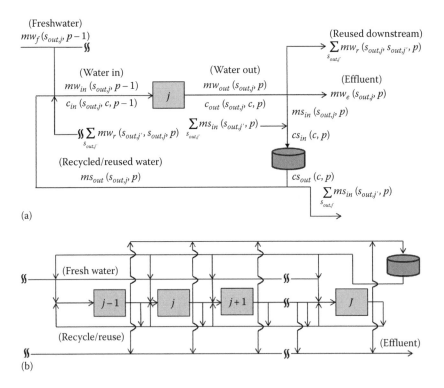

FIGURE 9.4 Superstructure for the mathematical formulation with reusable water storage (Majozi and Gouws, 2009).

nonlinear terms due to storage. These additional nonlinear terms can be linearized in the same manner as mentioned.

$$mw_{in}\left(s_{out,j}, p\right) = \sum_{s_{out,j'}} mw_r\left(s_{out,j'}, s_{out,j}, p\right) + mw_f\left(s_{out,j}, p\right) + ms_{out}\left(s_{out,j}, p\right),$$

$$\forall j, j' \in J, \quad s_{out,j} \in S_{out,j}, \quad p \in P \tag{9.11}$$

$$mw_{out}\left(s_{out,j}, p\right) = \sum_{s_{out,j'}} mw_r\left(s_{out,j}, s_{out,j'}, p\right) + mw_e\left(s_{out,j}, p\right) + ms_{in}\left(s_{out,j}, p\right),$$

$$\forall j, j' \in J, \quad s_{out,j} \in S_{out,j}, \quad p \in P \tag{9.12}$$

$$c_{in}\left(s_{out,j}, c, p\right) mw_{in}\left(s_{out,j}, p\right) = \sum_{s_{out,j'}} mw_r\left(s_{out,j'}, s_{out,j}, p\right) c_{out}\left(s_{out,j'}, c, p\right)$$

$$+ cs_{out}\left(c, p\right) ms_{out}\left(s_{out,j}, p\right), \quad \forall j, j' \in J, \quad s_{out,j} \in S_{out,j}, \quad p \in P, \quad c \in C \tag{9.13}$$

Besides the mass balance over a water-using operation, there also has to be mass balance over the storage unit. Constraint (9.14) is the mass balance over the storage unit. The water in the storage vessel at a certain time point is the difference between the water flowing into and from the storage vessel and the water possibly stored from the previous time point. Constraint (9.15) is a water mass balance at the beginning of the time horizon. Constraints (9.16) and (9.17) are the definition of the inlet and outlet concentration of the storage vessel, respectively. It should be noted that Constraint (9.17) is based on the assumption that the contaminant concentration inside the storage is the same as the concentration of the exit stream. Constraint (9.18) is the initial contaminant concentration of the water in the storage at the beginning of the time horizon, provided there is an initial amount of water in the storage tank. Constraint (9.19) ensures that the water in the storage vessel does not exceed the storage vessel's capacity. Constraint (9.20) ensures that water flowing from storage to unit j does not exceed unit j capacity.

$$qw_s\left(p\right) = qw_s\left(p-1\right) + \sum_{s_{out,j}} ms_{in}\left(s_{out,j}, p\right) - \sum_{s_{out,j}} ms_{out}\left(s_{out,j}, p\right)$$

$$\forall j \in J, \quad s_{out,j} \in S_{out,j}, \quad p \in P, \quad p > p_1 \tag{9.14}$$

$$qw_s\left(p_1\right) = Qw_s^o - \sum_{s_{out,j}} ms_{out}\left(s_{out,j}, p_1\right) \quad \forall j \in J, \quad s_{out,j} \in S_{out,j} \tag{9.15}$$

$$cs_{in}\left(c, p\right) = \frac{\displaystyle\sum_{s_{out,j}}\left(ms_{in}\left(s_{out,j}, p\right) c_{out}\left(s_{out,j}, c, p\right)\right)}{\displaystyle\sum_{s_{out,j}} ms_{in}\left(s_{out,j}, p\right)}$$

$$\forall j \in J, \quad s_{out,j} \in S_{out,j}, \quad p \in P \quad c \in C \tag{9.16}$$

$$cs_{out}(c,p) = \frac{qw_s(p-1)cs_{out}(c,p-1) + \left(\sum_{s_{out,j}} ms_{in}(s_{out,j},p)\right)cs_{in}(c,p)}{qw_s(p-1) + \sum_{s_{out,j}} ms_{in}(s_{out,j},p)},$$

$$\forall j \in J, \quad s_{out,j} \in S_{out,j}, \quad p \in P, \quad p > p_1, \quad c \in C \tag{9.17}$$

$$cs_{out}(c,p_1) = CS_{out}^o(c) \quad \forall c \in C \tag{9.18}$$

$$qw_s(p) \leq Qw_s^U \quad \forall p \in P \tag{9.19}$$

$$ms_{in}(s_{out,j},p) \leq Qw_s^U \, ys_{in}(s_{out,j},p) \quad \forall j \in J, \quad s_{out,j} \in S_{out,j}, \quad p \in P \tag{9.20}$$

As one would notice, there are nonlinear terms in Equations 9.16 and 9.17. One of the nonlinear terms can be eliminated by substituting Equations 9.16 and 9.17. The nonlinear terms in the resulting equation are then linearized in a similar fashion to that previously discussed. Due to the discontinuous nature of batch processes, the mass balances are not enough to fully describe the system. Sequencing constraints need to capture the discontinuous nature of the operation.

9.3.3.2 Sequencing and Scheduling Constraints

The sequencing and scheduling constraints pertaining to direct and indirect reuse/recycle are presented in this section. The constraints are presented for both cases where there is a central storage vessel and where there is none. The sequencing and scheduling constraints can be divided into three groups. The first group comprises of constraints pertaining to sequencing and scheduling of the washing operations. The second group comprises of constraints pertaining to direct recycle/reuse of reusable water, whilst the third group comprises of the constraints necessary for the scheduling of storage vessel. These groups are discussed here.

9.3.3.2.1 Scheduling Constraints Associated with the Washing Operation

Each water-using operation in the time horizon has to be scheduled accordingly, within the overall framework of operation scheduling. This is captured by Constraints (9.21) through (9.25). Constraints (9.21) and (9.22) ensure that unit j is washed immediately after a task that produced $s_{out,j}$. Constraint (9.23)) is the duration constraint, which defines the starting and the ending times of a washing operation. Constraint (9.24) stipulates that the washing operation can only commence at time point p if the task producing $s_{out,j}$ was active at the previous time point. Constraint (9.25) stipulates that only one washing operation can take place in a unit at any given time point.

$$tw_{in}(s_{out,j},p) \geq t_{out}(s_{out,j},p) - H^U(1 - yw(s_{out,j},p)), \quad \forall j \in J, \quad s_{out,j} \in S_{out,j}, \quad p \in P \tag{9.21}$$

$$tw_{in}(s_{out,j},p) \leq t_{out}(s_{out,j},p) + H^U(1 - yw(s_{out,j},p)), \quad \forall j \in J, \quad s_{out,j} \in S_{out,j}, \quad p \in P \tag{9.22}$$

$$tw_{out}\left(s_{out,j},p\right) = tw_{in}\left(s_{out,j},p-1\right) + \tau w\left(s_{out,j}\right) yw\left(s_{out,j},p-1\right),$$

$$\forall j \in J, \quad s_{out,j} \in S_{out,j}, \quad p \in P, \quad p > p_1 \tag{9.23}$$

$$yw\left(s_{out,j},p\right) = y\left(s_{in,j}^{*},p-1\right), \quad \forall j \in J, \quad s_{out,j} \in S_{out,j},$$

$$s_{in,j}^{*} \in S_{in,j}^{*}, \quad s_{in,j}^{*} \to s_{out,j}, \quad p \in P, \quad p > p_1 \tag{9.24}$$

$$\sum_{s_{out,j}} yw\left(s_{out,j},p\right) \le 1, \quad \forall j \in J, \quad s_{out,j} \in S_{out,j}, \quad p \in P_1 \tag{9.25}$$

9.3.3.2.2 Water Recycle/Reuse Sequencing Constraints

Scheduling of the recycle/reuse streams is important because of the discontinuous nature of the streams. Reusable water can only be directly recycled/reused if the unit producing water and the unit receiving water finish operating and begin operating at the same time, respectively. Constraint (9.26) stipulates that recycle/reuse between units can only take place when the unit receiving the reusable water start operating at that time point. The unit receiving the reusable water, however, does not necessarily need the recycled water to operate, that is, it can operate independently of the recycled water. Constraints (9.27) and (9.28) ensure that the time at which recycle/reuse of water occurs coincides with the time that the water is produced. Constraints (9.29) and (9.30) ensure that the starting time of the unit receiving water coincides with the time at which that water is recycled/reused.

$$yw_r\left(s_{out,j}, s_{out,j'}, p\right) \le yw\left(s_{out,j'}, p\right), \quad \forall j, j' \in J, \quad s_{out,j} \in S_{out,j}, \quad p \in P, \tag{9.26}$$

$$tw_r\left(s_{out,j}, s_{out,j'}, p\right) \le tw_{out}\left(s_{out,j}, p\right) + H^U\left(1 - yw_r\left(s_{out,j}, s_{out,j'}, p\right)\right),$$

$$\forall j, j' \in J, \quad s_{out,j} \in S_{out,j}, \quad p \in P, \tag{9.27}$$

$$tw_r\left(s_{out,j}, s_{out,j'}, p\right) \ge tw_{out}\left(s_{out,j}, p\right) - H^U\left(1 - yw_r\left(s_{out,j}, s_{out,j'}, p\right)\right),$$

$$\forall j, j' \in J, \quad s_{out,j} \in S_{out,j}, \quad p \in P \tag{9.28}$$

$$tw_r\left(s_{out,j}, s_{out,j'}, p\right) \le tw_{in}\left(s_{out,j'}, p\right) + H^U\left(1 - yw_r\left(s_{out,j}, s_{out,j'}, p\right)\right),$$

$$\forall j, j' \in J, \quad s_{out,j} \in S_{out,j}, \quad p \in P \tag{9.29}$$

$$tw_r\left(s_{out,j}, s_{out,j'}, p\right) \ge tw_{in}\left(s_{out,j'}, p\right) - H^U\left(1 - yw_r\left(s_{out,j}, s_{out,j'}, p\right)\right),$$

$$\forall j, j' \in J, \quad s_{out,j} \in S_{out,j}, \quad p \in P \tag{9.30}$$

9.3.3.2.3 Sequencing and Scheduling Constraints Associated with Storage

Constraints (9.31) and (9.32) ensure that the time at which reusable water goes to storage coincides with the time at which it is produced. Furthermore, reusable water can only be sent to storage at time point p if unit j has conducted a washing operation in the previous time point. This is captured by Constraint (9.33). However, the fact that unit j has conducted a washing operation in the previous time point does not mean that the resulting contaminated water needs to be sent to storage.

$$ts_{in}\left(s_{out,j},p\right) \geq tw_{out}\left(s_{out,j},p\right) - H^{U}\left(2 - ys_{out}\left(s_{out,j},p\right) - yw\left(s_{out,j},p\right)\right)$$
$$\forall j \in J, \quad s_{out,j} \in S_{out,j}, \quad p \in P \tag{9.31}$$

$$ts_{in}\left(s_{out,j},p\right) \leq tw_{out}\left(s_{out,j},p\right) + H^{U}\left(2 - ys_{out}\left(s_{out,j},p\right) - yw\left(s_{out,j},p\right)\right)$$
$$\forall j \in J, \quad s_{out,j} \in S_{out,j}, \quad p \in P \tag{9.32}$$

$$ys_{in}\left(s_{out,j},p\right) \leq yw\left(s_{out,j},p-1\right), \quad \forall j \in J, \quad s_{out,j} \in S_{out,j}, \quad p \in P, \quad p > p_1 \tag{9.33}$$

Constraints (9.34) and (9.35) ensure that the time at which water is used in unit j coincides with the time at which water is transferred from the storage to the unit. Constraint (9.36) ensures that unit j is indeed active when the unit is using recycled water from storage. The unit does not necessarily have to use the water from storage when it operates. A unit will not use the water in storage if there is a violation of the inlet concentration of the water into the unit or if there is simply no water in the storage vessel.

$$ts_{out}\left(s_{out,j},p\right) \geq tw_{in}\left(s_{out,j},p\right) - H^{U}\left(2 - ys_{out}\left(s_{out,j},p\right) - yw\left(s_{out,j},p\right)\right)$$
$$\forall j \in J, \quad s_{out,j} \in S_{out,j}, \quad p \in P \tag{9.34}$$

$$ts_{out}\left(s_{out,j},p\right) \leq tw_{in}\left(s_{out,j},p\right) + H^{U}\left(2 - ys_{out}\left(s_{out,j},p\right) - yw\left(s_{out,j},p\right)\right)$$
$$\forall j \in J, \quad s_{out,j} \in S_{out,j}, \quad p \in P \tag{9.35}$$

$$ys_{out}\left(s_{out,j},p\right) \leq yw\left(s_{out,j},p\right) \quad \forall j \in J, \quad s_{out,j} \in S_{out,j}, \quad p \in P \tag{9.36}$$

Constraint (9.37) ensures that when water is transferred to a unit at time point p, the time at which this happens is later in the time horizon than any previous time water was transferred to other units at previous time points. Constraints (9.38) and (9.39) ensure that the reusable water leaving the storage to different units at time point p leaves at the same time.

$$ts_{out}\left(s_{out,j},p\right) \geq ts_{out}\left(s_{out,j'},p'\right) - H^{U}\left(2 - ys_{out}\left(s_{out,j},p\right) - ys_{out}\left(s_{out,j'},p'\right)\right)$$
$$\forall j, j' \in J, \quad s_{out,j} \in S_{out,j}, \quad p, p' \in P, \quad p \geq p' \tag{9.37}$$

$$ts_{out}\left(s_{out,j},p\right) \geq ts_{out}\left(s_{out,j'},p\right) - H^{U}\left(2 - ys_{out}\left(s_{out,j},p\right) - ys_{out}\left(s_{out,j'},p\right)\right)$$
$$\forall j, j' \in J, \quad s_{out,j} \in S_{out,j}, \quad p \in P \tag{9.38}$$

$$ts_{out}\left(s_{out,j},p\right)\leq ts_{out}\left(s_{out,j'},p\right)+H^{U}\left(2-ys_{out}\left(s_{out,j},p\right)-ys_{out}\left(s_{out,j'},p\right)\right)$$

$$\forall j, j' \in J, \quad s_{out,j} \in S_{out,j}, \quad p \in P \tag{9.39}$$

Constraint (9.40) is similar to Constraint (9.37), but applies to inlet streams to the storage tank. Constraints (9.41) and (9.42) ensure that the time at which reusable water going to storage from unit j and unit j' corresponds to the same time at the same time point.

$$ts_{in}\left(s_{out,j},p\right)\geq ts_{in}\left(s_{out,j'},p'\right)-H^{U}\left(2-ys_{in}\left(s_{out,j},p\right)-ys_{in}\left(s_{out,j'},p'\right)\right)$$

$$\forall j, j' \in J, \quad s_{out,j} \in S_{out,j}, \quad p, p' \in P, \quad p \geq p' \tag{9.40}$$

$$ts_{in}\left(s_{out,j},p\right)\geq ts_{in}\left(s_{out,j'},p\right)-H^{U}\left(2-ys_{in}\left(s_{out,j},p\right)-ys_{in}\left(s_{out,j'},p\right)\right)$$

$$\forall j, j' \in J, \quad s_{out,j} \in S_{out,j}, \quad p \in P \tag{9.41}$$

$$ts_{in}\left(s_{out,j},p\right)\leq ts_{in}\left(s_{out,j'},p\right)+H^{U}\left(2-ys_{in}\left(s_{out,j},p\right)-ys_{in}\left(s_{out,j'},p\right)\right)$$

$$\forall j, j' \in J, \quad s_{out,j} \in S_{out,j}, \quad p \in P \tag{9.42}$$

Constraint (9.43) ensures that the outlet time of water from storage tank at a time point occurs later than the inlet time at the previous time point. Constraints (9.44) and (9.45) ensure that at time point p the time at which reusable water is moved to storage is the same as the time at which water leaves the storage.

$$ts_{out}\left(s_{out,j},p\right)\geq ts_{in}\left(s_{out,j'},p'\right)-H^{U}\left(2-ys_{out}\left(s_{out,j},p\right)-ys_{in}\left(s_{out,j'},p'\right)\right)$$

$$\forall j, j' \in J, \quad s_{out,j} \in S_{out,j}, \quad p, p' \in P, \quad p \geq p' \tag{9.43}$$

$$ts_{in}\left(s_{out,j},p\right)\geq ts_{out}\left(s_{out,j'},p\right)-H^{U}\left(2-ys_{in}\left(s_{out,j},p\right)-ys_{out}\left(s_{out,j'},p\right)\right)$$

$$\forall j, j' \in J, \quad s_{out,j} \in S_{out,j}, \quad p \in P \tag{9.44}$$

$$ts_{in}\left(s_{out,j},p\right)\leq ts_{out}\left(s_{out,j'},p\right)+H^{U}\left(2-ys_{in}\left(s_{out,j'},p\right)-ys_{out}\left(s_{out,j'},p\right)\right)$$

$$\forall j, j' \in J, \quad s_{out,j} \in S_{out,j}, \quad p \in P \tag{9.45}$$

9.3.3.3 Production Scheduling Constraints

As mentioned earlier, the intricate details of this sub-section are presented in another publication Majozi and Zhu (2001), but it is briefly presented here for continuity purposes. Emphasis will be on the new constraints added to cater for the concept of cyclic scheduling.

9.3.3.3.1 Capacity Constraint

The capacity constraint, given by Constraint (9.46), ensures that the amount of material processed in unit j at any time point p does not exceed the capacity of the unit.

$$V_j^L y\left(s_{in,j}^*, p\right) \le \sum_{s_{in,j}} m_{in}\left(s_{in,j}, p\right) \le V_j^U y\left(s_{in,j}^*, p\right),$$

$$\forall j \in J, \quad s_{in,j} \in S_{in,j}, \quad s_{in,j}^* \in S_{in,j}^*, \quad p \in P \qquad (9.46)$$

9.3.3.3.2 Material Balance Constraints

Constraints (9.47) through (9.53) are the mass balance constraints. These constraints ensure that the conservation of mass around each unit and the mass of each state involved in the production is maintained.

$$\sum_{s_{in,j}} m_{in}\left(s_{in,j}, p-1\right) = \sum_{s_{out,j}} m_{out}\left(s_{out,j}, p\right)$$

$$\forall j \in J, \quad s_{in,j} \in S_{in,j}, \quad s_{out,j} \in S_{out,j}, \quad p \in P, \quad p > p_1 \qquad (9.47)$$

$$q_s\left(s, p_1\right) = Q_s^0\left(s\right) - m_{in}\left(s, p_1\right), \quad s \ne product, \quad \forall s \in S \qquad (9.48)$$

$$q_s\left(s, p\right) = q_s\left(s, p-1\right) - m_{in}\left(s, p\right), \quad s = feed, \quad \forall s \in S, \quad p \in P, \quad p > p_1 \quad (9.49)$$

$$q_s\left(s, p\right) = q_s\left(s, p-1\right) + m_{out}\left(s, p\right) - m_{in}\left(s, p\right), \quad s \ne product, feed,$$

$$\forall s \in S, \quad p \in P, \quad p > p_1 \qquad (9.50)$$

$$q_s\left(s, p_1\right) = Q_s^0\left(s\right) - d\left(s, p_1\right), \quad s = product, \quad \forall s \in S \qquad (9.51)$$

$$q_s\left(s, p\right) = q_s\left(s, p-1\right) + m_{out}\left(s, p\right) - d\left(s, p\right), \quad s = product, byproduct,$$

$$\forall s \in S, \quad p \in P, \quad p > p_1 \qquad (9.52)$$

$$q_s\left(s, p\right) \le Q^U, \quad \forall s \in S, \quad p \in P \qquad (9.53)$$

9.3.3.3.3 Duration Constraints

The duration constraint is one of the most crucial constraints as it addresses the intrinsic aspects of time in batch plants. Constraint (9.54) simply states that the time at which a particular state is produced is dependent on the duration of the task that produces the same state.

$$t_{out}\left(s_{out,j}, p\right) = t_{in}\left(s_{in,j}^*, p-1\right) + \tau\left(s_{in,j}^*\right) y\left(s_{in,j}^*, p-1\right),$$

$$\forall j \in J, \quad s_{in,j}^* \in S_{in,j}^*, \quad s_{out,j} \in S_{out,j}, \quad p \in P, \quad p > p_1 \qquad (9.54)$$

9.3.3.3.4 Feasibility and Time Horizon Constraints

A washing operation starts immediately after a task, that is, if there is a task taking place in a unit at time point p, a washing operation will take place at the next time point. This is captured by Constraint (9.24). Constraint (9.55) ensures that a washing operation and a processing task do not coincide at the same time point. Constraint (9.56) ensures that only one task takes place in a given unit at a given time point.

$$yw\left(s_{out,j}, p\right) + y\left(s_{in,j}^*, p\right) \leq 1, \quad \forall j \in J, \quad s_{out,j} \in S_{out,j}, \quad s_{in,j}^* \in S_{in,j}^*, \quad p \in P \quad (9.55)$$

$$\sum_{s_{in,j}^*} y\left(s_{in,j}^*, p\right) \leq 1, \quad \forall j \in J, \quad s_{in,j}^* \in S_{in,j}^*, \quad p \in P \quad (9.56)$$

For cyclic scheduling, a task is allowed to cross the boundary of the cycle, as mentioned, but its duration must not be more than the length of two cycles ($2H$). Constraints (9.57) through (9.63) ensure that this is the case.

$$t_{in}\left(s_{in,j}, p\right) \leq 2H, \quad \forall j \in J, \quad s_{in,j} \in S_{in,j}, \quad p \in P \quad (9.57)$$

$$t_{out}\left(s_{out,j}, p\right) \leq 2H, \quad \forall j \in J, \quad s_{out,j} \in S_{out,j}, \quad p \in P \quad (9.58)$$

$$tw_{in}\left(s_{out,j}, p\right) \leq 2H, \quad \forall j \in J, \quad s_{out,j} \in S_{out,j}, \quad p \in P \quad (9.59)$$

$$tw_{out}\left(s_{out,j}, p\right) \leq 2H, \quad \forall j \in J, \quad s_{out,j} \in S_{out,j}, \quad p \in P \quad (9.60)$$

$$tw_r\left(s_{out,j}, s_{out,j'}, p\right) \leq 2H, \quad \forall j, j' \in J, \quad s_{out,j} \in S_{out,j}, \quad p \in P \quad (9.61)$$

$$ts_{in}\left(s_{out,j}, p\right) \leq 2H, \quad \forall j \in J, \quad s_{out,j} \in S_{out,j}, \quad p \in P \quad (9.62)$$

$$ts_{out}\left(s_{out,j}, p\right) \leq 2H, \quad \forall j \in J, \quad s_{out,j} \in S_{out,j}, \quad p \in P \quad (9.63)$$

9.3.3.3.5 Sequence Constraint within the Cycle

Before using a unit, all the tasks must be complete in the unit including their corresponding washing operations. Constraints (9.64) and (9.65) ensure that this is the case. For a unit that is suitable to conduct more than one task, Constraint (9.66) is necessary in the formulation.

$$t_{in}\left(s_{in,j}^*, p\right) \geq tw_{out}\left(s_{out,j}, p'\right) - H^U\left(2 - y\left(s_{in,j}^*, p\right) - yw\left(s_{out,j}, p'-1\right)\right),$$

$$\forall j \in J, \quad s_{in,j}^* \in S_{in,j}^*, \quad s_{out,j} \in S_{out,j}, \quad p, p' \in p, \quad p' > p_1, \quad p \geq p' \quad (9.64)$$

$$t_{in}\left(s_{in,j}, p\right) \geq t_{out}\left(s_{out,j}, p'\right) - H^U\left(2 - y\left(s_{in,j}, p\right) - y\left(s_{in,j}^*, p'-1\right)\right),$$

$$\forall j \in J, \quad s_{in,j}^* \in S_{in,j}^*, \quad s_{out,j} \in S_{out,j}, \quad p, p' \in p, \quad p' > p_1, \quad p \geq p', \quad s_{in,j}^* \to s_{out,j} \quad (9.65)$$

$$t_{in}\left(s_{in,j}, p\right) \geq \sum_{s_{out,j}} \left\{ \left(\tau\left(s_{in,j}^*\right) + \tau w\left(s_{out,j}\right) \right) yw\left(s_{out,j}, p'-1\right) \right\}, \quad \forall j \in J, \quad s_{in,j} \in S_{in,j},$$

$$s_{in,j}^* \in S_{in,j}^* \ s_{out,j} \in S_{out,j}, \quad p, p' \in P, \quad p \geq p', \quad p' \geq 2, \quad s_{in,j}^* \rightarrow s_{out,j} \tag{9.66}$$

9.3.3.3.6 Mass Balance between Two Cycles

Constraint (9.67) ensures that the amount of intermediate material needed to start the current cycle is stored at the last time point of the previous cycle in order to maintain smooth operation without any accumulation or shortage of material between cycles. This is the key constraint for cyclic scheduling.

$$q_s\left(s, p\right) = Q_s^0\left(s\right), \quad s \neq product, feed, \quad \forall s \in S, \quad p \in P, \quad p = |P| \tag{9.67}$$

9.3.3.3.7 Cycle Length Constraint

Constraint (9.68) stipulates that the sum of all the durations of the tasks and their corresponding washing operations in a processing unit must be less than the cycle length.

$$\sum_{s_{out,j}} \left\{ \left(\tau\left(s_{in,j}^*\right) + \tau w\left(s_{out,j}\right) \right) yw\left(s_{out,j}, p-1\right) \right\} \leq H,$$

$$\forall j \in J, \quad s_{in,j}^* \in S_{in,j}^*, \quad p \in P, \quad p > p_1, \quad s_{in,j}^* \rightarrow s_{out,j} \tag{9.68}$$

9.3.3.3.8 Sequence Constraints between Cycles

Constraints (9.69) and (9.70) express the relationship between the last task of the previous cycle and the first task of the current cycle in the same unit j to maintain continuity between cycles. Constraint (9.69) stipulates that the first task in unit j of the new cycle must start after the completion of the last task and its corresponding washing operation in the same unit j of the previous cycle. Constraint (9.70) holds for tasks which do not need washing after its completion.

$$t_{in}\left(s_{in,j}^*, p_1\right) \geq tw_{out}\left(s_{out,j}, p\right) - H^U\left(2 - y\left(s_{in,j}^*, p_1\right) - yw\left(s_{out,j}, p-1\right)\right) - H,$$

$$\forall j \in J, \quad s_{in,j}^* \in S_{in,j}^*, \quad s_{out,j} \in S_{out,j}, \quad p = |P| \tag{9.69}$$

$$t_{in}\left(s_{in,j}, p_1\right) \geq t_{out}\left(s_{out,j}, p\right) - H^U\left(2 - y\left(s_{in,j}, p_1\right) - y\left(s_{in,j}^*, p-1\right)\right) - H,$$

$$\forall j \in J, \quad s_{in,j}^* \in S_{in,j}^*, \quad s_{out,j} \in S_{out,j}, \quad p = |P|, \quad s_{in,j}^* \rightarrow s_{out,j} \tag{9.70}$$

9.3.3.3.9 Objective Function

The objective function takes the form of Equation 9.71 for cyclic scheduling, that is, performance index per unit time. The performance index can either be the maximization of profit or the minimization of effluent. This is dependent on the nature of the given data for the problem.

$$Z = \frac{\text{Performance index}}{H} \tag{9.71}$$

A typical objective function for the maximization of profit can take a form of Constraint (9.72), which is maximization of profit while taking into account freshwater and the effluent treatment cost.

Max Z

$$= \frac{\sum_s \sum_p CP(s)d(s,p) - CF \sum_{s_{out,j}} \sum_p mw_f(s_{out,j},p) - CE \sum_{s_{out,j}} \sum_p mw_e(s_{out,j},p)}{H} \tag{9.72}$$

The resulting mathematical formulation is an MINLP problem due to the presence of bilinear terms and a fractional term in the objective function. The linearization technique presented in Appendix 9A cannot be applied to the objective function. It is important to mention that for the presented model, global optimality cannot be guaranteed through the application of the technique presented in Appendix 9A. However, the technique can be used to provide a feasible starting point prior to solving the model.

9.3.3.3.10 Solution Procedure

The solution procedure is adopted from Wu and Ierapetritou (2004). The optimum cycle length is first determined from the methodology presented in this section with the objective function taking the form of Constraint (9.72). The results dictate the amount of intermediate products required to start the cyclic scheduling. With this data, the minimum duration of the initial period is determined through makespan minimization problem. This is done to ensure the existence of a feasible schedule to provide the required intermediates to start the cyclic scheduling period. The same problem is then solved with the objective of profit maximization with the time horizon obtained from the solution of the makespan minimization problem. The intermediate products from the main period, cyclic period, are consumed in the final period with the objective function being profit maximization. With the initial, the cyclic and the final period known, a wastewater minimization problem with a longer time horizon can be solved. Two illustrative examples are provided to demonstrate the effectiveness of the presented methodology.

9.4 ILLUSTRATIVE EXAMPLES

This section contains two literature examples to illustrate the applicability of the presented methodology.

9.4.1 EXAMPLE 1 (KONDILI ET AL., 1993)

The illustrative example considered is a well-published multipurpose facility which is commonly known as BATCH 1 in literature. It mainly consists of three chemical reactions which take place in two common reactors, Reactor 1 and 2, as depicted in Figure 9.5. In addition to the two common reactors, the flowsheet also entails the heater and the separator, before and after the reactors, respectively, as shown in Figure 9.5.

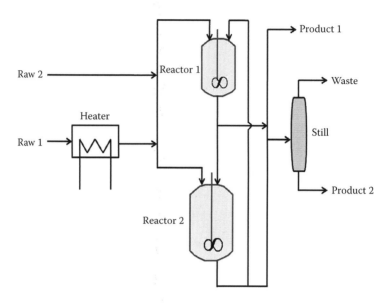

FIGURE 9.5 Flowsheet for BATCH 1 multipurpose plant.

The STN and SSN representations of the flowsheet are given in Figure 9.6a and b, respectively. The data for this example are given in Table 9.1.

This example was adapted to wastewater minimization in the work by Majozi and Gouws (2009). The philosophy is that the reactors need to be cleaned after each reaction in order to remove contaminants that are formed as by-products, so as to ensure product integrity. Data pertaining to cleaning tasks is shown in Table 9.2. It is important to note that the streams are characterized by multiple contaminants, which is a common occurrence in industry. The variation in performance in the reactors could be ascribed to the different designs, which is indeed a common encounter in practice. In addition to this data, it is known that the cost of freshwater is 2 c.u./kg water (c.u. – cost units), whilst the effluent treatment cost is 3 c.u./kg.

The method by Majozi and Gouws (2009) in the absence of a central storage vessel was applied to the illustrative example for different time horizons to demonstrate the common drawback of techniques currently available in literature and the results are given in Table 9.3. As mentioned earlier, when these techniques are applied to problems with a longer time horizons they present longer computational times. The model was formulated as an MINLP. The model was solved using GAMS 22.0, with CONOPT and CPLEX being the selected solvers for NLP and the MIP problems, respectively, in a DICOPT platform. All the results presented in this chapter were obtained using a Pentium 4, 3.2 GHz processor with a 512 MB RAM.

The results in Table 9.3 indicate that for a time horizon of 48 h, the solution could not be obtained. From these results, it is apparent that a methodology to solve scheduling problems for wastewater minimization in batch processes for longer time horizon is crucial, hence the proposed methodology.

The proposed methodology was applied to the illustrative example. From the solution procedure presented in Section 9.3.3.3.10, the first step is to determine the

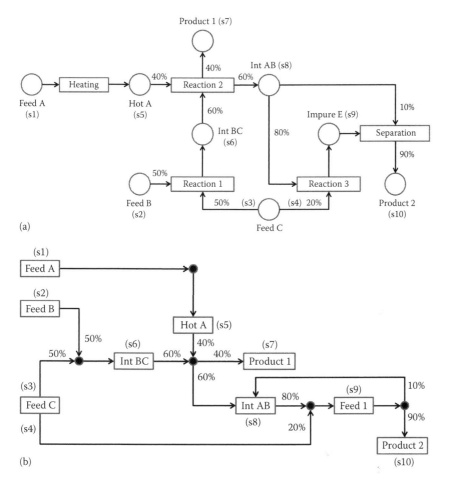

FIGURE 9.6 (a) STN and (b) SSN for the literature example.

optimal cycle length. The strategy used by Wu and Ierapetritou (2004) to determine the optimum cycle length was adopted. The time horizon range for determining the cycle length was 3–15 h. Instead of considering the whole time horizon range, the time horizon was subdivided into smaller intervals as indicated in Table 9.4. The objective function for the illustrative example is the maximization of profit per unit time, as given by Constraint (9.72), repeated for convenience.

$Max\,Z$

$$= \frac{\sum_s \sum_p CP(s) d(s,p) - CF \sum_{s_{out,j}} \sum_p mw_f\left(s_{out,j},p\right) - CE \sum_{s_{out,j}} \sum_p mw_e\left(s_{out,j},p\right)}{H}$$

The objective function comprises the product revenue, the freshwater and the wastewater treatment cost. For the objective function presented to be maximized, freshwater and effluent treatment cost must be minimized by minimizing the amount of freshwater used and wastewater generated per unit time.

TABLE 9.1

Scheduling Data for the Illustrative Example

Units	Capacity	Suitability	Mean Processing Time (τ)
Heater	100	Heating	1
Reactor 1	50	Reaction 1,2,3	2,2,1
Reactor 2	80	Reaction 1,2,3	2,2,1
Still	200	Separation	1 for product 2, 2 for IntAB
States	**Storage Capacity**	**Initial Amount**	**Price**
Feed A	Unlimited	Unlimited	0
Feed B	Unlimited	Unlimited	0
Feed C	Unlimited	Unlimited	0
Hot A	100	0	0
IntAB	200	0	0
IntBC	150	0	0
Impure E	200	0	0
Product 1	Unlimited	0	100
Product 2	Unlimited	0	100

9.4.1.1   Results and Discussions

9.4.1.1.1   Determination of the Optimum Cycle Length without Central Storage

The results for the determination of the optimum cycle length are presented in Table 9.4. The results presented in Table 9.4 are for a case where there is no central storage vessel for reusable water.

As shown in Table 9.4, the optimum cycle length was determined to be 5.75 h corresponding to a profit per unit time of 2785.507 c.u. per cycle. The schedule for the determined optimum cycle length is given in Figure 9.7. The values on top of the blocks represent the amount of material processed in the unit for production purposes. A washing operation starts after every material processing task, represented as the shaded blocks on the diagram. The values in the curly brackets represent the amount of freshwater supplied to the unit for the washing operation and the value in the square brackets represents the amount of reused water from one operation to another. In the absence of possible reuse and recycle, the amount of freshwater consumption is 527.78 kg per cycle. Applying reuse and recycle concept, the amount of effluent generated was reduced by 33.68% per cycle.

9.4.1.1.2   Determination of the Optimum Cycle Length with Central Storage

The cycle length for the illustrative example in the case where there is a central storage vessel for reusable water was also determined. It is important to note that to truly minimize wastewater generation in the presence of a storage vessel, the storage vessel must be empty at the end of every cycle. To ensure that this is the case, Constraint (9.73) was added to the model. The results for this case are given in Table 9.5.

$$qw_s(p) = 0 \quad \forall p = |P| \tag{9.73}$$

TABLE 9.2
Wastewater Minimization Data for the Illustrative Example

		Maximum Concentration (g contaminant/kg water) Contaminants		
		1	2	3
Reaction 1 (Reaction 1)	Max. inlet	0.5	0.5	2.3
	Max. outlet	1	0.9	3
Reaction 2 (Reaction 1)	Max. inlet	0.01	0.05	0.3
	Max. outlet	0.2	0.1	1.2
Reaction 3 (Reaction 1)	Max. inlet	0.15	0.2	0.35
	Max. outlet	0.3	1	1.2
Reaction 1 (Reaction 2)	Max. inlet	0.05	0.2	0.05
	Max. outlet	0.1	1	12
Reaction 2 (Reaction 2)	Max. inlet	0.03	0.1	0.2
	Max. outlet	0.075	0.2	1
Reaction 3 (Reaction 2)	Max. inlet	0.3	0.6	1.5
	Max. outlet	2	1.5	2.5

		Mass Load (g) Contaminants		
		1	2	3
Reaction 1	Reactor 1	4	80	10
	Reactor 2	15	24	358
Reaction 2	Reactor 1	28.5	7.5	135
	Reactor 2	9	2	16
Reaction 3	Reactor 1	15	80	85
	Reactor 2	22.5	45	36.5

	Duration of Washing (h)		
	Reaction 1	Reaction 2	Reaction 3
Reactor 1	0.25	0.5	0.25
Reactor 2	0.3	0.25	0.25

TABLE 9.3
Results for BATCH 1 Using Method by Majozi and Gouws (2009)

Time Horizon (h)	Objective Function (c.u.)	CPU Time (s)
10	11,537.5	0.31
13	19,587.5	2,044.44 (~0.57 h)
15	26,830.556	56,736.67 (~15.76 h)
48	—	Intractable

TABLE 9.4

Result for BATCH 1 Using the Presented Methodology without Storage

Cycle Time Range (h)	Number of Time Points	Objective Function Value (c.u.)	Optimal Cycle Time (h)	CPU Time (s)
3–6	7	2785.507	5.75	73.792
6–9	8	2669.444	6	106.634
9–12	10	2774.517	9.25	2056.567
12–15	11	2154.762	12	1353.127

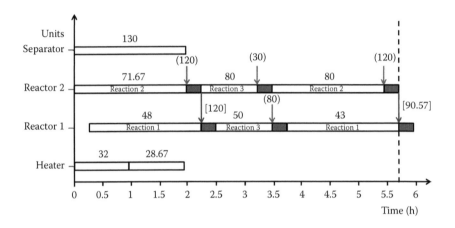

FIGURE 9.7 Schedule for the optimum cycle length without storage.

TABLE 9.5

Result for BATCH 1 Using the Developed Methodology with Storage

Cycle Time Range (h)	Number of Time Points	Objective Function Value (c.u.)	Optimal Cycle Time (h)	CPU Time (s)
3–6	7	2907.246	5.75	274.233
6–9	9	2669.444	8.05	345.485
9–12	11	2847.954	9.25	1125.784
12–15	12	2579.407	12	2157.469

As shown in Table 9.5, the optimum cycle length for this case is also 5.75 h corresponding to an average profit of 2907.246 c.u. per cycle. The schedule for the determined cycle length in the presence of a central storage vessel is given in Figure 9.8. The values on the diagram are as described in Figure 9.7. In the absence of possible indirect and direct water reuse/recycle, the amount of freshwater consumed is 518.89 kg per cycle. Applying reuse and recycle in the presence of a central storage vessel, the amount of effluent generated was reduced by 59.53% per cycle.

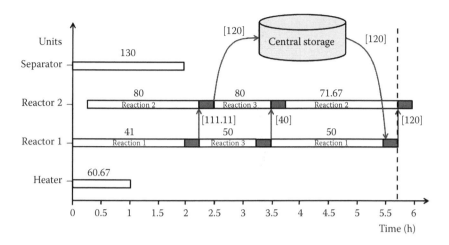

FIGURE 9.8   Schedule for the optimum cycle length without storage.

It is important to note the margin of difference in reduction/savings for the two presented cases with and without central storage vessel is quite big. This is due to the flexibility in the recycling and reusing time between the units, introduced by the presence of the storage vessel. In the case where the storage vessel is absent, only direct reuse is possible which decreases the degree of flexibility in the recycling and reusing opportunities.

9.4.1.1.3   Wastewater Minimization without Central Storage for 48 h Time Horizon

The illustrative example could not be solved for a time horizon of 48 h when the methodology by Majozi and Gouws (2009) was used, as demonstrated in Table 9.3. Therefore, the proposed methodology was applied for 48 h time horizon. The time horizon was divided into three periods, as mentioned in the solution procedure in Section 9.3.3.3.10. The optimum cycle length in the absence of central storage is as determined in Table 9.4, 5.75 h, with the average profit of 2785.507 c.u. per cycle.

Six cycles were determined to provide enough time for the initial period, to produce the necessary intermediates, and the final period to wrap up the intermediates. The CPU time for the initial period problem was 321.276 s. The duration of the initial period was determined to be 10.35 h with an objective value of 6201.786 c.u. using 11 time points. The schedule for the initial period is shown in Figure 9.9. The freshwater usage without exploiting reuse and recycle opportunities is 1103.889 kg. Exploiting these opportunities, the freshwater consumption and wastewater generated was reduced by 9.185%.

The duration of the final period was determined to be 3.15 h. This duration was determined from the duration of the total time horizon and the duration of the other two periods. The CPU time for the final period problem was 65.73 s. The objective value for this period is 18237.5 c.u., which was obtained using 5 time points. The schedule for the final period is shown in Figure 9.10. The freshwater

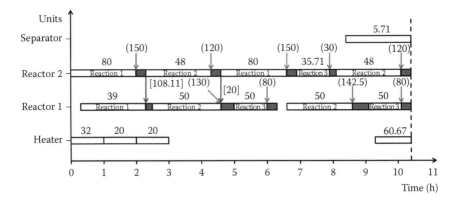

FIGURE 9.9 Initial period without a central storage vessel.

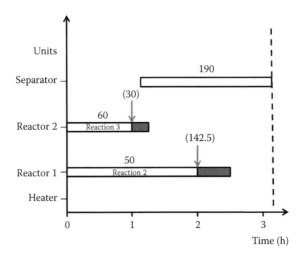

FIGURE 9.10 Final period without storage for a time horizon of 48 h.

consumption for this period was determined to be 172.5 kg. There were no reuse and recycle opportunities for this period as shown in Figure 9.10.

The overall objective value representing the total profit, including the profit of the initial and the final period, for the time horizon of 48 h is 41,152.33 c.u. Disregarding the principle of reuse and recycle, the total freshwater consumption is 4443.069 kg. Exploiting all the possible reuse and recycle opportunities, the total freshwater consumption was reduced by 26.287%.

9.4.1.1.4 Wastewater Minimization with Central Storage for 48 h Time Horizon

The same problem of 48 h was solved in the presence of a central storage vessel. The optimum cycle length in the presence of the storage vessel was determined to be 5.75 h with an average profit of 2907.246 c.u. per cycle, as mentioned in Table 9.5. The procedure used to determine the initial and the final period for the case without central storage vessel was also used for the case with central storage vessel and the results are presented in Table 9.6.

The corresponding Gantt charts for the initial and the final periods are given Figures 9.11 and 9.12, respectively.

The overall objective value representing the total profit for the time horizon of 48 h in the presence of a central storage vessel is 41,990.7 c.u. The total freshwater consumption is 4089.579 kg in the absence of both direct and indirect water reuse/recycle. Exploiting all the available direct and indirect water reuse/recycle opportunities, the total freshwater consumption was reduced by 47.088%.

To facilitate understanding of the presented results, the freshwater savings for the time horizon of 48 h in the presence and absence of a central storage vessel are given in Table 9.7. The savings for the cyclic period are given per cycle and six operating cycles were obtained for both cases.

TABLE 9.6

Results for Initial and Final Period with Storage for the Time Horizon of 48 h

Period	Duration (h)	Objective Function Value (c.u.)	Water Usage without Storage (kg)	CPU (s)	Freshwater Reduction
Initial	9.5	7,197.222	866.239	254.675	8.35%
Final	4	17,350	110	93.756	0%

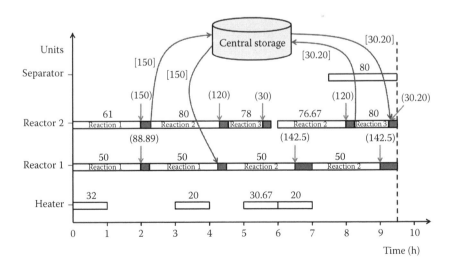

FIGURE 9.11 Initial period with storage.

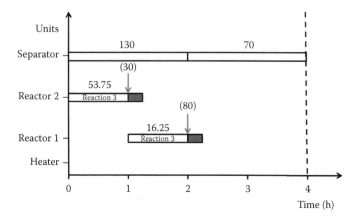

FIGURE 9.12   Final period with storage.

TABLE 9.7
Summary of Freshwater Savings for Example 1

	With Storage		Without Storage	
Period	Duration (h)	Savings (%)	Duration (h)	Savings (%)
Initial	9.5	8.35	10.35	9.185
Cyclic	5.75	59.53	5.75	33.68
Final	4	0	3.15	0
Overall	48	47.008	48	26.287

9.4.2 EXAMPLE 2 (SUNDARAMOORTHY AND KARIMI, 2005)

This second illustrative example was presented in the work by Sundaramoorthy and Karimi (2005). It is similar to the first example but considerably more complex. In this example, two products are produced through six processing units and seven processing tasks. It consists of three reactions which take place in two common reactors and two heating tasks which take place in one common heater. In addition, it involves one separation task and two mixing tasks which take place in two different mixers. The STN and SSN representations of the example are given in Figures 9.13 and 9.14, respectively. The scheduling data for this example is given in Table 9.8.

This example is adapted to wastewater minimization in this chapter to illustrate the applicability of the presented methodology to a more complex multipurpose problem. The philosophy is that the reactors and the heater must be washed after each production operation. This is to remove contaminants that are formed as by-products in the units, so as to ensure product integrity. Data pertaining to cleaning tasks is given in Table 9.9.

The method by Majozi and Gouws (2009) could not solve the illustrative example when a time horizon of 168 h was considered. The proposed methodology was applied to the illustrative example with the same objective function considered for

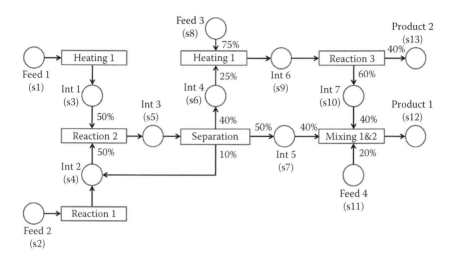

FIGURE 9.13 STN representation of the second illustrative example.

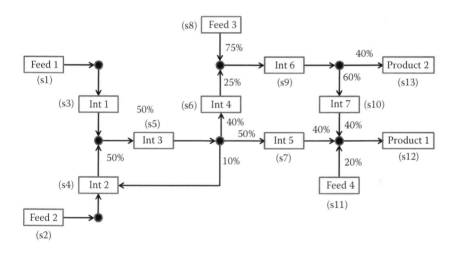

FIGURE 9.14 SSN representation of the second illustrative example.

the first illustrative example for a time horizon of 168 h. The results are presented in the following sub-section.

9.4.2.1   Results and Discussions

9.4.2.1.1   *Wastewater Minimization without Central*
Storage for 168 h Time Horizon

The procedure used in the first illustrative example for determining the optimal cycle is used in this example. The time horizon range for determining the optimal cycle length and the results are given in Table 9.10. The results presented in Table 9.10 are for case where a central storage vessel is absent.

TABLE 9.8

Scheduling Data for the Second Illustrative Example

Units	Capacity Min	Capacity Max	Suitability	Mean Processing Time
Heater	0	100	Heating 1,2	1, 1.5
Reactor 1	0	100	Reaction 1,2,3	2,1,2
Reactor 2	0	150	Reaction 1,2,4	2,1,2
Still	0	300	Separation	3
Mixer 1	20	200	Mixing	2
Mixer 2	20	200	Mixing	2

State	Price	Initial Amount	Storage Capacity
Feed 1	0	10,000	Unlimited
Feed 2	0	10,000	Unlimited
Feed 3	0	10,000	Unlimited
Feed 4	0	10,000	Unlimited
Int1	0	0	100
Int2	0	0	100
Int3	0	0	300
Int4	0	50	150
Int5	0	50	150
Int6	0	0	150
Int7	0	0	150
Product 1	100	0	Unlimited
Product 2	100	0	Unlimited

The optimum cycle length was determined to be 6.45 h with an average profit of 5167.557 c.u. per cycle. The schedule for the optimum cycle length for this case is given in Figure 9.15. In the absence of possible reuse and recycle, the amount of freshwater consumed is 501.675 kg per cycle. Applying the concept of reuse and recycle, the amount of freshwater consumption for the determined cycle length was reduced by 12.53%.

Considering the scheduling problem for the time horizon of 168 h, the proposed methodology determines 24 cycles of operation. This leaves enough time for the initial and the final period. The results for the initial and the final schedules are given in Table 9.11.

The corresponding schedules for the initial and final period are given in Figures 9.16 and 9.17, respectively.

The objective value corresponding to the total profit for the time horizon of 168 h is 179,386.596 c.u. The total freshwater consumption for this time horizon is 12,902.914 kg in the absence of water reuse and recycle. Exploiting all the available reuse and recycle opportunities, the total freshwater consumption is reduced by 16.363%.

TABLE 9.9

Wastewater Minimization Data for the Second Illustrative Example

		Maximum Concentration (g Contaminant/kg Water) Contaminant		
		1	2	3
Heating 1 (Heater)	Max. inlet	0.5	0.35	1
	Max. outlet	1.15	0.65	1.5
Heating 2 (Heater)	Max. inlet	0.65	0.2	1.35
	Max. outlet	1.5	0.7	2.5
Reaction 1 (Reaction 1)	Max. inlet	0.5	0.5	2.3
	Max. outlet	1	0.9	3
Reaction 2 (Reaction 1)	Max. inlet	0.01	0.05	0.3
	Max. outlet	0.2	0.1	1.2
Reaction 3 (Reaction 1)	Max. inlet	0.15	0.2	0.35
	Max. outlet	0.3	1	1.2
Reaction 1 (Reaction 2)	Max. inlet	0.05	0.2	0.05
	Max. outlet	0.1	1	12
Reaction 2 (Reaction 2)	Max. inlet	0.03	0.1	0.2
	Max. outlet	0.075	0.2	1
Reaction 3 (Reaction 2)	Max. inlet	0.3	0.6	1.5
	Max. outlet	2	1.5	2.5

		Mass Load (g) Contaminant		
		1	2	3
Heating 1	Heater	7.5	13	20
Heating 2	Heater	11	15	25
Reaction 1	Reactor 1	4	80	10
	Reactor 2	15	24	358
Reaction 2	Reactor 1	28.5	7.5	135
	Reactor 2	9	2	16
Reaction 3	Reactor 1	15	80	85
	Reactor 2	22.5	45	36.5

	Duration of Washing (h)		
	Heating 1	**Heating 2**	
Heater	0.2	0.25	
	Reaction 1	**Reaction 2**	**Reaction 3**
Reactor 1	0.25	0.5	0.25
Reactor 2	0.3	0.25	0.25

TABLE 9.10

Result for the Example without a Central Storage Vessel

Cycle Time Range (h)	Number of Time Points	Objective Function Value (c.u.)	Optimal Cycle Time (h)	CPU Time (s)
3–6	8	4720.253	4.7	2,391.904
6–9	9	5167.557	6.45	5,042.156
9–12	10	3880.959	9.2	3,845.292
12–15	12	3008.294	12	29,239.092

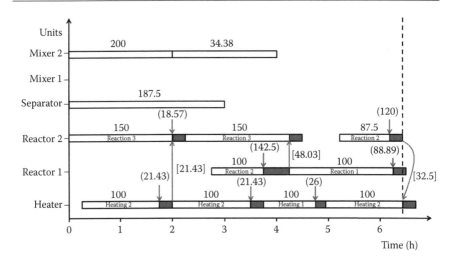

FIGURE 9.15 Schedule for the optimum cycle length without central storage.

TABLE 9.11

Results for Initial and Final Period for 168 h Time Horizon

Period	Duration (h)	Objective Function Value (c.u.)	Water Usage without Storage (kg)	CPU Time (s)	Freshwater Reduction
Initial	6.7	12,897.371	559.857	2923.03	8.00%
Final	6.5	42,467.857	302.857	166.439	7.08%

9.4.2.1.2 Wastewater Minimization with Central Storage for 168 h Time Horizon

The illustrative example for the time horizon of 168 h in the presence of central storage for reusable water was also considered. The optimum cycle length for the example for case where a central storage was presented was also determined and the results are shown in Table 9.12.

The optimum cycle length for this case was also determined to be 6.45 h, corresponding to the average production of 5326.345 c.u. per cycle as shown in Table 9.12.

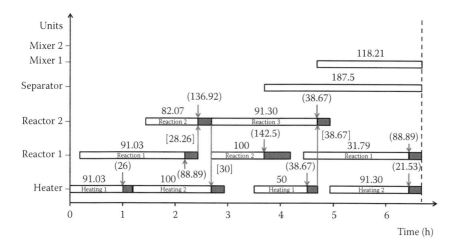

FIGURE 9.16 Initial period without storage for a time horizon of 168 h.

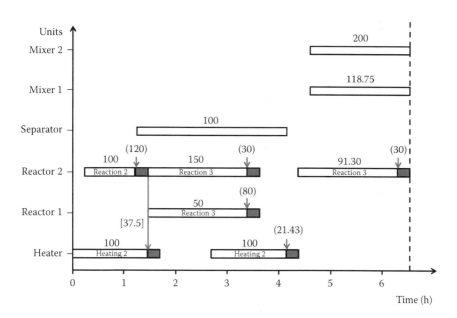

FIGURE 9.17 Final period without storage for a time horizon of 168 h.

The schedule for the optimum cycle length for this case is given in Figure 9.18. The freshwater consumption for the cycle with exploiting the concept of reuse and recycle is 372.22 kg per cycle. Considering this concept, the freshwater consumption for the cycle was reduced by 41.83% per cycle.

Considering the illustrative example for the time horizon of 168 h in the presence of a central storage, the proposed methodology determines 24 cycles of operation

TABLE 9.12

Result for Example with a Central Storage Vessel Present

Cycle Time Range (h)	Number of Time Points	Objective Function Value (c.u.)	Optimal Cycle Time (h)	CPU Time (s)
3–6	7	4885.904	4.7	2,019.64
6–9	9	5326.345	6.45	14,963.811
9–12	10	4006.984	9	8,401.495
12–15	11	3025.014	12	22,937.718

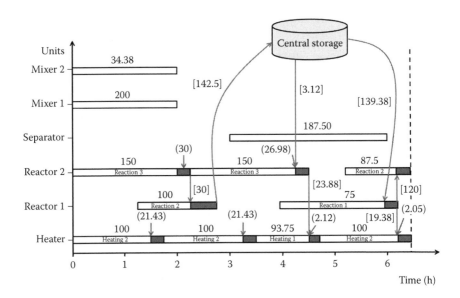

FIGURE 9.18   Schedule for the optimum cycle length for the case with central storage.

which leaves enough time for the initial and the final period. The results for the initial and the final period are given in Table 9.13.

The corresponding Gantt charts for the initial and final period are given in Figures 9.19 and 9.20, respectively.

The objective value representing the total profit for the time horizon of 168 h is 184,731.05 c.u. The total freshwater consumption corresponding for this time horizon in the absence of direct and indirect water reuse/recycle is 9799.845 kg. Exploiting all possible direct and indirect water reuse/recycle, the total freshwater consumption is reduced by 39.498%.

The freshwater savings results for the time horizon of 168 h in the presence and absence of a central storage vessel are given in Table 9.14. The savings for the cyclic period are given per cycle and 24 operating cycles were obtained for both cases.

TABLE 9.13

Result for the Initial and the Final Period in the Presence of a Central Storage

Period	Duration (h)	Objective Function Value (c.u.)	Water Usage without Storage (kg)	CPU Time (s)	Freshwater Reduction
Initial	8.85	14,111.27	592.635	2728.143	19.39%
Final	4.35	42,787.5	273.93	98.235	6.94%

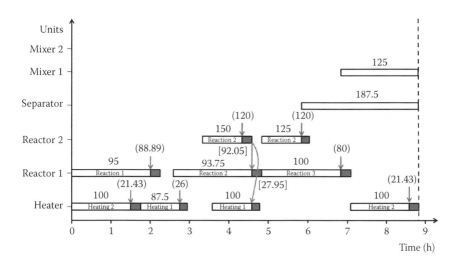

FIGURE 9.19   Initial period with storage.

9.4.2.2   Remarks

From the results presented, it is evident that the developed methodology can be used to reduce problems with longer time horizon to a smaller problem which can be solved with ease. The first illustrative example could not be solved for 48 h using the usual technique for wastewater minimization for short-term time horizon. With the application of the proposed methodology, the 48 h time horizon in the absence of central storage was reduced to six cycles with a time horizon of 5.75 h, initial period with a duration of 10.35 h and a final period with a duration of 3.15 h. The second illustrative example with the time horizon 168 h was reduced 24 h cycles, each cycle with the length of 6.45 h, initial period of 6.7 h and final period of 6.5 h in the absence of central storage vessel. The global optimality of the presented results cannot be guaranteed since the developed methodology is nonlinear. The nonlinearity of the objective function cannot be linearized with the technique presented in

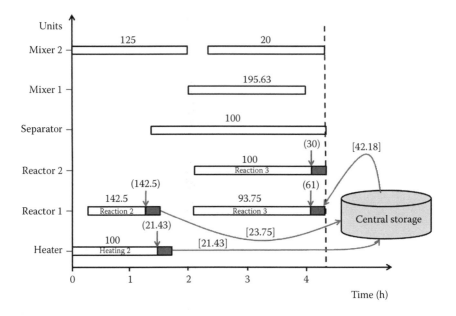

FIGURE 9.20   Final period with central storage present.

TABLE 9.14
Summary of Example 2 for 168 h Time Horizon

	With Storage		Without Storage	
Period	**Duration (h)**	**Savings (%)**	**Duration (h)**	**Savings (%)**
Initial	8.85	19.39	6.7	8
Cyclic	6.45	41.83	6.45	16.985
Final	4.35	6.94	6.5	7.08
Overall	168	39.498	168	16.363

the Appendix. The developed technique is further applied to an industrial case study in the following section to demonstrate the practical application.

9.5   INDUSTRIAL CASE STUDY

In this section, the developed methodology is applied to an industrial case study to demonstrate practical application. The case study is taken from a pharmaceuticals facility that produces four types of products, namely, shampoos, deodorants, lotions and creams. All the products involve mixing and there are four mixing units dedicated to each product, that is, one unit is suitable for mixing of one product. It is mandatory to wash the unit after every mixing task to ensure product integrity.

TABLE 9.15
Data for the Case Study

Mixer	Product	Residue Mass (kg)	Max. Outlet Conc (kg/kg Water)	Mass Water (kg)	Duration (h)
1	Shampoos	15	0.040	576.9	7
2	Deodorants	15	0.045	361.4	5.5
3	Lotions	30	0.050	697.6	11
4	Creams	70	0.060	1238.9	11

TABLE 9.16
Maximum Inlet Concentrations for Cleaning Operation for the Case Study

Mixer	Shampoos (kg/kg Water)	Deodorants (kg/kg Water)	Lotions (kg/kg Water)	Creams (kg/kg Water)
1	0.014	0	0.007	0.0035
2	0.014	0.0035	0.007	0.007
3	0.014	0	0.007	0.0035
4	0.014	0	0.007	0.0035

The residue mass left in the unit after a mixing operation is given in Table 9.15. Also given in Table 9.15 is the necessary information to conduct wastewater minimization and the duration of each mixing operation. Because of different designs of the stirrers in the mixing vessels, mixing durations vary according to the vessel used. All the mixers have a capacity of 2 tons.

The maximum inlet concentration of each mixer is given in Table 9.16. It is important to note that the inlet concentration for the deodorant is zero in mixers 1, 3 and 4 (due to incompatibility of this residue with other residues). Thus, the reuse of contaminated water resulting from unit 2 to any other unit is forbidden. The duration of the washing operation is 30 min. A 10 ton central storage vessel for water reuse is available.

In addition to the given data, the cost of freshwater is 0.2 c.u./kg of water whilst the effluent treatment cost is 0.3 c.u./kg.

9.5.1 RESULTS AND DISCUSSIONS

The methodology by Majozi and Gouws (2009) was applied to the industrial case study to demonstrate the computational problem short-term scheduling techniques for wastewater minimization have when they are applied to industrial scale problems. The same objective function used for the illustrative example was used for the industrial case study, that is, maximization of profit per unit time. The results are

TABLE 9.17

Demonstration of Computational Problems

Time Horizon (h)	Objective Function	CPU (s)
24	41,758.062	10.625
48	105,167.217	5,724.550
72	229,163.475	50,786.576
168	—	—

shown Table 9.17. The methodology encountered a computational difficulty for the time horizon of 168 h, and the computational time for the time horizon of 72 h is quite long.

9.5.1.1 Wastewater Minimization without Central Storage for 168 h Time Horizon

The developed methodology was applied to the presented industrial case study. The formulated problem was an MINLP for both cases, with and without central storage tank. It is important to note that for the case study the raw materials needed for mixing are readily available, hence the initial and the final period in this case are not required. The same procedure followed for the illustrative example to determine the optimal cycle length is followed for the case study. The time horizon range for determining the optimal cycle length for the case where there is no central storage vessel was chosen to be between 12 and 72 h, as shown in Table 9.18. The results for industrial case study without storage tank are presented in Table 9.18.

As shown in Table 9.18, the optimum cycle length was determined to be 23 h with a profit per unit time of 1740.910 c.u. per cycle. The Gantt chart corresponding to the optimum cycle length showing both the production operations and the washing operations is depicted in Figure 9.21. The values in the curly brackets represent the amount of freshwater supplied to the unit and the value in the square brackets represents the amount of water reused from one operation to another. In the absence of possible reuse and recycle, the amount of freshwater used is 5658.33 kg per cycle which corresponds to a profit per unit time of 1718.623 c.u. per cycle. Applying the

TABLE 9.18

Results for the Case Study without Central Storage

Cycle Time Range (h)	Number of Time Points	Objective Function Value (c.u.)	Optimal Cycle Time (h)	CPU Time (s)
12–24	7	1740.910	23	55.202
24–36	9	1730.886	24	71.538
36–48	11	1718.852	37.5	984.048
48–60	12	1663.541	48	771.453
60–72	14	1595.436	64	2036.576

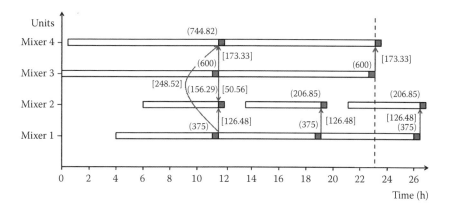

FIGURE 9.21   Optimum schedule for the case study without a storage vessel.

concept of reuse and recycle, the amount of effluent generated was reduced by 18.12% per cycle. Consequently, the average profit per cycle is increased by 1.28%. The overall objective value which represents the total profit for the time horizon of 168 h is 12,716.21 c.u. This corresponds to a total freshwater consumption of 41,330.41 kg in the absence of reuse and recycle. The reduction in the overall freshwater consumption is the same as the reduction per cycle for the case study since the initial and the final period are not present, as mentioned.

9.5.1.2   Wastewater Minimization with Central Storage for 168 h Time Horizon

In a case where a central storage was present, the cycle length was also obtained to be 23 h with the objective function of 1787.278 c.u. per cycle, as shown in Table 9.19. In the absence of possible reuse and recycle, the amount of freshwater required is 5658.33 kg per cycle which corresponds to a profit per unit time of 1726.775 c.u. per cycle. Applying the concept of reuse and recycle, the amount of effluent generated was reduced by 45.40% per cycle and the schedule for these results is given in Figure 9.22. Consequently, the average profit per cycle is increased by 3.39%. The objective value

TABLE 9.19

Results for the Case Study with Central Storage Present

Cycle Time Range (h)	Number of Time Points	Objective Function Value (c.u.)	Optimal Cycle Time (h)	CPU Time (s)
12–24	8	1787.278	23	64.53
24–36	9	1750.281	24	87.435
36–48	11	1734.563	37.5	882.361
48–60	13	1683.342	51.5	561.453
60–72	14	1695.436	60	1045.583

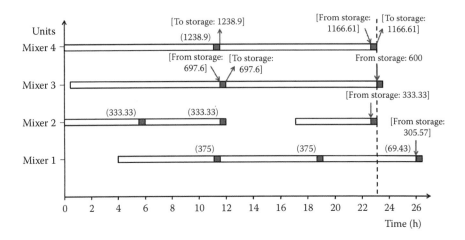

FIGURE 9.22 Optimum schedule with the presence of a central storage vessel.

for the time horizon of 168 h in the presence of a central storage vessel is 13,054.9 c.u. This corresponds to a total freshwater consumption of 41,330.41 kg in the absence of direct and indirect water reuse/recycle. The reduction in the overall freshwater consumption is the same as the reduction per cycle for the case study since the initial and the final period are not present, as mentioned earlier.

For the case of 168 h time horizon, the short-term scheduling methodology could not find a solution. Applying the developed methodology, the time horizon of 168 h can be reduced to 7.304 (≈7 cycles) with the length of 23 h. The CPU time for the optimum cycle length is 64.53 s. The reduction in freshwater usage and wastewater generation for the determined cycle will be 18.12% per cycle in the absence of central storage vessel and 45.40% per cycle in the presence of the central storage vessel.

From the presented results of the case study, the developed methodology proves to be capable of reducing an industrial scale problem to a problem with a smaller time horizon. As mentioned for the illustrative example, the global optimality of the presented results cannot be proven due to the nonlinearities present in the formulation of the methodology, particularly in the objective function.

9.6 CONCLUSION

A long-term scheduling methodology for wastewater minimization in multipurpose batch facilities is presented in this chapter. The main advantage of the presented methodology is the ability to reduce industrial scale problems to a problem with a smaller time horizon, which can be solved within reasonable CPU time. This is achieved through the exploitation of cyclic scheduling concepts. The concept of water reuse and recycle is used to minimize wastewater generation. The proposed methodology optimises both the production schedule and wastewater generation simultaneously. It is applicable to operations with streams characterized by multiple contaminants, which is more prevalent in industry. The application of the methodology to an industrial case study has proven its effectiveness. A time horizon of 168 h of the case study was reduced to seven cycles with the length of 23 h. The total freshwater consumption for the time horizon of 168 h was reduced by 18.12% in the absence of central storage vessel and 45.40% in the presence of the central storage vessel.

9A APPENDIX

9A.1 LINEARIZATION

To demonstrate the linearization technique presented in the work by Majozi and Gouws (2009), Constraints (9.4) and (9.5) in the presented model will be used. The two constraints contain bilinear terms. These nonlinear terms can be linearized according to the linearization of bilinear terms proposed by Quesada and Grossmann (1995). The linearization is given below.
Let

$$c_{in}\left(s_{out,j}, c, p\right)mw_{in}\left(s_{out,j}, p\right) = \Gamma_1\left(s_{out,j}, c, p\right)$$

$$c_{out}\left(s_{out,j}, c, p\right)mw_{out}\left(s_{out,j}, p\right) = \Gamma_2\left(s_{out,j}, c, p\right)$$

$$c_{out}\left(s_{out,j'}, c, p\right)mw_r\left(s_{out,j'}, s_{out,j}, p\right) = \Gamma_3\left(s_{out,j'}, s_{out,j}, c, p\right)$$

with each variable having the following bounds

$$0 \le c_{in}\left(s_{out,j}, c, p\right) \le C_{in}^U\left(s_{out,j}, c\right)$$

$$0 \le mw_{in}\left(s_{out,j}, p\right) \le Mw^U\left(s_{out,j}\right)$$

$$0 \le c_{out}\left(s_{out,j}, c, p\right) \le C_{out}^U\left(s_{out,j}, c\right)$$

$$0 \le mw_{out}\left(s_{out,j}, p\right) \le Mw^U\left(s_{out,j}\right)$$

$$0 \le mw_r\left(s_{out,j'}, s_{out,j}, p\right) \le Mw^U\left(s_{out,j}\right)$$

then the following constraints are true for Γ_1:

$$\Gamma_1\left(s_{out,j}, c, p\right) \ge 0, \quad \forall j \in J, \quad s_{out,j} \in S_{out,j}, \quad p \in P, \quad c \in C \tag{9A.1}$$

$$\Gamma_1\left(s_{out,j}, c, p\right) \ge Mw^U\left(s_{out,j}\right)c_{in}\left(s_{out,j}, c, p\right) + C_{in}^U\left(s_{out,j}, c\right)mw_{in}\left(s_{out,j}, p\right)$$
$$- Mw^U\left(s_{out,j}\right)c_{in}\left(s_{out,j}, c, p\right), \quad \forall j \in J, \quad s_{out,j} \in S_{out,j}, \quad p \in P, \quad c \in C \tag{9A.2}$$

$$\Gamma_1\left(s_{out,j}, c, p\right) \geq Mw^U\left(s_{out,j}\right) c_{in}\left(s_{out,j}, c, p\right), \quad \forall j \in J, \quad s_{out,j} \in S_{out,j}, \quad p \in P, \quad c \in C$$
(9A.3)

$$\Gamma_1\left(s_{out,j}, c, p\right) \leq mw_{in}\left(s_{out,j}, p\right) C_{in}^U\left(s_{out,j}, c\right), \quad \forall j \in J, \quad s_{out,j} \in S_{out,j}, \quad p \in P, \quad c \in C$$
(9A.4)

and the following constraints are true for Γ_2:

$$\Gamma_2\left(s_{out,j}, c, p\right) \geq 0, \quad \forall j \in J, \quad s_{out,j} \in S_{out,j}, \quad p \in P, \quad c \in C$$
(9A.5)

$$\Gamma_2\left(s_{out,j}, c, p\right) \geq Mw^U\left(s_{out,j}\right) c_{out}\left(s_{out,j}, c, p\right) + C_{out}^U\left(s_{out,j}, c\right) mw_{in}\left(s_{out,j}, p\right)$$
$$-Mw^U\left(s_{out,j}\right) C_{out}^U\left(s_{out,j}, c\right), \quad \forall j \in J, \quad s_{out,j} \in S_{out,j}, \quad p \in P, \quad c \in C$$
(9A.6)

$$\Gamma_2\left(s_{out,j}, c, p\right) \leq mw_{out}\left(s_{out,j}, p\right) C_{out}^U\left(s_{out,j}, c\right), \quad \forall j \in J, \quad s_{out,j} \in S_{out,j}, \quad p \in P, \quad c \in C$$
(9A.7)

and the following constraints are true for Γ_3:

$$\Gamma_3\left(s_{out,j'}, s_{out,j}, c, p\right) \geq 0, \quad \forall j \in J, \quad s_{out,j} \in S_{out,j}, \quad p \in P, \quad c \in C \quad (9A.8)$$

$$\Gamma_3\left(s_{out,j'}, s_{out,j}, c, p\right) \geq Mw^U\left(s_{out,j}\right) c_{out}\left(s_{out,j'}, c, p\right) + C_{out}^U\left(s_{out,j'}, c\right) mw_r\left(s_{out,j'}, s_{out,j}, p\right)$$
$$-Mw^U\left(s_{out,j}\right) C_{out}^U\left(s_{out,j}, c\right), \quad \forall j, j' \in J, \quad s_{out,j} \in S_{out,j}, \quad p \in P, \quad c \in C \quad (9A.9)$$

$$\Gamma_3\left(s_{out,j}, s_{out,j}, c, p\right) \leq Mw^U\left(s_{out,j}\right) c_{out}\left(s_{out,j'}, c, p\right),$$
$$\forall j, j' \in J, \quad s_{out,j} \in S_{out,j}, \quad p \in P, \quad c \in C$$
(9A.10)

$$\Gamma_3\left(s_{out,j'}, s_{out,j}, c, p\right) \leq mw_r\left(s_{out,j}, p\right) C_{out}^U\left(s_{out,j'}, c\right),$$
$$\forall j, j' \in J, \quad s_{out,j} \in S_{out,j}, \quad p \in P, \quad c \in C$$
(9A.11)

Substituting these linearized variables into Constraints (9.4) and (9.5) gives Constraints (9A.12) and (9A.13):

$$\Gamma_2\left(s_{out,j}, c, p\right) = \Gamma_1\left(s_{out,j}, c, p-1\right) + M\left(s_{out,j}, c\right) yw\left(s_{out,j}, p-1\right),$$
$$\forall j \in J, \quad s_{out,j} \in S_{out,j}, \quad p \in P, \quad c \in C$$
(9A.12)

$$\Gamma_1\left(s_{out,j}, c, p\right) = \sum_{s_{out,j'}} \Gamma_3\left(s_{out,j'}, s_{out,j}, c, p\right), \quad \forall j, j' \in J, \quad s_{out,j} \in S_{out,j}, \quad p \in P, \quad c \in C$$
(9A.13)

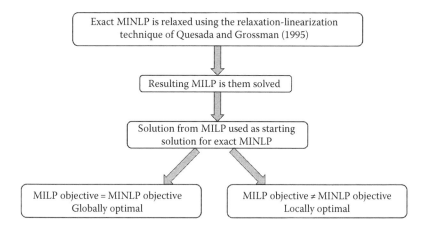

FIGURE 9A.1 Solution procedure from Majozi and Gouws (2009).

The same linearization procedure can be followed to linearize any source of nonlinearity in the model with the exception of the nonlinearity present in the objective function.

The solution procedure presented in Figure 9A.1 is followed when the nonlinear terms in the model are linearized. It is important to mention that for the presented model, global optimality cannot be guaranteed through the application of this solution procedure. However, this procedure can be used to provide a feasible starting point prior to solving the exact model.

REFERENCES

Kondili, E., Pantelides, C.C., Sargent, R.W.H., 1993. A general algorithm for short-term scheduling of batch operations I. MILP formulation. *Computers and Chemical Engineering.* 17(2), 211–227.

Majozi, T., Gouws, J., 2009. A mathematical optimization approach for wastewater minimization in multiple contaminant batch plants. *Computers and Chemical Engineering.* 33, 1826–1840.

Majozi, T., Zhu, X.X., 2001. A novel continuous time MILP formulation for multipurpose batch plants. 1. Short-term scheduling. *Industrial and Engineering Chemical Research.* 40(25), 5935–5949.

Quesada, I., Grossmann, I.E., 1995. Global optimization of bilinear process networks with multi-component flows. *Computers and Chemical Engineering.* 19, 1219.

Shah, N., Pantelides, C.C., Sargent, R.W.H., 1993. Optimal periodic scheduling of multipurpose batch plants. *Annals of the Operation Research.* 42(1), 193–228.

Sundaramoorthy, A., Karimi, I.A., 2005. A simple better slot-based continuous-time formulation for short-term scheduling in multipurpose batch plants. *Chemical Engineering Science.* 60, 2679–2702.

Wu, D., Ierapetritou, M., 2004. Cyclic short-term scheduling of multiproduct batch plants using continuous-time representation. *Computers and Chemical Engineering.* 28, 2271–2286.

10 Long-Term Heat Integration in Multipurpose Batch Plants Using Heat Storage

10.1 INTRODUCTION

Most scheduling methods are limited to the short-term scheduling case, and solution of problems over long time horizons may prove challenging or impossible with these methods. Inclusion of additional considerations such as heat integration further complicates the problem. A model for the simultaneous optimization of the schedule and energy usage in heat-integrated multipurpose batch plants operated over long time horizons is presented in this chapter. The method uses a cyclic scheduling solution procedure. The proposed model includes indirect heat integration via heat storage, rather than just direct heat integration. This has largely not been considered in long-term heat integration models in current literature. Both the heat storage size and initial heat storage temperature are also optimized. The solution obtained over 24 h using the proposed cyclic scheduling model with direct heat integration for a multipurpose example is compared to the result obtained from the direct solution and an error of less than 6% is achieved.

10.2 NECESSARY BACKGROUND

In a batch process, discrete tasks follow a specific sequence or recipe, whereby raw materials are transformed to final products. The recipe includes the amounts of materials to be processed as well as the processing times of the various tasks (Majozi, 2010). Batch processes are commonly used for the manufacture of products required in small quantities or for specialty and complex products of high value. Approximately half of all production facilities make use of batch processes (Stoltze et al., 1995). Batch plants are also popular due to their flexible and adaptable nature, which is particularly important in volatile markets.

Heating and cooling are required in most processing facilities. The objective of heat integration is to optimize the use of energy. This becomes a possibility if a process includes both heat generating and heat consuming operations. Heat integration may be accomplished in two ways in a batch process. If the operating schedule allows an overlap in time of hot and cold units, direct heat integration may be used if both units are active. However, due to the time-dependent nature of batch processes it may be necessary to store heat from a hot unit using an intermediate heat storage fluid and reuse this heat at a later time when it is required, resulting in indirect heat integration. The inclusion of heat storage instead of only direct heat integration leads to more flexibility in the process and therefore improved energy usage.

Heat integration in batch plants has in the past been largely disregarded as utility requirements were considered less significant due to the smaller scale of batch operations compared to continuous plants. However, utility requirements in some batch plants, such as in the food and drink industries, dairies, meat processing facilities, biochemical plants and agrochemical facilities, contribute largely to their overall costs. Knopf et al. (1982) analysed a non-continuous cottage cheese process and concluded that capital costs were far outweighed by the energy costs in the plant. Boyadjiev et al. (1996) applied a sequential analysis for direct heat integration in an existing antibiotics plant. The effluent cooling water was used as makeup for the hot water, which reduced the freshwater consumption and wastewater generated. The overall energy costs of the plant decreased by 39%. Rašković et al. (2010) identified an opportunity for significant waste heat recovery in a yeast and ethanol production plant using pinch analysis. Majozi (2009) combined both direct and indirect heat integration with scheduling and applied the model to an agrochemical facility. Savings of more than 75% in external steam consumption were achieved.

Optimal scheduling and equipment use and decreased energy requirements can have a significant effect on the efficiency and revenue of a batch plant. Minimizing energy usage is also influenced by the need to comply with stricter environmental regulations, reduce the effects of higher energy prices and conserve scarce environmental resources.

Early techniques for heat integration in batch processes were based on pinch analysis, originally developed for heat integration in continuous plants at steady state. Variations of the technique for batch processes still appear in literature (Kemp, 1990; Wang and Smith, 1995; Foo et al., 2008; Wang et al., 2014). However, their reliance on a predefined schedule and averaging energy requirements over time intervals may lead to suboptimal results.

For many mathematical heat integration techniques presented in published literature, the processing schedule also tends to be predefined, leading to suboptimal results. Some methods may include heuristic approaches which also cannot guarantee optimality (Vaselenak et al., 1986; Vaklieva-Bancheva et al., 1996; Chen and Ciou, 2008; Halim and Srinivasan, 2009).

Methods specifically applicable to batch plants are required as they capture the essence of time. Scheduling and heat integration may be combined into a single problem for a more optimal solution (Papageorgiou et al., 1994; Pinto et al., 2003; Majozi, 2006, 2009; Chen and Chang, 2009; Stamp and Majozi, 2011; Seid and Majozi, 2014; Lee et al., 2015). However, models may then need to be simplified in order to avoid excessive solution times.

Most heat integration methods are limited to the short-term case and solution of problems over long time horizons may prove challenging or impossible with these current methods. The model of Stamp and Majozi (2011) has been extended in this chapter for the simultaneous optimization of the schedule and energy usage in multipurpose batch plants operated over long time horizons. The proposed model uses the cyclic scheduling concepts and solution procedure presented by Wu and Ierapetritou (2004). These concepts were also used in a similar technique for the solution of long-term wastewater minimization problems in multipurpose batch plants by Nonyane and Majozi (2012). Rather than just considering direct heat integration, the proposed method also includes the concept of indirect heat integration via heat storage, which has not been considered in long-term heat integration models in current literature. The initial heat storage temperature and heat storage capacity are also optimized.

A brief discussion on the concepts used in cyclic scheduling is given in the next section. The problem statement and objectives are then presented. The mathematical model and solution procedure are then discussed. The model is applied to a literature example and an industrial case study, and conclusions are then drawn.

10.3 CYCLIC SCHEDULING CONCEPTS

Shah et al. (1993) proposed an axiom which stated that, in the case where the time horizon of interest is much longer than the individual task durations, there exists a sub-schedule of much shorter duration which, when repeated, can achieve production close to the optimal production achievable for the original long time horizon. This was the basis for the development of cyclic scheduling techniques. It involves obtaining an optimal cyclic schedule which is then repeated and this therefore reduces the problem size. Although cyclic scheduling sacrifices some accuracy when compared to the direct solution of the scheduling problem, this is balanced by the easier and quicker solution of a less complex scheduling problem. Cyclic scheduling methods may be preferred in order to obtain a feasible solution where a solution with short-term methods might not be possible or practical otherwise.

In the cyclic scheduling approach, the length of the cycle is an optimization variable. The unit schedule is then also optimized. For the unit schedule to be repeatable, a certain amount of each intermediate state must be available to start the cyclic period. It is also required that the intermediates be produced and stored at the end of the period to be available for the next period.

In cyclic scheduling, each unit has an individual cycle with cycle time equal to the unit period cycle duration. The units therefore do not need to share the same starting and ending times. This leads to the concept of the "wrapping around" of tasks as proposed by Shah et al. (1993) which allows tasks to cross a unit schedule boundary for improved unit utilization. This concept is demonstrated in Figures 10.1 and 10.2. The cyclic scheduling model of Wu and Ierapetritou (2004) was based on the previous discrete time model of Shah et al. (1993) and uses the continuous time scheduling constraints of Ierapetritou and Floudas (1998) based on the state task network (STN) representation.

The overall long time horizon is divided into three periods (Wu and Ierapetritou, 2004). The optimal unit schedule with optimal cycle length is solved for first, where the objective is to maximize the average profit over the cycle. This is the main period which will be repeated. Sub-ranges for the cycle time may be used rather than considering the entire cycle time at once. As the cycle time is a variable, this makes it

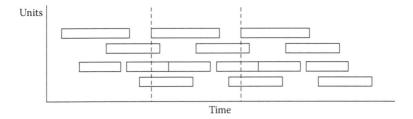

FIGURE 10.1   Representation of a cyclic schedule. (From Wu, D. and Ierapetritou, M., *Comp. Chem. Eng.*, 28, 2271, 2004.)

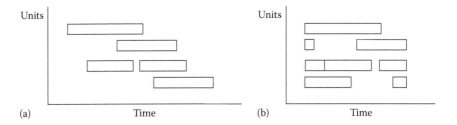

FIGURE 10.2   (a) Unit schedule. (b) With tasks wrapping around. (From Wu, D. and Ierapetritou, M., *Comp. Chem. Eng.*, 28, 2271, 2004.)

easier to determine the required number of time points. These sub-problems may also be solved in parallel and generate alternative schedules with different cycle lengths which may be useful depending on the application.

The initial period, the start-up phase, is then solved. The results from this period ensure that the intermediates required to start the cyclic period are produced. The first objective is to minimize the makespan which ensures there is a feasible solution and the required intermediates are produced in the shortest possible time. The problem is then solved to maximize profit over the time horizon obtained from the makespan minimization problem.

A profit maximization problem is then solved for the final period, which is the remainder of the long time horizon. The final period, shut-down phase, uses up all the remaining intermediates to form products. The initial period and final period profit maximization problems are both mixed integer linear programming (MILP) short-term scheduling problems with known time horizons. The combined lengths of the three different periods add up to the overall long time horizon.

This method can easily be applied to any long time horizon as the initial period is not affected by the overall length of the time horizon considered. The number of cycles can be increased or decreased and the final period solved for again if the overall time horizon changes. The solution procedure and computational complexity are also not affected by a change in the overall time horizon. The method provides another option for the solution of long-term scheduling problems rather than solving them directly with short-term scheduling methods and its usefulness will therefore depend on the application.

10.4 PROBLEM STATEMENT AND OBJECTIVES

For the problem addressed in this chapter, the following is given.

10.4.1 SCHEDULING DATA

1. Production recipe for each product.
2. Available units and their capacities.
3. Maximum storage capacity for each material.
4. Task durations.
5. Time horizon of interest.
6. Costs of raw materials.
7. Selling price of final products.

10.4.2 HEAT INTEGRATION DATA

1. Hot duties for tasks require heating and cold duties for tasks that require cooling.
2. Operating temperatures of heat sources and heat sinks.
3. Minimum allowable temperature differences.
4. Heat capacities of materials.
5. Costs of hot and cold utilities.
6. Design limits on heat storage.

The objectives are then to determine the optimum cycle time and cyclic operating schedule as well as the schedules for the initial and final periods in order to optimize the profit over the given long time horizon.

10.5 MATHEMATICAL MODEL

The mathematical formulation comprises the following sets, continuous variables, binary variables and parameters.

10.5.1 SETS

J $\{j|j = \text{processing unit}\}$

J_c $\{j_c|j_c = \text{processing unit which may conduct tasks requiring heating}\} \subseteq J$

J_h $\{j_h|j_h = \text{processing unit which may conduct tasks requiring cooling}\} \subseteq J$

P $\{p|p = \text{time point}\}$

S $\{s|s = \text{any state}\}$

S_{in} $\{s_{in}|s_{in} = \text{input state into any unit}\}$

$S_{in,j}$ $\{s_{in,j}|s_{in,j} = \text{input state to a processing unit}\} \subseteq S$

$S_{in,j}^*$ $\{s_{in,j}^*|s_{in,j}^* = \text{effective state into a processing unit}\} \subseteq S_{in,j}$

$S_{in,j}^{sc}$ $\{s_{in,j}^{sc}|s_{in,j}^{sc} = \text{task consuming state } s\} \subseteq S_{in,j}$

$S_{in,j}^{sp}$ $\{s_{in,j}^{sp}|s_{in,j}^{sp}$ = task which produces state s, other than a product$\} \subseteq S_{in,j}$

S^p $\{s^p|s^p$ = a state which is a final product$\}$

U $\{u|u$ = heat storage unit$\}$

10.5.2   Continuous Variables

$B\left(s_{in,j}, p\right)$	Batch size, either fixed or variable
$CL\left(s_{in,j_h}, p\right)$	Cooling load for hot state
$\dot{CL}\left(s_{in,j_h}, p\right)$	Cooling load per time, for hot state
$cw\left(s_{in,j_h}, p\right)$	External cooling required by unit j_h conducting the task corresponding to state s_{in,j_h} at time point p
$dur\left(s_{in,j}, p\right)$	Duration of task, dependent on batch size
$extra\_cw\left(u\right)$	Additional cooling required in heat storage vessel
$extra\_st\left(u\right)$	Additional heating required in heat storage vessel
G_{cw}	Glover transformation variable
G_{st}	Glover transformation variable
H	Single cycle length for cyclic scheduling problem
$HL\left(s_{in,j_c}, p\right)$	Heating load for cold state
$\dot{HL}\left(s_{in,j_c}, p\right)$	Heating load per time, for cold state
K_{cw}	Reformulation–linearization variable
K_{st}	Reformulation–linearization variable
$m_u\left(s_{in,j}, p\right)$	Amount of material processed in a unit at time point p
$Q\left(s_{in,j}, u, p\right)$	Heat exchanged with heat storage unit u at time point p
$q\left(s_{in,j_h}, s_{in,j_c}, p\right)$	Amount of heat exchanged during direct heat integration
$q_s\left(s, p\right)$	Amount of state s stored at time point p
$Q_s^0\left(s\right)$	Initial amount of intermediate state s stored (cyclic scheduling)
$st\left(s_{in,j_c}, p\right)$	External heating required by unit j_c conducting the task corresponding to state s_{in,j_c} at time point P
$T_0\left(u, p\right)$	Initial temperature in heat storage unit u at time point p
$T_f\left(u, p\right)$	Final temperature in heat storage unit u at time point p
ΔT_{cw}	Temperature change required in heat storage vessel to return to the starting temperature, when additional cooling is required
ΔT_{st}	Temperature change required in heat storage vessel to return to the starting temperature, when additional heating is required
T_{start}	Temperature of heat storage at beginning of cycle
T_{end}	Temperature of heat storage at end of cycle
$t_0\left(s_{in,j}, u, p\right)$	Time at which heat storage unit u commences activity
$t_f\left(s_{in,j}, u, p\right)$	Time at which heat storage unit u ends activity
$t_u\left(s_{in,j}, p\right)$	Time at which a task starts or state is used in unit j

$t_p\left(s_{in,j}, p\right)$ Time at which a task ends in unit j

$W(u)$ Capacity of heat storage unit u

10.5.3 BINARY VARIABLES

$tt(j,p)$ Binary variable associated with usage of state produced by unit j at time point p

$x\left(s_{in,j_c}, s_{in,j_h}, p\right)$ Binary variable associated with heat integration between unit j_c conducting the task corresponding to state s_{in,j_c} and unit j_h conducting the task corresponding to state s_{in,j_h}, at time point p

x_{cw} Binary variable signifying that extra cooling is required in the heat storage vessel

x_{st} Binary variable signifying that extra heating is required in the heat storage vessel

$y\left(s_{in,j}, p\right)$ Binary variable associated with usage of state s in unit j at time point p

$z\left(s_{in,j}, u, p\right)$ Binary variable associated with heat integration between unit j conducting the task corresponding to state $s_{in,j}$ with heat storage unit u at time point p

10.5.4 PARAMETERS

β Variable coefficient of processing time

$CL\left(s_{in,j_h}\right)$ Fixed cooling load for hot state

cp_{fluid} Specific heat capacity of heat storage fluid

$cp_{state}\left(s_{in,j}\right)$ Specific heat capacity of state

$Cost\_cw$ Cost of cooling water

$Cost\_st$ Cost of steam

$CP(s)$ Selling price of product s, s = product

$HL\left(s_{in,j_c}\right)$ Fixed heating load for cold state

H^U Upper bound for cycle length

MM Any large number

$Q^{max}\left(s_{in,j}\right)$ Maximum possible heating load or cooling load for a cold state or hot state, respectively

$T\left(s_{in,j}\right)$ Operating temperature for processing unit j conducting the task corresponding to state $s_{in,j}$, for constant temperature processes

$T_{in}\left(s_{in,j}\right)$ Inlet temperature for state $s_{in,j}$

$T_{out}\left(s_{in,j}\right)$ Outlet temperature for state $s_{in,j}$

T^L Lower bound for heat storage temperature

T^U Upper bound for heat storage temperature

ΔT^L Lower bound for temperature difference in heat storage vessel

ΔT^U	Upper bound for temperature difference in heat storage vessel
$\Delta T^{\min}$	Minimum allowable thermal driving force
$\tau\left(s_{in,j}\right)$	Duration of the task corresponding to state $s_{in,j}$ conducted in unit j
W^L	Lower bound for heat storage capacity
W^U	Upper bound for heat storage capacity

The mathematical formulation consists of the scheduling constraints, with modifications for cyclic scheduling, as well as the heat integration constraints of Stamp and Majozi (2011). The model uses the scheduling framework of Seid and Majozi (2012), which is based on the state sequence network (SSN) recipe representation and an uneven discretization of the time horizon (Majozi and Zhu, 2001). This scheduling framework has proven to result in fewer time points required compared to models based on other representations, with improvements in computational time as well as objective function in short-term scheduling problems. It also provides for the correct sequencing of tasks for the fixed intermediate storage (FIS) policy.

The mathematical model is based on the superstructure in Figure 10.3 and includes opportunities for both direct and indirect heat integration. Direct heat integration refers to the use of heat generated from a processing unit to supply a processing unit requiring heat without the use of heat storage. Indirect heat integration refers to the use of heat previously stored in a heat storage vessel to supply a processing unit requiring heat. Processing units may also operate in standalone mode, using only external utilities. This may be required for control reasons or when thermal driving forces or time do not allow for heat integration. If either direct or indirect heat integration is not sufficient to satisfy the required duty, external utilities may make up for any deficit.

Constraints (10.1) and (10.2) are active simultaneously and ensure that one hot unit will be integrated with one cold unit when direct heat integration takes place in order to simplify operation of the process. Also, if two units are to be heat integrated at a given time point, they must both be active at that time point. However, if a unit is active, it may operate in either integrated or standalone mode.

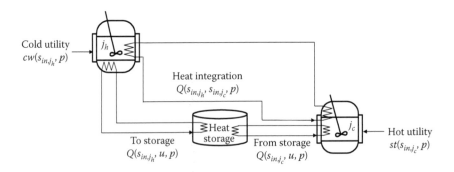

FIGURE 10.3 Superstructure for mathematical formulation when units perform heating/cooling tasks.

$$\sum_{s_{in,jc}} x\left(s_{in,j_c}, s_{in,j_h}, p\right) \le y\left(s_{in,j_h}, p\right), \quad \forall p \in P, \quad s_{in,j_h} \in S_{in,j} \tag{10.1}$$

$$\sum_{s_{in,jh}} x\left(s_{in,j_c}, s_{in,j_h}, p\right) \le y\left(s_{in,j_c}, p\right), \quad \forall p \in P, \quad s_{in,j_c} \in S_{in,j} \tag{10.2}$$

Constraints (10.3) and (10.4) ensure that heat integration between a unit and heat storage may occur only if the unit is active at that time point. However, if a unit is active, it will not necessarily integrate with heat storage.

$$z\left(s_{in,j_c}, u, p\right) \le y\left(s_{in,j_c}, p\right), \quad \forall p \in P, \quad s_{in,j_c} \in S_{in,j}, \quad u \in U \tag{10.3}$$

$$z\left(s_{in,j_h}, u, p\right) \le y\left(s_{in,j_h}, p\right), \quad \forall p \in P, \quad s_{in,j_h} \in S_{in,j}, \quad u \in U \tag{10.4}$$

Constraint (10.5) ensures that heat storage is heat integrated with either one hot unit or one cold unit at any point in time. This is to simplify and improve operational efficiency in the plant.

$$\sum_{s_{in,jc}} z\left(s_{in,j_c}, u, p\right) + \sum_{s_{in,jh}} z\left(s_{in,j_h}, u, p\right) \le 1, \quad \forall p \in P, \quad u \in U \tag{10.5}$$

Constraints (10.6) and (10.7) ensure that a unit cannot simultaneously undergo direct and indirect heat integration. This condition simplifies the operation of the process.

$$\sum_{s_{in,jh}} x\left(s_{in,j_c}, s_{in,j_h}, p\right) + z\left(s_{in,j_c}, u, p\right) \le 1, \quad \forall p \in P, \quad s_{in,j_c} \in S_{in,j}, \quad u \in U \tag{10.6}$$

$$\sum_{s_{in,jc}} x\left(s_{in,j_c}, s_{in,j_h}, p\right) + z\left(s_{in,j_h}, u, p\right) \le 1, \quad \forall p \in P, \quad s_{in,j_h} \in S_{in,j}, \quad u \in U \tag{10.7}$$

Constraints (10.8) and (10.9) quantify the amount of heat received from or transferred to the heat storage unit, respectively. There will be no heat received or transferred if the binary variable signifying usage of the heat storage vessel, $z\left(s_{in,j}, u, p\right)$, is zero. These constraints are active over the entire time horizon, where P is the current time point and $p-1$ is the previous time point.

$$Q\left(s_{in,j_c}, u, p-1\right) = W\left(u\right) cp_{fluid}\left(T_0\left(u, p-1\right) - T_f\left(u, p\right)\right) z\left(s_{in,j_c}, u, p-1\right),$$

$$\forall p \in P, \quad p > p0, \quad s_{in,j_c} \in S_{in,j}, \quad u \in U \tag{10.8}$$

$$Q\left(s_{in,j_h}, u, p-1\right) = W\left(u\right) cp_{fluid}\left(T_f\left(u, p\right) - T_0\left(u, p-1\right)\right) z\left(s_{in,j_h}, u, p-1\right),$$

$$\forall p \in P, \quad p > p0, \quad s_{in,jk} \in S_{in,j}, \quad u \in U \tag{10.9}$$

Constraint (10.10) quantifies the heat transferred to the heat storage vessel at the beginning of the time horizon. The initial temperature of the heat storage fluid is $T_0(u, p0)$.

$$Q(s_{in,j_h}, u, p0) = W(u) cp_{fluid}(T_f(u, p1) - T_0(u, p0)) z(s_{in,j_h}, u, p0),$$

$$\forall s_{in,j_h} \in S_{in,j}, \quad u \in U \tag{10.10}$$

Constraint (10.11) ensures that the final temperature of the heat storage fluid at any time point becomes the initial temperature of the heat storage fluid at the next time point. This condition will hold regardless of whether or not there was heat integration at the previous time point.

$$T_0(u, p) = T_f(u, p-1), \quad \forall p \in P, \quad u \in U \tag{10.11}$$

Constraints (10.12) and (10.13) ensure that temperature of heat storage does not change if there is no heat integration with the heat storage unit, unless there is heat loss from the heat storage unit. MM is any large number, thereby resulting in an overall "Big M" formulation. If either $z(s_{in,j_c}, u, p-1)$ or $z(s_{in,j_h}, u, p-1)$ is equal to one, Constraints (10.12) and (10.13) will be redundant. However, if these two binary variables are both zero, the initial temperature at the previous time point will be equal to the final temperature at the current time point.

$$T_0(u, p-1) \leq T_f(u, p) + MM\left(\sum_{s_{in,j_c}} z(s_{in,j_c}, u, p-1) + \sum_{s_{in,j_h}} z(s_{in,j_h}, u, p-1)\right),$$

$$\forall p \in P, \quad p > p0, \quad u \in U \tag{10.12}$$

$$T_0(u, p-1) \geq T_f(u, p) - MM\left(\sum_{s_{in,j_c}} z(s_{in,j_c}, u, p-1) + \sum_{s_{in,j_h}} z(s_{in,j_h}, u, p-1)\right),$$

$$\forall p \in P, \quad p > p0, \quad u \in U \tag{10.13}$$

Constraint (10.14) ensures that minimum thermal driving forces are obeyed when there is direct heat integration between a hot and a cold unit. This constraint holds when both hot and cold units operate at constant temperature, which is commonly encountered in practice. An example is when there is heat integration between an exothermic and an endothermic reaction.

$$T(s_{in,j_h}) - T(s_{in,j_c}) \geq \Delta T^{min} - MM(1 - x(s_{in,j_c}, s_{in,j_h}, p-1)),$$

$$\forall p \in P, \quad p > p0, \quad s_{in,j_c}, s_{in,j_h} \in S_{in,j} \tag{10.14}$$

Constraints (10.15) and (10.16) ensure that minimum thermal driving forces are obeyed when there is heat integration with the heat storage unit. Constraint (10.15) applies for heat integration between heat storage and a heat sink, while constraint (10.16) applies for heat integration between heat storage and a heat source. In Constraints (10.15) and (10.16), the units operate at fixed temperatures. For units not operating at fixed temperatures, both inlet and outlet minimal thermal driving forces between the two integrated tasks need also to be enforced.

$$T_f\left(u,p\right)-T\left(s_{in,j_c}\right)\geq \Delta T^{\min}-MM\left(1-z\left(s_{in,j_c},u,p-1\right)\right),$$

$$\forall p\in P,\quad p>p0,\quad s_{in,j_c}\in S_{in,j},\quad u\in U \tag{10.15}$$

$$T\left(s_{in,j_h}\right)-T_f\left(u,p\right)\geq \Delta T^{\min}-MM\left(1-z\left(s_{in,j_h},u,p-1\right)\right),$$

$$\forall p\in P,\quad p>p0,\quad s_{in,j_c}\in S_{in,j},\quad u\in U \tag{10.16}$$

Constraints (10.17) and (10.18) give the heating load for a cold state and cooling load for a hot state, respectively, for variable batch size and changing temperature.

$$HL\left(s_{in,j_c},p\right)=B\left(s_{in,j_c},p\right)cp_{state}\left(s_{in,j_c}\right)\left(T_{out}\left(s_{in,j_c}\right)-T_{in}\left(s_{in,j_c}\right)\right),\quad \forall p\in P,\quad s_{in,j_c}\in S_{in,j} \tag{10.17}$$

$$CL\left(s_{in,j_h},p\right)=B\left(s_{in,j_h},p\right)cp_{state}\left(s_{in,j_h}\right)\left(T_{in}\left(s_{in,j_h}\right)-T_{out}\left(s_{in,j_h}\right)\right),\quad \forall p\in P,\quad s_{in,j_h}\in S_{in,j} \tag{10.18}$$

Constraint (10.19) ensures that the heating of a cold state will be satisfied by either direct or indirect heat integration as well as external utility if required.

$$HL\left(s_{in,j_c},p\right)=Q\left(s_{in,j_c},u,p\right)+st\left(s_{in,j_c},p\right)+\sum_{s_{in,j_h}}q\left(s_{in,j_c},s_{in,j_h},p\right),$$

$$\forall p\in P,\quad s_{in,j_c}\in S_{in,j},\quad u\in U \tag{10.19}$$

Constraint (10.20) states that the cooling of a hot state will be satisfied by either direct or indirect heat integration as well as external utility if required.

$$CL\left(s_{in,j_h},p\right)=Q\left(s_{in,j_h},u,p\right)+cw\left(s_{in,j_h},p\right)+\sum_{s_{in,j_c}}q\left(s_{in,j_c},s_{in,j_h},p\right),$$

$$\forall p\in P,\quad s_{in,j_k}\in S_{in,j},\quad u\in U \tag{10.20}$$

The upper bounds of the heating load of a cold state, the cooling load of a hot state and the amount of heat exchanged during direct heat integration are given in Constraints (10.21) through (10.23).

$$HL\left(s_{in,j_c},p\right)\leq y\left(s_{in,j_c},p\right)Q^{\max}\left(s_{in,j_c}\right),\quad \forall p\in P,\quad s_{in,j_c}\in S_{in,j} \tag{10.21}$$

$$CL\left(s_{in,j_h},p\right)\leq y\left(s_{in,j_h},p\right)Q^{\max}\left(s_{in,j_h}\right),\quad \forall p\in P,\quad s_{in,j_k}\in S_{in,j} \tag{10.22}$$

$$q\left(s_{in,j_c}, s_{in,j_h}, p\right) \le x\left(s_{in,j_c}, s_{in,j_h}, p\right) \min\left\{Q^{max}\left(s_{in,j_c}\right), Q^{max}\left(s_{in,j_h}\right)\right\}$$

$$\forall p \in P, \quad s_{in,j_c}, s_{in,j_h} \in S_{in,j} \tag{10.23}$$

For the specific case where the heating and cooling loads are fixed, Constraints (10.24) and (10.25) are used instead of Constraints (10.19) and (10.20).

$$HL\left(s_{in,j_c}\right) y\left(s_{in,j_c}, p\right) = Q\left(s_{in,j_c}, u, p\right) + st\left(s_{in,j_c}, p\right) + x\left(s_{in,j_c}, s_{in,j_h}, p\right)$$
$$\times \sum_{s_{in,jh}} \min_{s_{in,jc}, s_{in,jh}} \left\{HL\left(s_{in,j_c}\right), CL\left(s_{in,j_h}\right)\right\},$$
$$\forall p \in P, \quad s_{in,j_c} \in S_{in,j}, \quad u \in U \tag{10.24}$$

$$CL\left(s_{in,j_h}\right) y\left(s_{in,j_h}, p\right) = Q\left(s_{in,j_h}, u, p\right) + cw\left(s_{in,j_h}, p\right) + x\left(s_{in,j_c}, s_{in,j_h}, p\right)$$
$$\times \sum_{s_{in,jc}} \min_{s_{in,jc}, s_{in,jh}} \left\{HL\left(s_{in,j_c}\right), CL\left(s_{in,j_h}\right)\right\},$$
$$\forall p \in P, \quad s_{in,j_h} \in S_{in,j}, \quad u \in U \tag{10.25}$$

The amount of heat transferred through direct heat integration will be limited by the smaller heating or cooling requirement of the heat integrated tasks. Constraints (10.26) and (10.27) express this. Constraint (10.26) calculates the heat load of the cold task, while Constraint (10.27) calculates the cooling load of the hot task.

$$q\left(s_{in,j_c}, s_{in,j_h}, p\right) \le B\left(s_{in,j_c}, p\right) cp_{state}\left(s_{in,j_c}\right)\left(T_{out}\left(s_{in,j_c}\right) - T_{in}\left(s_{in,j_c}\right)\right) x\left(s_{in,j_c}, s_{in,j_h}, p\right),$$

$$\forall p \in P, \quad s_{in,j_c}, s_{in,j_h} \in S_{in,j} \tag{10.26}$$

$$q\left(s_{in,j_c}, s_{in,j_h}, p\right) \le B\left(s_{in,j_h}, p\right) cp_{state}\left(s_{in,j_h}\right)\left(T_{in}\left(s_{in,j_h}\right) - T_{out}\left(s_{in,j_h}\right)\right) x\left(s_{in,j_c}, s_{in,j_h}, p\right),$$

$$\forall p \in P, \quad s_{in,j_c}, s_{in,j_h} \in S_{in,j} \tag{10.27}$$

Furthermore, it is possible that a given pair of tasks cannot be heat integrated or that a possible ΔT^{min} violation may occur. The possibility of heat integration between pairs of tasks as well as possible ΔT^{min} violations should be investigated for each pair of hot and cold tasks beforehand. If ΔT^{min} violations occur, the temperatures in Constraints (10.26) and (10.27) should be adjusted for this.

The amount of heat transferred through direct heat integration can also be limited by the duration of the shorter task if the tasks have different durations. Constraints (10.28) through (10.33) capture this. Constraints (10.28) and (10.29) calculate the heating load per time and cooling load per time, of the cold task and hot task, respectively.

$$\dot{HL}\left(s_{in,j_c}, p\right) = \frac{HL\left(s_{in,j_c}, p\right)}{dur\left(s_{in,j_c}, p\right)}, \quad \forall p \in P, \quad s_{in,j_c} \in S_{in,j} \tag{10.28}$$

$$\dot{CL}\left(s_{in,j_h},p\right)=\frac{CL\left(s_{in,j_h},p\right)}{dur\left(s_{in,j_h},p\right)}, \quad \forall p \in P, \quad s_{in,j_h} \in S_{in,j} \tag{10.29}$$

Constraint (10.30) calculates the heat load of the cold task based on the duration of the same cold task. Constraint (10.31) calculates the heat load of the cold task based on the duration of the hot task. Constraint (10.32) calculates the cooling load of the hot task based on the duration of the same hot task. Constraint (10.33) calculates the cooling load of the hot task based on the duration of the cold task. The amount of heat integrated directly will effectively be the minimum of these four quantities.

$$q\left(s_{in,j_c},s_{in,j_h},p\right)\le \dot{HL}\left(s_{in,j_c},p\right)dur\left(s_{in,j_c},p\right), \quad \forall p \in P, \quad s_{in,j_c},s_{in,j_h} \in S_{in,j} \tag{10.30}$$

$$q\left(s_{in,j_c},s_{in,j_h},p\right)\le \dot{HL}\left(s_{in,j_c},p\right)dur\left(s_{in,j_h},p\right), \quad \forall p \in P, \quad s_{in,j_c},s_{in,j_h} \in S_{in,j} \tag{10.31}$$

$$q\left(s_{in,j_c},s_{in,j_h},p\right)\le \dot{CL}\left(s_{in,j_h},p\right)dur\left(s_{in,j_h},p\right), \quad \forall p \in P, \quad s_{in,j_c},s_{in,j_h} \in S_{in,j} \tag{10.32}$$

$$q\left(s_{in,j_c},s_{in,j_h},p\right)\le \dot{CL}\left(s_{in,j_h},p\right)dur\left(s_{in,j_c},p\right), \quad \forall p \in P, \quad s_{in,j_c},s_{in,j_h} \in S_{in,j} \tag{10.33}$$

In Constraints (10.28) through (10.33), the duration is a function of batch size. If the duration is fixed, $\tau\left(s_{in,j}\right)$ is used and these constraints are then linear.

Constraints (10.34) and (10.35) ensure that the times at which units are active are synchronized when direct heat integration takes place. Starting times for the tasks in the integrated units are the same. This constraint may be relaxed for operations requiring preheating or precooling and is dependent on the process.

$$t_u\left(s_{in,j_h},p\right)\ge t_u\left(s_{in,j_c},p\right)-MM\left(1-x\left(s_{in,j_c},s_{in,j_h},p\right)\right) \quad \forall p \in P, \quad s_{in,j_c},s_{in,j_h} \in S_{in,j}$$
$$\tag{10.34}$$

$$t_u\left(s_{in,j_h},p\right)\le t_u\left(s_{in,j_c},p\right)+MM\left(1-x\left(s_{in,j_c},s_{in,j_h},p\right)\right) \quad \forall p \in P, \quad s_{in,j_c},s_{in,j_h} \in S_{in,j}$$
$$\tag{10.35}$$

Constraints (10.36) and (10.37) ensure that if indirect heat integration takes place, the time at which a heat storage unit starts either to transfer or receive heat will be equal to the time a unit is active.

$$t_u\left(s_{in,j},p\right)\ge t_0\left(s_{in,j},u,p\right)-MM\left(y\left(s_{in,j},p\right)-z\left(s_{in,j},u,p\right)\right)$$
$$\forall p \in P, \quad u \in U, \quad s_{in,j} \in S_{in,j} \tag{10.36}$$

$$t_u\left(s_{in,j},p\right)\le t_0\left(s_{in,j},u,p\right)+MM\left(y\left(s_{in,j},p\right)-z\left(s_{in,j},u,p\right)\right)$$
$$\forall p \in P, \quad u \in U, \quad s_{in,j} \in S_{in,j} \tag{10.37}$$

Constraints (10.38) and (10.39) state that the time when heat transfer to or from a heat storage unit is finished will coincide with the time the task transferring or receiving heat has finished processing.

$$t_u\left(s_{in,j}, p-1\right) + \tau\left(s_{in,j}\right) y\left(s_{in,j}, p-1\right) \geq t_f\left(s_{in,j}, u, p\right)$$
$$- MM\left(y\left(s_{in,j}, p-1\right) - z\left(s_{in,j}, u, p-1\right)\right) \quad \forall \in P, \quad p > p0, \quad u \in U, \quad s_{in,j} \in S_{in,j}$$

$$(10.38)$$

$$t_u\left(s_{in,j}, p-1\right) + \tau\left(s_{in,j}\right) y\left(s_{in,j}, p-1\right) \leq t_f\left(s_{in,j}, u, p\right)$$
$$+ MM\left(y\left(s_{in,j}, p-1\right) - z\left(s_{in,j}, u, p-1\right)\right) \quad \forall p \in P, \quad p > p0, \quad u \in U, \quad s_{in,j} \in S_{in,j}$$

$$(10.39)$$

The necessary constraints for determining the optimal cyclic schedule are now described. The scheduling model of Seid and Majozi (2012) was used. Some constraints were modified to accommodate cyclic scheduling and are shown here. The value for "Big M" in the cyclic scheduling constraints and heat integration constraints is the value of the upper bound of the cycle length, H^U.

10.5.5 Mass Balance between Two Cycles

Constraint (10.40) ensures that the amount of intermediate state s stored at the end of a cycle is the same as that stored at the beginning of the cycle. There is therefore no accumulation or shortage of intermediate states when the cycle is repeated.

$$q_s\left(s, p\right) = Q_s^0\left(s\right), \quad \forall s \in S, \quad s \neq product, feed, \quad p = |P| \tag{10.40}$$

10.5.6 Sequence Constraints between Cycles: Completion of Previous Tasks

Constraint (10.41) defines the relationship between the last task of the previous cycle and the first task of the current cycle when tasks occur in different units. This maintains continuity of tasks between cycles.

$$t_u\left(s_{in,j'}, p0\right) \geq t_p\left(s_{in,j}, p\right) - H^U\left(1 - y\left(s_{in,j}, p-2\right)\right) - H,$$
$$\forall j \in J, \quad p = |P|, \quad s_{in,j} \in S_{in,j}^{sp}, \quad s_{in,j'} \in S_{in,j'}^{sc} \tag{10.41}$$

Constraint (10.42) is similar to Constraint (10.41), but is applicable when intermediate state s is produced from one unit.

$$t_u\left(s_{in,j'}, p0\right) \geq t_p\left(s_{in,j}, p\right) - H^U\left(2 - y\left(s_{in,j}, p-1\right) - tt\left(j, p\right)\right) - H,$$
$$\forall j \in J, \quad p = |P|, \quad s_{in,j} \in S_{in,j}^{sp}, \quad s_{in,j'} \in S_{in,j}^{sc} \tag{10.42}$$

Constraint (10.43) ensures that units are available when the same task is performed in the same unit (the same state is used in the same unit). A state can only be used in a unit after all preceding tasks in the unit have been completed.

$$t_u\left(s_{in,j}, p0\right) \geq t_p\left(s_{in,j}, p\right) - H, \quad \forall j \in J, \quad p = |P|, \quad s_{in,j} \in S_{in,j}^* \qquad (10.43)$$

Constraint (10.44) is similar to Constraint (10.43), but pertains to different tasks being performed in the same unit (different states used in the same unit). A task can only start in a unit after all previous tasks in the unit are complete.

$$t_u\left(s_{in,j}, p0\right) \geq t_p\left(s'_{in,j}, p\right) - H, \quad \forall j \in J, \quad p = |P|, \quad s_{in,j} \neq s'_{in,j}, \quad s_{in,j}, s'_{in,j} \in S_{in,j}^*$$

$$(10.44)$$

10.5.7 Tightening Constraints

This is the same constraint as in the scheduling model, but the time horizon is now the cycle length. Constraint (10.45) is used to tighten the model. The sum of the durations of all tasks in a unit must be within one cycle length.

$$\sum_{s_{in,j} \in S_{in,j}^*} \sum_p \left(\tau\left(s_{in,j}\right) y\left(s_{in,j}, p\right) + \beta\left(s_{in,j}\right) m_u\left(s_{in,j}, p\right)\right) \leq H, \quad \forall p \in P, \quad j \in J \qquad (10.45)$$

Constraints (10.46) and (10.47) ensure processing tasks take place within two cycles.

$$t_u\left(s_{in,j}, p\right) \leq 2H, \quad \forall s_{in,j} \in S_{in,j}, \quad p \in P, \quad j \in J \qquad (10.46)$$

$$t_p\left(s_{in,j}, p\right) \leq 2H, \quad \forall s_{in,j} \in S_{in,j}, \quad p \in P, \quad j \in J \qquad (10.47)$$

Constraints (10.48) and (10.49) ensure that the times the heat storage unit is active are within two cycles.

$$t_0\left(s_{in,j}, u, p\right) \leq 2H, \quad \forall s_{in,j} \in S_{in,j}, \quad p \in P, \quad j \in J, \quad u \in U \qquad (10.48)$$

$$t_f\left(s_{in,j}, u, p\right) \leq 2H, \quad \forall s_{in,j} \in S_{in,j}, \quad p \in P, \quad j \in J, \quad u \in U \qquad (10.49)$$

Constraints (10.50) through (10.73) cater for the wrapping of the heat storage temperature and ensure the temperature in heat storage is the same at the beginning and end of the cycle in order for the cycle to be repeated.

Constraints (10.50) and (10.51) are defined, where T_{start} is the initial heat storage temperature at the beginning of the cycle and T_{end} is the heat storage temperature at the end of the cycle.

$$\Delta T_{cw} = T_{end} - T_{start} \qquad (10.50)$$

$$\Delta T_{st} = T_{start} - T_{end} \qquad (10.51)$$

If the heat storage temperature at the end of the cycle is higher than the temperature at the beginning of the cycle, cooling will be required to bring the storage temperature back to the starting temperature—essential for repetition of the cycle. ΔT_{cw} will then be positive. If the heat storage temperature at the end of the cycle is lower than the temperature at the beginning of the cycle, heating will be required and ΔT_{st} will be positive.

Constraint (10.52) is used if the heat storage temperature at the end of the cycle is too high and cooling is required to bring it back to the temperature at the beginning of the cycle. Constraint (10.53) is similar, but will be used if extra heating is required at the end of the cycle to bring the heat storage temperature back to the starting temperature.

$$extra\_cw(u) = W(u)\ cp_{fluid}\Delta T_{cw}x_{cw} \tag{10.52}$$

$$extra\_st(u) = W(u)\ cp_{fluid}\Delta T_{st}x_{st} \tag{10.53}$$

Both Constraints (10.52) and (10.53) contain trilinear terms, where a binary variable and two continuous variables are multiplied. This results in a non-convex mixed integer nonlinear programming (MINLP) formulation. The bilinearity resulting from the multiplication of a continuous variable with a binary variable may be handled effectively with the Glover transformation (Glover, 1975). This is an exact linearization technique and as such will not compromise the accuracy of the model. Constraints (10.54) and (10.55) are then obtained.

$$extra\_cw(u) = W(u)\ cp_{fluid}G_{cw} \tag{10.54}$$

$$extra\_st(u) = W(u)\ cp_{fluid}G_{st} \tag{10.55}$$

Constraints (10.54) and (10.55) still contain bilinear terms where two continuous variables are multiplied. A method to handle this is a reformulation–linearization technique (Sherali and Alameddine, 1992) as discussed by Quesada and Grossmann (1995). This is an inexact linearization technique and increases the size of the model by an additional type of continuous variable and four types of continuous constraints. These constraints correspond to the convex and concave envelopes of the bilinear terms over the given bounds. Constraint (10.54) is linearized using Constraints (10.56) through (10.62) and Constraint (10.55) is linearized using Constraints (10.63) through (10.69). Constraints (10.70) and (10.71) are then obtained.

$$W(u)G_{cw} = K_{cw} \tag{10.56}$$

$$W^L \le W(u) \le W^U \tag{10.57}$$

$$\Delta T^L \le G_{cw} \le \Delta T^U \tag{10.58}$$

$$K_{cw} \le W^U G_{cw} + \Delta T^L W(u) - W^U \Delta T^L \tag{10.59}$$

$$K_{cw} \leq W^L G_{cw} + \Delta T^U W(u) - W^L \Delta T^U \tag{10.60}$$

$$K_{cw} \geq W^L G_{cw} + \Delta T^L W(u) - W^L \Delta T^L \tag{10.61}$$

$$K_{cw} \geq W^U G_{cw} + \Delta T^U W(u) - W^U \Delta T^U \tag{10.62}$$

$$W(u) G_{st} = K_{st} \tag{10.63}$$

$$W^L \leq W(u) \leq W^U \tag{10.64}$$

$$\Delta T^L \leq G_{st} \leq \Delta T^U \tag{10.65}$$

$$K_{st} \leq W^U G_{st} + \Delta T^L W(u) - W^U \Delta T^L \tag{10.66}$$

$$K_{st} \leq W^L G_{st} + \Delta T^U W(u) - W^L \Delta T^U \tag{10.67}$$

$$K_{st} \geq W^L G_{st} + \Delta T^L W(u) - W^L \Delta T^L \tag{10.68}$$

$$K_{st} \geq W^U G_{st} + \Delta T^U W(u) - W^U \Delta T^U \tag{10.69}$$

$$extra\_cw(u) = cp_{fluid} K_{cw} \tag{10.70}$$

$$extra\_st(u) = cp_{fluid} K_{st} \tag{10.71}$$

Constraints (10.8) through (10.10) also have trilinear terms where a binary variable and two continuous variables are multiplied, while Constraints (10.28) through (10.33) have bilinear terms where two continuous variables are multiplied. These constraints are linearized similarly.

Constraint (10.72) is also necessary, as either extra cooling or extra heating will be required for the heat storage vessel at the end of the cycle, or neither will be required.

$$x_{cw} + x_{st} \leq 1 \tag{10.72}$$

If neither extra heating nor extra cooling are required to bring the heat storage temperature back to its initial temperature, T_{start} will be equal to T_{end} and the values for $extra\_cw(u)$ and $extra\_st(u)$ will both be zero.

In order for the heat storage temperature to return to its starting temperature, the total energy into the heat storage vessel must be equal to the energy out of the heat storage vessel within the cycle. Constraint (10.73) is used to ensure this.

$$\sum_{s_{in.jh}} \sum_p Q(s_{in.jh}, u, p) + extra\_st(u) = \sum_{s_{in.jc}} \sum_p Q(s_{in.jc}, u, p) + extra\_cw(u) \tag{10.73}$$

The profit is given by Equation 10.74.

$$\text{Profit} = \sum_{s} CP\left(s^{p}\right) q_{s}\left(s^{p}, p\right) - Cost\_cw \sum_{s_{in,jh}} \sum_{p} cw\left(s_{in,jh}, p\right) - Cost\_cw * extra\_cw\left(u\right)$$

$$- Cost\_st \sum_{s_{in,jc}} \sum_{p} st\left(s_{in,jc}, p\right) - Cost\_st * extra\_st\left(u\right),$$

$$\forall p = P, \quad s_{in,jh} \in S_{in,j}, \quad s_{in,jc} \in S_{in,j}, \quad s^{p} \in S^{p} \tag{10.74}$$

The objective for the cyclic period is then to maximize the average profit per cycle, as given by Constraint (10.75).

$$\max \frac{\text{Profit}}{H} \tag{10.75}$$

The same solution procedure as used by Stamp and Majozi (2011) is applied. The overall MINLP model is linearized and solved as a MILP. The solution obtained is then used as a starting point for the exact MINLP model. If the solutions from the two models are equal, the solution is globally optimal, as global optimality can be proven for MILP problems. If the solutions differ, the MINLP solution is locally optimal. The possibility also exists that no feasible starting point is found.

The objective function for the cyclic portion of the problem, Constraint (10.75) is, however, nonlinear and the MINLP can therefore not be linearized completely to a MILP due to the nonlinear objective function. All of the nonlinear constraints apart from the objective function are therefore linearized to produce a relaxed MINLP problem, the solution of which provides a starting point for the exact MINLP problem. The resulting solution to the exact MINLP cannot then be guaranteed to be a global optimum.

Constraints (10.40) through (10.51) and Constraints (10.56) through (10.75) are used in the linearized model, while Constraints (10.40) through (10.51), Constraints (10.54) and (10.55) and Constraints (10.72) through (10.75) are used in the exact model. This is in addition to the scheduling and heat integration constraints and applies to the determination of the optimal cyclic schedule. Once the optimal cycle length has been determined, the initial and final periods can be solved. The overall objective is the maximization of profit over the entire long time horizon.

10.6   MULTIPURPOSE EXAMPLE

Figure 10.4 shows the STN of a multipurpose batch plant, while Figure 10.5 shows the SSN(Seid and Majozi, 2012). This example is commonly referred to as "BATCH1" in literature and has been modified to include heating and cooling requirements for certain tasks to allow for the possibility of heat integration. For this example, the batch sizes were not fixed and the energy requirements varied linearly with batch size.

The required scheduling data may be found in Tables 10.1 and 10.2, while the necessary heat integration data and heating and cooling requirements may be found in Tables 10.3 and 10.4, respectively.

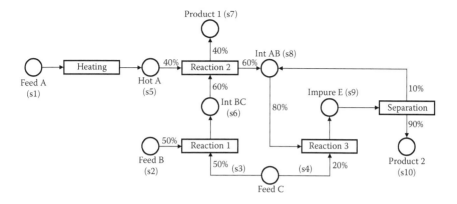

FIGURE 10.4   STN for multipurpose example.

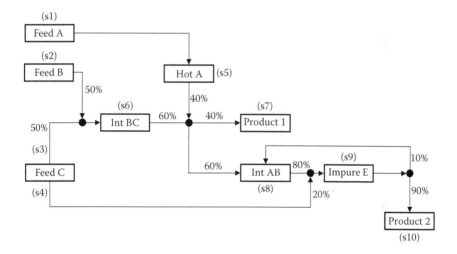

FIGURE 10.5   SSN for multipurpose example.

TABLE 10.1
Scheduling Data for Multipurpose Example

Unit	Capacity	Suitability	Mean Processing Time (h)
Heater	100	Heating	1
Reactor 1	50	RX1, RX2, RX3	2, 2, 1
Reactor 2	80	RX1, RX2, RX3	2, 2, 1
Still	200	Separation	1 (Product 2)
			2 (Int AB)

TABLE 10.2
Scheduling Data for Multipurpose Example

State	Storage Capacity (ton)	Initial Amount (ton)	Revenue (c.u./ton)
Feed A (s1)	Unlimited	Unlimited	0
Feed B (s2)	Unlimited	Unlimited	0
Feed C (s3/s4)	Unlimited	Unlimited	0
Hot A (s5)	100	0	0
Int AB (s8)	200	0	0
Int BC (s6)	150	0	0
Impure E (s9)	200	0	0
Product 1	Unlimited	0	100
Product 2	Unlimited	0	100

TABLE 10.3
Heat Integration Data for Multipurpose Example

Parameter	Value
Specific heat capacity, cp_{fluid} (kJ/kg °C)	4.2
Product selling price (c.u./ton)	100
Steam cost (c.u./kWh)	10
Cooling water cost (c.u./kWh)	2
ΔT^{min} (°C)	10
T^L (°C)	20
T^U (°C)	250
ΔT^L (°C)	0
ΔT^U (°C)	250
W^L (ton)	0.1
W^U (ton)	4

TABLE 10.4
Heating/Cooling Requirements for Multipurpose Example

Task	Type	Max Heating/Cooling Requirement (kWh)	Operating Temperature (°C)
RX1	Exothermic	60 (cooling)	100
RX2	Endothermic	80 (heating)	60
RX3	Exothermic	70 (cooling)	140

TABLE 10.5

Results of Cyclic Portion for Multipurpose Example

Cycle Range (h)	Optimal Cycle Time (h)	Objective (c.u./h)	Time Points	CPU Time (s)
3–6	5	3581.920	6	204.55
6–9	6	3478.086	7	1,781.50
9–12	12	3560.000	9	280,048.74

TABLE 10.6

Results of Optimal Cyclic Length for Multipurpose Example

Starting and ending amount of intermediate state (ton)	
s5	0
s6	0
s8	48
s9	150
Starting and ending storage temperature (°C)	70
Heat storage size (ton)	0.357
Cooling water (kWh)	95.2
Steam (kWh)	0
Extra cooling water (kWh)	0
Extra steam (kWh)	0

10.6.1 CYCLIC PORTION

As the model used to solve the cyclic portion contains nonlinear terms, the solution procedure as described earlier this chapter is used. The objective function will, however, still contain a nonlinearity which cannot be linearized and as such both the linearized and exact models will be MINLP problems. The problem was solved with GAMS 24.1.3 using DICOPT for the MINLP with CPLEX as the MIP solver and MINOS5 as the NLP solver. All models were solved using a computer with an Intel Core i3, 3.1 GHz processor with 2.0 GB RAM.

The results obtained for the cyclic portion can be seen in Table 10.5. The optimal cycle length is 5 h and an average profit of 3581.920 c.u./h is achieved. The CPU times in the table are the sum of the CPU times for both the linearized and the exact models.

The results obtained for the optimal schedule for the optimal cycle length of 5 h are shown in Table 10.6. The optimal schedule obtained is shown in Figure 10.6.

10.6.2 INITIAL PERIOD

In order to solve for the initial period of the time horizon, the amounts of intermediates at the end of the initial period are fixed to the values required to start the

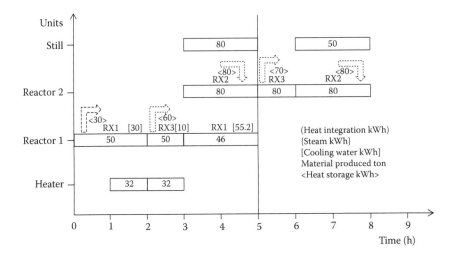

FIGURE 10.6   Gantt chart for optimal cycle length (5 h) for multipurpose example.

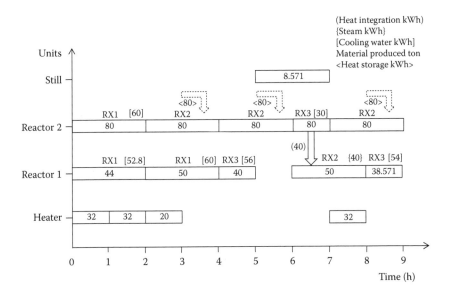

FIGURE 10.7   Gantt chart for maximization of profit over initial period (9 h) for multipurpose example.

cyclic period. These values were obtained from the solution of the cyclic period. This ensures there will be sufficient of the intermediate states produced in the initial period to be available to start the cyclic scheduling period. First, a makespan minimization problem is solved to determine a feasible time horizon for the initial period. A profit maximization problem is then solved in order to maximize the profit in the

TABLE 10.7

Results for Maximization of Profit over Initial Period for Multipurpose Example

Time period (h)	9
Objective (c.u.)	11,345.829
Time points	6
Cooling water (kWh)	321.8
Steam (kWh)	40
Initial heat storage temperature (°C)	230.064
CPU time (s)	0.983

initial period. The optimum heat storage size of 0.357 ton which was solved for in the cyclic portion is now fixed in the initial period. Since this is the case, both the makespan minimization and profit maximization problems will be MILP problems. The final temperature for the heat storage vessel is set at 70°C.

From the makespan minimization problem, a feasible time horizon of 9 h was determined for the initial period, in a CPU time of 0.593 s and required 6 time points. The Gantt chart for the profit maximization problem over 9 h is shown in Figure 10.7 and the results are shown in Table 10.7.

10.6.3 FINAL PERIOD

A total time horizon of 24 h was chosen. Repeating the cyclic portion once and accounting for the initial period of 9 h, a final period of 10 h remains. A profit maximization problem was solved for the final period and an objective value of 46,605.928 c.u. was determined in a CPU time of 1.794 s and required 6 time points.

The Gantt chart for the maximization of profit over the final period is shown in Figure 10.8 and the results are shown in Table 10.8.

10.6.4 COMPARISON WITH DIRECT SOLUTION

The problem was solved directly over a time horizon of 24 h for the case where only direct heat integration was used as well as the case where only utilities were available for heating and cooling. In order to have a long enough time horizon for a fair comparison, the problem becomes too complex to solve directly for the heat storage case. A fairly long final period is required in the cyclic scheduling model as most of the products are formed during this time and this makes the overall time horizon lengthy. However, the results for the direct solution of the problem for the other two cases are shown in Table 10.9 as well as a summary of the results for the cyclic scheduling heat integration model.

As seen from Table 10.9, an error of less than 6% is achieved when the solution obtained using the proposed cyclic scheduling model with direct heat integration is compared to the result obtained from the direct solution over 24 h.

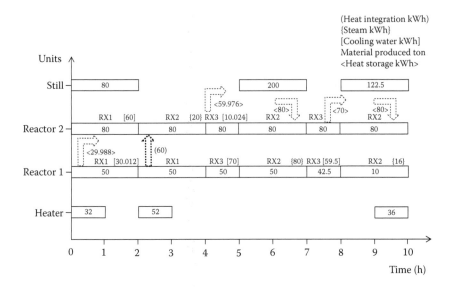

FIGURE 10.8 Gantt chart for maximization of profit over final period (10 h) for multipurpose example.

TABLE 10.8

Results for Maximization of Profit over Final Period for Multipurpose Example

Time period (h)	10
Objective (c.u.)	46,605.928
Time points	6
Cooling water (kWh)	229.536
Steam (kWh)	116
CPU time (s)	1.794

10.7 INDUSTRIAL CASE STUDY

Figure 10.9 shows the STN of an industrial case study, while Figure 10.10 shows the SSN (Chen and Chang, 2009).

The scheduling data for the problem may be found in Tables 10.10 and 10.11 with the stoichiometric data available in Table 10.12. The heat integration data and heating and cooling requirements may be found in Tables 10.13 and 10.14, respectively.

10.7.1 CYCLIC PORTION

The results obtained for the cyclic portion can be seen in Table 10.15. The optimal cycle length is 9 h and an average profit of 20,528.395 c.u./h is achieved. The CPU times in the table are the sum of the CPU times for both the linearized and the exact models.

TABLE 10.9

Comparison between Direct Solution and Cyclic Scheduling Solution over 24 h

Period	Duration (h)	Profit (c.u.)	Direct Solution (c.u.)	% Error from Exact Solution
Heat integration with storage				
Initial	9	11,345.829		
Cyclic	5	17,909.600		
Final	10	46,605.928		
Overall	24	75,861.357	—	—
Direct heat integration				
Initial	6	5,547.2		
Cyclic	5 (×2)	34,744.0		
Final	8	31,840.0		
Overall	24	72,131.2	76,580	5.809
Utilities only				
Initial	11	10,312.8		
Cyclic	5	15,989.6		
Final	8	40,840.0		
Overall	24	67,142.4	70,790	5.153

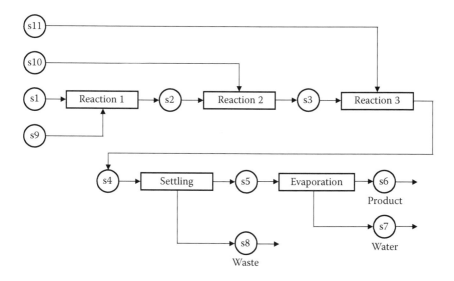

FIGURE 10.9 STN for industrial case study.

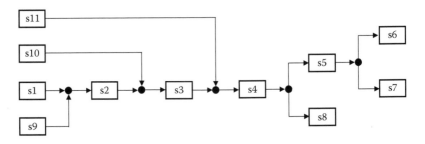

FIGURE 10.10 SSN for industrial case study.

TABLE 10.10
Scheduling Data for Industrial Case Study

Unit	Capacity	Suitability	Mean Processing Time (h)
R1	10	RX1	2
R2	10	RX1	2
R3	10	RX2, RX3	3, 1
R4	10	RX2, RX3	3, 1
SE1	10	Settling	1
SE2	10	Settling	1
SE3	10	Settling	1
EV1	10	Evaporation	3
EV2	10	Evaporation	3

TABLE 10.11
Scheduling Data for Industrial Case Study

State	Storage Capacity (ton)	Initial Amount (ton)	Revenue (c.u./ton)
s1	Unlimited	Unlimited	0
s2	100	0	0
s3	100	0	0
s4	100	0	0
s5	100	0	0
s6	100	0	10,000
s7	100	0	0
s8	100	0	0
s9	Unlimited	Unlimited	0
s10	Unlimited	Unlimited	0
s11	Unlimited	Unlimited	0

TABLE 10.12
Stoichiometric Data for Industrial Case Study

State	Ton/Ton Output	Ton/Ton Product
s1	0.20	
s9	0.25	
s10	0.35	
s11	0.20	
s7		0.7
s8		1

TABLE 10.13
Heat Integration Data for Industrial Case Study

Parameter	Value
Specific heat capacity, cp_{fluid} (kJ/kg °C)	4.2
Product selling price (c.u./ton)	10,000
Steam cost (c.u./kWh)	20
Cooling water cost (c.u./kWh)	8
ΔT^{min} (°C)	5
T^L (°C)	20
T^U (°C)	180
ΔT^L (°C)	0
ΔT^U (°C)	180
W^L (ton)	0.1
W^U (ton)	1.5

TABLE 10.14
Heating/Cooling Requirements for Industrial Case Study

Task	Type	Max Heating/Cooling Requirement (kWh)	Operating Temperature (°C)
RX2	Exothermic	100 (cooling)	150
Evaporation	Endothermic	110 (heating)	90

TABLE 10.15

Results of Cyclic Portion for Industrial Case Study

Cycle Range (h)	Optimal Cycle Time (h)	Objective (c.u./h)	Time Points	CPU Time (s)
3–6	4	18,475.556	3	1.079
6–9	9	20,528.395	6	123.734
9–12	9	20,528.395	6	42.137

TABLE 10.16

Results of Optimal Cyclic Length for Industrial Case Study

Starting and ending amount of intermediate state (ton)	
s2	5.625
s3	30
s4	10
s5	4.882
Starting and ending storage temperature (°C)	95
Heat storage size (ton)	0.22
Cooling water (kWh)	53.704
Steam (kWh)	0
Extra cooling water (kWh)	0
Extra steam (kWh)	0

The results obtained for the optimal schedule for the optimal cycle length of 9 h are shown in Table 10.16. The optimal schedule obtained is shown in Figure 10.11.

10.7.2 INITIAL PERIOD

The amounts of intermediate states at the end of the initial period are fixed to the values required to start the cyclic period to ensure there will be sufficient of the intermediate states produced in the initial period available to start the cyclic scheduling period. The optimum heat storage size of 0.22 ton, which was solved for in the cyclic period, is now fixed in the initial period. The final temperature for the heat storage vessel is set at 95°C.

From the makespan minimization problem, a feasible time horizon of 11 h was determined for the initial period, in a CPU time of 0.73 s and required 6 time points. The Gantt chart for the profit maximization problem over 11 h is shown in Figure 10.12 and the results are shown in Table 10.17.

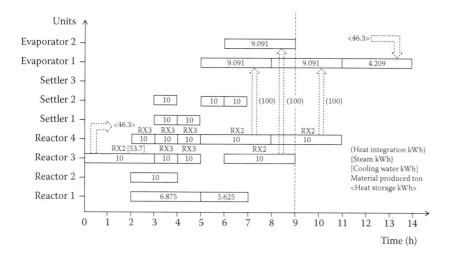

FIGURE 10.11 Gantt chart for optimal cycle length (9 h) for industrial case study.

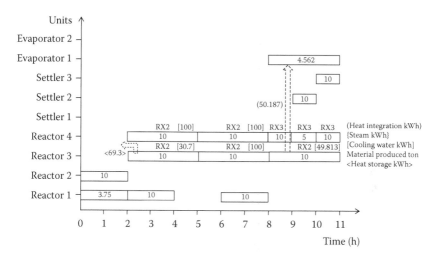

FIGURE 10.12 Gantt chart for maximization of profit over initial period (11 h) for industrial case study.

10.7.3 FINAL PERIOD

A total time horizon of 48 h was chosen. Repeating the cyclic portion three times and accounting for the initial period of 11 h, a final period of 10 h remains. A profit maximization problem was solved for the final period and an objective value of 338,884.348 c.u. was determined in a CPU time of 1,515.2 s and required 9 time points.

TABLE 10.17

Results for Maximization of Profit over Initial Period for Industrial Case Study

Time period (h)	11
Objective (c.u.)	23,793.804
Time points	7
Cooling water (kWh)	380.513
Steam (kWh)	0
Initial heat storage temperature (°C)	20
CPU time (s)	81.89

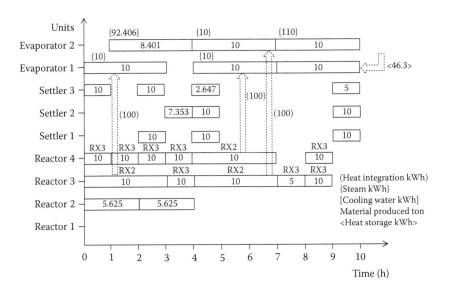

FIGURE 10.13 Gantt chart for maximisation of profit over final period (10 h) for industrial case study.

The Gantt chart for the profit maximization over the final period is shown in Figure 10.13 and the results are shown in Table 10.18.

The problem was also solved for the case with direct heat integration only as well as the case where only utilities are available. A summary of results for each case is given in Table 10.19. In each case, the cyclic period was repeated three times.

10.7.4 COMPARISON WITH EXISTING LITERATURE MODEL

This same example was considered by Chen and Chang (2009). In solving for the optimal cycle length, a series of fixed values were used and the cycle time giving the best objective, which is the average profit per time, was considered optimal.

TABLE 10.18

Results for Maximization of Profit over Final Period for Industrial Case Study

Time period (h)	10
Objective (c.u.)	338,884.348
Time points	9
Cooling water (kWh)	0
Steam (kWh)	232.406
CPU time (s)	1,515.2

TABLE 10.19

Results of Cyclic Scheduling for Different Cases over 48 h for Industrial Case Study

Period	Duration (h)	Profit (c.u.)
Heat integration with storage		
Initial	11	23,793.804
Cyclic	9 (×3)	554,266.665
Final	10	338,884.348
Overall	48	916,944.817
Direct heat integration		
Initial	13	11,625.325
Cyclic	9 (×3)	554,266.668
Final	8	234,494.118
Overall	48	800,386.111
Utilities only		
Initial	14	6,457.583
Cyclic	9 (×3)	525,177.777
Final	7	226,494.118
Overall	48	758,129.478

The model proposed in this chapter, however, includes the optimal cycle length as an optimization variable which is solved for between a given upper and lower bound. Also, the initial and final periods of the time horizon were not considered in the model of Chen and Chang (2009).

To compare the models, the heat integration and cost data as used by Chen and Chang (2009) were used. These values may be obtained from Table 10.20. In the example, batch sizes were not fixed and the maximum batch sizes were given as 8 ton for all units. Only direct heat integration was considered.

The model proposed in this chapter, with direct heat integration only, was used to solve for the optimal cycle time between 6 and 9 h and the same optimal objective value of $161.05 was obtained. The proposed model was also applied over an interval

TABLE 10.20

Data for Comparison of Industrial Case Study

Product	Selling price ($/ton)
s6	100
Utility	Unit cost ($/ton)
Cooling water	8
Steam	15
Task	Cooling/ Heating duty (ton)
Reaction 2	5
Evaporation	4

of 9–12 h. The optimal cycle time was found to be 11 h and gave a better objective value of $164.71 compared to $161.05 for the reported optimal cycle time of 9 h.

10.8 CONCLUSIONS

A model for the cyclic scheduling of multipurpose batch plants using direct and indirect heat integration has been presented. Cyclic scheduling constraints were incorporated into the heat integration model of Stamp and Majozi (2011). The presented cyclic scheduling model can be used for any chosen long time horizon. Once the cyclic and initial portions have been solved, the flexibility of the model allows the solution over any long time horizon, as only the profit maximization problem of the final period must be solved for again. The time horizon therefore does not affect the computational complexity required to solve the model.

The model may give better results when compared to an existing method as the cycle time is an optimization variable rather than a fixed value and both the initial and final periods are also accounted for.

The model may be preferred for the solution of problems over long time horizons as direct solution using short-term methods may not be feasible. The solution obtained over 24 h using the proposed cyclic scheduling model with direct heat integration for a multipurpose example was compared to the result obtained from the direct solution and an error of less than 6% was achieved.

REFERENCES

Boyadjiev, C.H.R., Ivanov, B., Vaklieva-Bancheva, N., Pantelides, C.C., Shah, N., 1996. Optimal energy integration in batch antibiotics manufacture. *Computers and Chemical Engineering*. 20, S31–S36.

Chen, C.L., Chang, C.Y., 2009. A resource-task network approach for optimal short-term/periodic scheduling and heat integration in multipurpose batch plants. *Applied Thermal Engineering*. 29, 1195–1208.

Chen, C.L., Ciou, Y.J., 2008. Design and optimization of indirect energy storage systems for batch process plants. *Industrial and Engineering Chemical Research*. 47(14), 4817–4829.

Foo, D.C.Y., Chew, Y.H., Lee, C.T., 2008. Minimum units targeting and network evolution for batch heat exchanger network. *Applied Thermal Engineering.* 28, 2089–2099.

Glover, F., 1975. Improved linear integer programming formulations of nonlinear integer problems. *Management Science.* 22(4), 455–460.

Halim, I., Srinivasan, R., 2009. Sequential methodology for scheduling of heat-integrated batch plants. *Industrial and Engineering Chemical Research.* 48(18), 8551–8565.

Ierapetritou, M.G., Floudas, C.A., 1998. Effective continuous-time formulation for short-term scheduling. 1. Multipurpose batch processes. *Industrial and Engineering Chemical Research.* 37(11), 4341–4359.

Kemp, I.C., 1990. Applications of the time-dependent cascade analysis in process integration. *Heat Recovery System and CHP.* 10(4), 423–435.

Knopf, F.C., Okos, M.R., Reklaitis, G.V., 1982. Optimal design of batch/semicontinuous processes. *Industrial and Engineering Chemical Process Design and Development.* 21(1), 79–86.

Lee, J.Y., Seid, E.R., Majozi, T., 2015. Heat integration of intermittently available continuous streams in multipurpose batch plants. *Computers and Chemical Engineering.* 74, 100–114.

Majozi, T., 2006. Heat integration of multipurpose batch plants using a continuous-time framework. *Applied Thermal Engineering.* 26, 1369–1377.

Majozi, T., 2009. Minimization of energy use in multipurpose batch plants using heat storage: An aspect of cleaner production. *Journal of Cleaner Products.* 17, 945–950.

Majozi, T., 2010. *Batch Chemical Process Integration—Analysis, Synthesis and Optimization,* Springer, New York.

Majozi, T., Zhu, X.X., 2001. A novel continuous-time MILP formulation for multipurpose batch plants. 1. Short-term scheduling. *Industrial and Engineering Chemical Research.* 40(25), 5935–5949.

Nonyane, D.R., Majozi, T., 2012. Long term scheduling technique for wastewater minimisation in multipurpose batch processes. *Applied Mathematical Model.* 36, 2142–2168.

Papageorgiou, L.G., Shah, N., Pantelides, C.C., 1994. Optimal scheduling of heat-integrated multipurpose plants. *Industrial and Engineering Chemical Research.* 33(12), 3168–3186.

Pinto, T., Novais, A.Q., Barbosa-Póvoa, A.P.F.D., 2003. Optimal design of heat-integrated multipurpose batch facilities with economic savings in utilities: A mixed integer mathematical formulation. *Annals of the Operating Research.* 120, 201–230.

Quesada, I., Grossmann, I.E., 1995. Global optimization of bilinear process networks with multicomponent flows. *Computers and Chemical Engineering.* 19(12), 1219–1242.

Rašković, P., Anastasovski, A., Markovska, L.J., Meško, V., 2010. Process integration in bioprocess indystry: Waste heat recovery in yeast and ethyl alcohol plant. *Energy.* 35, 704–717.

Seid, E.R., Majozi, T., 2014. Heat integration in multipurpose batch plants using a robust scheduling framework. *Energy.* 71, 302–320.

Seid, R., Majozi, T., 2012. A robust mathematical formulation for multipurpose batch plants. *Chemical Engineering Science.* 68, 36–53.

Shah, N., Pantelides, C.C., Sargent, R.W.H., 1993. Optimal periodic scheduling of multipurpose batch plants. *Annals of the Operating Research.* 42(1), 193–228.

Sherali, H.D., Alameddine, A., 1992. A new reformulation–linearization technique for bilinear programming problems. *Journal of Global Optimum.* 2(4), 379–410.

Stamp, J.D., Majozi, T., 2011. Optimum heat storage design for heat integrated multipurpose batch plants. *Energy.* 36, 5119–5131.

Stoltze, S., Mikkelsen, J., Lorentzen, B., Petersen, P.M., Qvale, B., 1995. Waste-heat recovery in batch processes using heat storage. *Journal of Energy Resources and Technology.* 117, 142–149.

Vaklieva-Bancheva, N., Ivanov, B.B., Shah, N., Pantelides, C.C., 1996. Heat exchanger network design for multipurpose batch plants. *Computers and Chemical Engineering.* 20(8), 989–1001.

Vaselenak, J.A., Grossmann, I.E., Westerberg, A.W., 1986. Heat integration in batch processing. *Industrial and Engineering Chemical Process Design and Development.* 25(2), 357–366.

Wang, Y., Wei, Y., Feng, X., Chu, K.H., 2014. Synthesis of heat exchanger networks featuring batch streams. *Applied Energy.* 114, 30–44.

Wang, Y.P., Smith, R., 1995. Time pinch analysis. *Trans IChemE.* 73(A), 905–914.

Wu, D., Ierapetritou, M., 2004. Cyclic short-term scheduling of multiproduct batch plants using continuous-time representation. *Computers and Chemical Engineering.* 28, 2271–2286.

Index